WIND ENERGY CONVERSION 1986

Conference Organizing Committee

M.B. Anderson
S.J.R. Powles

J.A. Halliday
S. Wilmshurst

Council of the British Wind Energy Association 1985 – 1986

Chairman
N.H. Lipman
  Rutherford Appleton Laboratory
Vice Chairman
R.R. Wilson
  James Howden and Company
Secretary
P.J. Musgrove
  University of Reading

Treasurer
M.B. Anderson
  Sir Robert McAlpine and Sons Ltd
Membership Secretary
J.A. Halliday
  Rutherford Appleton Laboratory

Council Members

W.S. Bannister
  Napier College
R. Clare
  Sir Robert McAlpine and Sons Ltd
B. Clayton
  University College, London
P. Fraenkel
  Intermediate Technology Power Limited
J. Galt
  James Howden and Company
A.J. Garside
  Cranfield Institute of Technology
G. Elliott
  National Engineering Laboratory
Administrator
F.J. Low
  Royal Aeronautical Society

D. Lindley
  Taylor Woodrow Construction
J.C. Riddell
  J.C. Riddell and Associates
W.M. Somerville
  IRDC
D.T. Swift-Hook
  CERL
R.H. Taylor
  CEGB
J.W. Twidell
  University of Strathclyde

Co-opted Member
A.L. Challis

Readers wishing to contact members of the Organizing Committee or officers of the British Wind Energy Association should write to 4 Hamilton Place, London W1V 0BQ.

WIND ENERGY CONVERSION 1986

Proceedings of the 8th British Wind Energy Association Conference
Cambridge 19–21 March 1986

Edited by M.B. Anderson and S.J.R. Powles, Sir Robert McAlpine and Sons Ltd

Published by
Mechanical Engineering Publications Limited
LONDON

Printed by Waveney Print Services Ltd, Beccles, Suffolk

CONTENTS

The Proceedings of the 8th British Wind Energy Conference
are dedicated to
Professor Sir Martin Ryle FRS
who was a pioneer in the field of Wind Energy Research

A review of wind energy in the United Kingdom

L A W BEDFORD
Energy Technology Support Unit, Harwell
J HALLIDAY, SERC
Rutherford Appleton Laboratory, Chilton
D MILBORROW
Central Electricity Generating Board, London

SYNOPSIS

The paper presents an up to date view of progress in the development of wind energy in the United Kingdom, covering activities sponsored by Central Government, the electricity generating boards and industrial companies.

1   INTRODUCTION

Wind energy activities in the United Kingdom were last reviewed to BWEA in 1984 (1) and 1985 (2). Since then wind energy has become widely accepted as the most promising renewable energy resource option for electricity generation in the United Kingdom (3). The past year has seen further important progress highlighted by:

- increasing commitments and achievements being made by British companies.
- substantial progress being made with major construction projects.
- an expansion of generic research based on a growing understanding of problems and indentification of the uncertainties remaining to be resolved.
- the announcement of some important new projects.

2. FUNDING

Funding support for UK wind energy activities is provided from a number of sources. Firstly from Government bodies including Department of Energy/ETSU through its substantial R&D programme, Department of Trade and Industry through its support of new product development, and the Science and Engineering Research Council with its support for academic research. Secondly from the electricity generating boards who wish to maintain their position as informed users through in-house research and purchase of equipment for evaluation. Thirdly from industry in support of commercial ventures into emerging markets. Finally, opportunites for funding support from the European Commission have opened up through the Demonstration Programme (since 1983) and the Energy R&D Programme (since 1985).

3. ACTIVITIES SUPPORTED BY CENTRAL GOVERNMENT

3.1 Department of Energy

This programme, which forms the major R&D activity within the UK, was established jointly by the Department of Energy, which as the funding department is responsible for policy and setting aims and objectives, and the Energy Technology Support Unit, Harwell which is responsible for defining and setting up individual projects, and their technical and financial management. A wind energy steering committee (WISC), comprising senior representatives of Government bodies, electricity boards and industrial companies provides advice to D Energy and ETSU on broad strategy and technical and commercial aspects of proposals.

The Department's major long term interest is in the development of wind turbines for the generation of electricity which, if needed, could be deployed in numbers sufficient to contribute a substantial amount of energy towards UK needs. It is currently believed that for central grid applications, the optimum size of machine lies in the range of 1MW upwards for land-based sites and probably larger for offshore sites where the tower, foundation installation and maintenance costs have a comparatively greater influence on the economics.

The Department's approach has, up to recently, followed two main routes. Firstly to explore the more conventional horizontal-axis type by gaining experience with a 60m diameter, 3MW rated wind turbine generator (WTG). Secondly to examine the new concept of a vertical axis machine which utilises a variable geometry configuration, in order to assess the claims made about inherent advantages over other machine types. Within the past year, a third route has been identified and is now included within the programme. This consists of exploring the extent to which medium scale machines, experience of which is rapidly increasing, may be scaled-up, thus stretching demonstrated technology up to the megawatt range in gradual steps which minimise technical risk.

The Department's programme, underway since 1977, can be conveniently sub-divided under two main headings, with total expenditure to 31 March 1986 by D Energy roughly as follows:

1

		£M	
Large Wind Turbines:	Horizontal Axis	10.3	13.1
	Vertical Axis	2.8	
Generic Research:	Basic Studies	1.6	
	Wind Data	1.5	3.9
	Offshore	0.8	

Total: £17.0 Million

Expenditure in 1985/86 is expected to be about £6.0 Million.

Progress on projects within the D Energy programme is described below:

3.1.1 Large Wind Turbines

The Orkney Project

In July 1983 a 20m diameter 250kW HAWTG was installed by the Wind Energy Group (WEG) at the same site selected for the 60m machine on Burgar Hill, Orkney to act as a development prototype (4). The cost of this machine was shared by WEG and the Department of Trade and Industry and the North of Scotland Hydro-Electric Board (NSHEB). Up to the middle of December 1985 the machine had operated for over 5400 hours and generated more than 470MWh. The availability of the machine has steadily improved over the past year, with figures of over 90% being achieved over the past few months.

Data acquisition and analysis for monitoring wind conditions together with performance and loadings of the 20m machine on Burgar Hill are being carried out by the Wind Energy Group under contract to the Department of Energy. Reports covering specific technical areas have been produced. The equipment is also currently in use measuring wake interactions between the 20m machine and a 22m Howden machine on the same site. The data acquisition and analysis equipement will be utilised for the 60m machine once it is in service.

The largest single project in the Department's programme is the 3MW rated, 60m diameter HAWTG being built on Burgar Hill, Orkney. Funding of this project is provided jointly by D Energy and NSHEB, with the latter contributing over £1 million towards the total cost of over £10.5 million. This Project (5) has now moved firmly into the manufacture and construction stage. The concrete tower (Figure 1) was completed in late 1985 and all major components eg rotor, gearbox/transmission, nacelle, were at an advanced stage of manufacture by January 1986 (See Figure 2). The 33kV overhead line to the site was completed during 1985,, and the associated sub-station is currently being built. The rotor is being fabricated in five sections – hub/inner blades, outer blades (2), and variable pitch tips (2), and will be assembled at Hatfield during the summer 1986. Sensors including strain gauges and accelerometers will be fitted to structural elements at this stage and the teeter bearings and hydraulic control systms will be checked out. A modal analysis will also be carried out to check the calculated vibrational modes.

The upper tower frustum, containing the generator, secondary gearbox and yawing ring, and the nacelle are due for delivery to Burgar Hill by late summer, and lifting on to the tower will follow soon after. The Primary Gear Box, including the bevel stage, is expected to be lifted to the nacelle using the built-in lifting jib during the Autumn. On present plans it is expected that the rotor will reach the site in time for lifting in late November – early December. If the planned dates are achieved it is probable that commissioning will be well underway and first rotation achieved before the end of 1986. The timing of first synchronisation of the machine to the local electrical grid is dependent on the NSHEB being satisfied that all systems have passed the appropriate acceptance tests. It is currently expected that this stage will be reached by the end of 1986.

Following first sychronisation, and over a period of several months, the machine will be subjected to a series of on load acceptance tests. During these preliminary tests, data from several hundred sensors will be evaluated using a comprehensive data acquisition system which has already been extensively used over the past two and half years with the 20m HAWT prototype. When the acceptance tests are satisfactorily completed the machine will be monitored over an extended period during which a comprehensive experimental programme will be carried out.

The Carmarthen Bay Project

By January 1986, construction of the 135kW, 25m diameter VAVGWTG (6) by Vertical Axis Wind Turbines Ltd (VAWTL) had reached the point where rotor blades, struts and cross-arm were approaching the final stages of fabrication. It was expected that a factory assembly of the complete rotor would be underway by April 1986, at which time monitoring sensors would be fitted and a modal survey carried out (See Figure 3). The tower, together with the control room, were in the process of being fitted out. Other major components – pintle shaft bearings, upper gearbox and generator – were all close to completion. Hydraulic power control system and the data handling system were all complete. The computerised control system has been successfully commissioned and tested at the manufacturer's works. Anemometry masts and equipment have now been installed on site and data logging will start shortly.

Assembly of major components at the Carmarthen Bay Power Station site is expected to start in May 1986 followed soon after by commissioning and first rotation by the end of June. At the completion of commissioning a period of intensive testing will proceed for about six months during which the machines performance will be assessed against agreed acceptance criteria. Beyond these tests it is expected that the machine will be operated for a further period of about eighteen months, during which time it will be extensively monitored.

The Richborough Project

It was announced in January 1986 (7) that agreement had been reached for the construction of a 1MW rated 55m diameter HAWTG on a site at the CEGB power station at Richborough.

The machine, with three wood/epoxy rotor blades mounted on a 45m high turbular steel tower, is based on the technology developed at medium scale by James Howden Ltd who will be responsible for design and construction. A photomontage of the machine is shown in Figure 4. Subject to contract, funding for the project will be shared by the European Commission, the Department of Energy, the CEGB, and James Howden Ltd. Based on the current planned timetable, it is expected that the design will be finished by the end of 1986, and the machine up and running by the end of 1987. A test period of about 12 months to the end of 1988 is planned. The main features of the wind turbine are shown in Figure 5 together with those of the Orkney 60m machine for comparison.

3.1.2 Generic Research

The objective of the generic programme is to ensure that the major wind turbine development projects are supported by an adequate level of underlying research. Appropriate areas of R&D have been identified and significant progress has been made in getting work underway in these areas. An important contribution towards the identification of research needs is made through the joint D Energy/BWEA workshops which are now becoming a regular feature within the UK.

A one-day meeting to discuss "R&D Needs in Wind Energy" was held at Harwell in September 1985. Over 80 people with a diversity of interests heard accounts of the D Energy, DTI, CEGB and SERC R&D programmes, as well as the views of manufacturers and utilities. There was general agreement that the topics identified as important by ETSU were correct. The proceedings of the meeting have been published (8).

A second more detailed one day meeting to discuss aerodynamics and control, attended by 60 people, was held in February 1986.

The general programme has expanded significantly over the past year, and work is now underway in the following areas:

Materials -

Fatigue testing of welded steel, wood composites and glass reinforced plastics (grp)

Aerodynamics -

Aerodynamic characteristics of thick aerofoil sections.
Variable aerodynamic loads due to unsteady air flow
Control methods.

Cost Forecasting -

Development of a cost forecasting methodology.

Structures and Aeroelasticity -

Wake measurements on two medium size HAWTGs on Burgar Hill Orkney.

Modal Analysis

Turbulent wind fatigue damage predictions. Dynamic analysis by the application of finite element structural modelling techniques.

Environment -

Measurements of noise generated by WTG blades.

Component Development -

Hydraulic transmission as alternative to mechanical gearbox.

Comparison of different types of generators.

Wind Data and the Resource -

Collaboration within the IEA and collection of offshore wind data.

Offshore Potential -

Feasibility study of a 100m diameter offshore version of the 60m Orkney machine. Collaborative studies within the IEA.

3.2 Department of Trade and Industry

The DTI is providing support to the evolving UK wind turbine industry in several important ways:

(i) by means of funding grants for projects selected under the Support for Innovation Schemes; to date approximately £1 Million has been provided to wind turbine manufacturers in support of projects costing nearly £4 Million in total.

(ii) by providing financial support in collaboration with the Scottish Development Agency, for the setting up of a National Wind Turbine Test Centre (NWTC) (9) for commercial wind turbines up to medium size by the National Engineering Laboratory (NEL); this facility is intended to act as a focal point for UK industry for testing, development and accreditation of machines.

(iii) by providing representation on behalf of UK industry in a number of areas of international collaboration; for example in the consideration and preparation of recommended practises which may lead to establishing EEC, IEC and ISO standards.

(iv) by managing specific international collaborative projects such as that on Autonomous Wind Turbine Systems being examined under Annex VIII of the International Energy Agency R&D Agreement (Wind Energy Systems).

The commercial development of wind energy in the UK has continued to make steady progress; currently about 25 companies (10) have been set up to supply wind turbines covering sizes ranging from about 50 Watts up to about 750 kilowatts. Success is being achieved by a number of manufacturers in both home and export markets. Notable examples being that by Marlec

3

Ltd who have manufactured and sold in excess of 6,500 battery charging machines and James Howden Ltd who have succeeded in gaining a substantial foothold in the highly competitive Californian wind farm market referred to later in Section 5. It is important to bear in mind that not only should such success be measured in overseas earnings but also in terms of jobs created. For example James Howden claimed that this project led to the creation of 70 new staff posts and provided 600 person years of work.

The NWTC, as well as providing a wind range of technical support and test facilities at NEL to help manufacturers research and develop their products, has created test sites both at NEL and at a new site located 10km South West of NEL. This test site was completed in Autumn 1985 and is now undergoing commissioning with individual machines set up on universal test stands. A wide range of meteorological equipment as well as a sophisticated real-time data analysis system is available for application to machines under test at either site. It is expected that the facilities will be expanded and that new equipment will be introduced as required to meet the needs of users.

The DTI is responsible for the development of British Standards and together with NWTC and in collaboration with D Energy/ETSU has instituted work aimed at examining the need for standards in wind energy, especially in the important area of structural safety, against the background of developing standards in other countries and existing relevant standards already applicable in the UK. During the past year meetings have been held with representatives of both industry and potential customers, aimed at pulling together a unified UK view which can be put forward to the various international agencies.

3.3   Science and Engineering Research Council

The remit of SERC is to fund research in UK universities and polytechnics in the fields of science and engineering. It does this in a number of ways including awarding 2 or 3 year research grants, awarding studentships and maintaining central facilities for the general use and benefit of the academic community. SERC's budget for wind energy research is currently about £370k/year. The majority of this is spent in the form of direct grants and studentships and the rest in support of the Energy Research Group (ERG) at its Rutherford Appleton Laboratory (RAL). The Group, which is led by Prof N H Lipman, not only maintains and develops facilities at its test site at RAL but is actively involved in several national collaborative research projects and assists in the co-ordination of wind energy research in UK academic institutions.

The SERC funding of wind energy has increased in recent years (reflecting the greater interest) from £150k/year in 1980 to about £370k/year in 1986. During 1985 5 major research grants have been awarded; and work on another 7 has been continued. It is particularly noteworthy that 5 of these 12 are in collaboration with an industrial body –

thereby confirming wind energy research to be particularly relevant. SERC has funded a comprehensive research programme on wind systems intregration, which has examined integration for large utility systems (Cambridge and previously Reading), meso-scale island systems (Strathclyde) and isolated community/island systems (Reading and Imperial College). Other areas funded have included electrical engineering – dynamics (Imperial College); variable speed drives (Leicester), and single phase generators (Imperial College); aerodynamics (City and Cranfield); materials (Bath); novel concepts – the Vee type vertical axis turbine (Open University); the vertical axis shrouded turbine (Kingston Polytechnic), and displacement control of high torque pumps (Edinburgh).

During 1985 SERC hosted two meetings of relevance to wind energy. In October it organised an "Energy Round Table" under the chairmanship of Dr A A L Challis which reviewed funding of energy-related research. The meeting broadly confirmed that SERC's level of funding to all areas (including wind energy) was correct. In August the Energy Research Group hosted the first of a series of meetings for users and potential users of its test site. A need for improved monitoring equipment was expressed by the potential users and has been met by recent purchases.

Finally, at the end of 1985, agreement was reached with ETSU that a project which was of relevance to both D Energy's generic programme and SERC's academic programme, should be jointly funded. This project consisted of a feasibility, study for compiling a UK Wind Energy Atlas including climatological wind variations. This development is significant, as in the past there has been a risk of projects falling between the remits of the two bodies.

4.   PROJECTS SUPPORTED BY THE UK GENERATING BOARDS

4.1   Central Electricity Generating Board

The Central Electricity Generating Board has its own programme of research into wind energy which is co-ordinated with that of the Department of Energy and provides a valuable input to it. The CEGB wishes to be in a position to exploit wind energy for large-scale electricity generation, when and if wind turbines are sufficiently reliable, economic and environmentally acceptable (11). To this end the CEGB has a step-by-step strategy to test these factors and gain early experience of the technology. In the light of recent developments, this strategy is being further developed to assist the UK industry.

4.1.1   Medium Size Wind Turbine

As a first step, the CEGB purchased in 1982 a 25m 200kW HAWTG of US design from James Howden. This is sited at Carmarthen Bay Power Station in South Wales. Considerable effort has been put into obtaining satisfactory operation of the machine but there have been a number of technical difficulties which have severely limited its operation. Nevertheless a considerable amount of useful scientific and engineering data has been obtained. The machine will shortly be dismantled and negotiations for its replacement by a machine representative of current design trends have reached an advanced stage.

4.1.2 MW Size Wind Turbines

The CEGB hoped that if progress in the development of multi-megawatt machines is satisfactory, it will be possible to purchase a machine of demonstrated capability as a second step. Planning consent for such a machine at Richborough in Kent has been obtained, a specification produced and preparatory measurements at the site carried out. However, progress in developing such machines has been slower than expected by manufacturers in the early 1980s and most designs have experienced teething problems. The third step in the Board's strategy - to develop an array of about ten machines - remains an objective for the longer term.

The CEGB wishes to leave open the option of purchasing a multi-megawatt wind turbine when it is satisfied that such machines can achieve acceptable reliability. In the meantime, however, the Board is anxious to assist Government and UK industry to demonstrate wind turbines of interest to the CEGB. It has recently been approached by the Department of Energy and agreed to assist the development of the successful range of medium-size wind turbines built by James Howden. CEC part funding for this project is anticipated, in addition to the contributions from CEGB, the Department of Energy and the manufacturer. The Board will site the machine at Richborough and will collaborate in carrying out R&D on the wind turbine.

4.1.3 Hosted Machines

The CEGB will continue to offer the facility of 'hosting' UK wind turbines, thereby acting as a shop-window for UK designs. The vertical axis wind turbine being constructed by Vertical Axis Wind Turbines Ltd with Department of Energy funding at Carmarthen Bay Power Station is an example of this approach. In addition to providing a site, the CEGB has agreed to assist, if required, in data analysis and to purchase the machine when tested. Recently a small prototype machine of novel design has been hosted on the same site for Balfour Beatty.

4.1.4 Offshore Wind Energy

The siting of wind turbines in shallow waters offshore, to take advantage of the greater wind speeds and reduced environmental impact, has also been investigated by CEGB in conjunction with the Department of Energy (12). The potential resource is very large (13), but the construction costs would, of course, be higher. The CEGB currently leads an international study, IEA R&D WECS Annex VII, of offshore wind energy and has indicated its willingness to host a demonstration machine built by international collaboration.

4.1.5 Supporting Research

The maximum output of a large wind turbine is only a few megawatts and CEGB would need to build sizeable arrays of clusters to provide a significant contribution to electricity needs. It is necessary, therefore, to decide on suitable spacings and quantify the energy losses and increased blade loading due to interactive effects with various designs of arrays and machines. The Board's Research Laboratory (CERL) is studying these wake and cluster problems and is leading an International Energy Agency research programme (Annex IX). The studies involve wind tunnel tests to examine aerodynamics in detail at model scale, field measurements on wind turbines to obtain data under realistic operating conditions, and theoretical work to provide an understanding of the flow processes. CERL is also leading an international collaborative wake measurements project on the two 630 kW, 40m rotor machines at Nibe in Denmark, which are ideally spaced at only 200m apart. Measurement masts and equipment were installed in 1983, assisted by financial contributions from the CEC and the Department of Energy. Valuable full-scale performance and wake data have been obtained with the machines operating alone and interactively. Results so far show that the power loss of a machine operating in the wake of the other at this spacing is about 50% at low wind speeds, falling as rated power is approached (14). There is moderate agreement with calculated values which will help with the design and assessment of proposed arrays.

Site assessments are an important element in wind energy studies. Accurate statistics of wind are needed to estimate the energy yield at a prospective wind turbine site since available power is proportional to the cube of the wind speed. The selection of the Richborough site for the CEGB's first large wind turbine was made after making wind measurements with a 50m tower for at least a year and making short and long term comparisons with measurements from nearby meteorological stations (15). Other meteorological studies cover Carmarthen Bay in support of the CEGB's medium sized machine, offshore sites to determine how greater windspeeds might offset extra construction costs, and the comparison of Meteorological Office upper wind data with those from surface stations.

The feasibility of making remote measurements of windspeed is being studied by CERL to overcome the limitations of mast mounted anemometers. SODAR (Sonic detection and ranging) uses the Doppler frequency shift in sound waves scattered from temperature and velocity fluctuations and allows wind profiles to be obtained from ground based transmitters and receivers. Inexpensive, easily portable equipment is being developed. The Kite Wind Indicator (KIWI) is an alternative method for measuring winds using simple, lightweight equipment and relies on the aerodynamic force on a tethered kite.

Safety and environmental studies are also being carried out, as an essential part of site assessments. A particularly detailed study has been made of the consequences of the unlikely occurrence of fracture of the main rotor, throwing a blade or fragments in the vicinity (16). Simulations, following the motion of fragments as they are acted on by gravity and drag have been employed to estimate the chance of an impact at a particular location. Risk levels and safety zones have now been defined.

The need to take account of noise levels also influence siting and resource assessments and measurements are now being made in the vicinity of the medium size machine at Carmarthen Bay. These data are being assessed, compared with other data from operational machines and predictive techniques are being developed which take both aerodynamic and mechanical sources into account. Background noise measurements are already being made at the Richborough site and a study of measurement techniques is also in progress.

4.2 North of Scotland Hydro Electric Board

In addition to its support for the large and medium size wind turbine projects discussed in sections 3 and 5, the NSHEB has supported other projects as part of its active interest in the development of alternative sources of energy (17). A small 22kW wind turbine was installed Orkney in 1980 and its performance has been monitored to obtain data for future developments, some of which might be of particular value in helping to contain generating costs in the islands powered by diesel generators.

The Board are currently co-funding with the SERC a 3 year project to assess the wind energy potential on Shetland, both by taking on-site measurements and by examining the impact that wind turbines would have on the operation of the power station and grid, with computer models. They have also provided technical support and advice for the installation of the 55kW wind turbine on Fair Isle (18). The Board have also participated in an evaluation of a small 15kW wind turbine, installed near Thurso, suitable for providing lighting and heating to premises isolated from the supply network.

4.3 South of Scotland Electricity Board

The SSEB have acquired two wind turbines to provide practical experience of the operation and economics of small machines associated with agricultural or small industrial operations in the Board's district. A 15kW wind turbine, supplied by NEI, was commissioned in August 1984 at the West of Scotland Agricultural College, Auchincruive and a 60kW, 15m machine, made by James Howden has since been installed in Mid-Lothian.

5 PROGRESS BY UK INDUSTRIAL COMPANIES

Following on from their successful 20m 250kW wind turbine project on Orkney, the Wind Energy Group with funding support from the CEC as part of the demonstration scheme designed and built near Ilfracombe a commercial machine at 25m diameter, 200/250kW (19). Project authorisation was given in March 1984 and the wind turbine first generated power on New Year's Eve that year. This achievement was commended by the

Commission of the Economic Communities DG XVII as the first of the demonstration projects they had supported to generate electricity.

The wind turbine (See Figure 6) has three wood/epoxy composite blades which are fully variable pitch to give a high quality of power output. Pitching is achieved using an electromechanical system which is failsafe in the event of a loss of power. A "soft" steel tower carries the nacelle which houses the epicyclic transmission and a synchronous generator with provision for an induction generator as an alternative. The wind turbine is connected to the South West Area Electricity Board's distribution grid in Ilfracombe. Its measured output has exceeded designer's expectations which bodes well for the anticipated sales to both wind farms and stand-alone installations.

James Howden & Co Ltd, Glasgow, through an arrangement with the North of Scotland Hydro-Electric Board installed, in July 1983, a 22m 300kW 3 bladed HAWTG at Burgar Hill, Orkney (20). Up to the end of October 1985 the unit generated for 6,414 hours and produced 751,980 kWh. The success of the design and the experience gained has been instrumental in Howden being awarded a contract to build a similar machine on Barbados. In addition Howden have installed a 26m 330kW version of this design for Southern California Edison at their Palm Springs Test Centre.

Howden achieved further success in January 1986 (21) when it was announced that an agreement valued at 48 million dollars had been reached covering the sale of 75 machines each 31m diameter 330kW rated, all of which are now operational on a wind farm site at Altamont Pass, Northern California (See Figure 7).

Howden have now extended its design range up to 45m diameter (750kW). NSHEB plan to buy a machine of this type for installation, subject to statutory consent, in 1987, on the mainland island of Shetland. This project will receive support through the European Commission's demonstration project. A machine of similar size has already been installed in California for the Pacific Gas and Electric Company.

As part of the Department of Energy's "technology stretching" strategy referred to earlier, it was also announced in January 1986 (7) that a 55m diameter 1MW HAWTG was to be designed and built by Howden for the CEGB at its power station site at Richborough in Kent.

In parallel with their current work on the 25m VAVGWT at Carmarthen Bay, VAWTL have produced designs for their first commercial VA machine. Under a European Commission supported demonstration project, VAWTL are building a 17m diameter 100kW rated machine on the Isles of Scilly (22).

Over the past few years there has been increasing interest in applying the useful properties of wood laminates to the design requirements of wind turbine rotor blades. Gifford Technology has made a very successful input to this area and as a result of its design, development and prototype testing, this company has supplied rotor blades for all the Howden designs as well as the WEG MS-2 25m diameter machine.

Northern Engineering Industries plc also have an active interest in horizontal axis wind turbines. This activity is centred at International Research & Development Co Ltd, Newcastle upon Tyne, and covers both network connected machines and stand alone wind turbine systems for the augmentation of diesel generation schemes for isolated and remote communities. The first network connected machine rated 22kW was installed for the North of Scotland Hydro Electric Board on South Ronaldsay in December 1980 and is still operating satisfactorily requiring only routine maintenance after 5 years' service.

The company's major achievement has been in the development of a successful and practical load management system for stand alone applications which maximises the utilisation of energy captured by the wind turbine and thereby reduces the consumption of diesel for generation. Such systems have now been operating for more than three years on the Islands of Fair Isle and Lundy. The last year of operation on Fair Isle shows a demand increased to 325% being met by the system with diesel consumption reduced to 38.5% of its former annual rate. Approximately 90% of the energy was provided by the wind turbine, in a year when the mean wind speed was only 85% of the norm.

The company has made careful study of the operational problems encountered and has evolved a new machine design for greater reliability and easier servicing which was erected on their own test site in Northumberland last July. Testing on the company test site is being conducted in open dialogue with the National Wind Turbine Test Centre at East Kilbride. A new scheme, with an offer of EEC demonstration programme support, is under consideration for the remote Shetland island of Foula which has a microhydro system as a pumped storage element of 2,500 kWhr capacity operating in conjunction with a stand alone wind turbine.

There is also a substantial wind energy effort being mounted by a number of companies who are concerned with the smaller size range. The world-wide activities of the Northumbrian Energy Workshop are an example of this important area of work. This company's main sphere of work during the year has been with applied industrial and domestic hybrid systems.

Wind, coupled with either diesel or photovoltaic power, has proved more cost effective, and offers higher continous power than any of the three in isolation. As international consultants, NEW has found a sustained global interest in this form of remote power production with recent contracts in Central American, Africa, the Artic, South Atlantic, Pacific and, of course, the UK. A good example of small machines being exported is that of Marlec Engineering Ltd referred to earlier in Section 3.2.

Finally in this section it is important to note the significant level of collaboration between industrial companies and the utilities. A good example of this is the project covering wake measurements and interactions on Burgar Hill close liaison and cooperation between Howden, WEG, NSHEB and CEGB.

6. THE FUTURE

The development of wind energy in the UK is expected over the next few years to reach a number of important milestones which will have crucial implications for future activities.

(a) the testing during 1987 of the 60m HAWT on Orkney will, for the first time in the UK, provide experience and data on the performance and reliability of multi-MW machines of this type.

(b) the testing during 1986 of the 25m VAWT at Carmarthen Bay will provide experience of this unique variable geometry design, which together with the results of other design work and assessment studies should enable many important questions to be clarified regarding the potential of this type compared with HAWT's and other designs of VAWT.

(c) the 55m HAWT at Richborough should by 1988 be providing valuable information on the prospects for stretching the proven medium scale technology up to and perhaps beyond the MW rating.

(d) the amount of progress made by UK industry with their products in the various overseas markets, and how much the size of those markets are affected by the current uncertainty over oil prices will clearly have a major influence on the commitment of resources by those companies.

Progress in other countries during this period will also provide valuable insights into the questions of optimum size and preferred designs. Trends towards larger wind turbines in the wind farm developments will be watched with interest as well as progress with other MW sized machines in North America and in Europe.

All the above will be particularly important to the Department of Energy and the CEGB's development programmes. If the results are encouraging, they will reinforce the case for continuing support of further developments aimed at building on the progress already made. The Department of Energy has already made it clear (3) that it is important to examine renewable energy options and classify them as either (a) economically attractive , (b) promising but uncertain, or (c) long shots. Land based wind energy is currently in category (b) and it is important to determine whether or not the technology can be pushed into the higher classification or should be relegated to the lower level. It is also worth bearing in mind that if land based wind energy can be moved upwards towards category (a) then offshore wind energy could also be moved from being currently regarded as a long shot up to category (b), thus providing a stronger case for further R&D.

Academic research of the type funded by SERC will continue, reflecting the greater interest in, and application of, wind energy. However, it is likely that research will be even more related to the needs of industry and to the problems associated with the application of wind energy. For example, it is likely that more research into materials, understanding the wind resource and structure, electrical intregration aspects and ways of improving component lifetimes and reliability will be needed.

The future for commercial machines largely depends on UK manufacturer's success in promoting sales and hence creating a home market with cost effective products. In the overseas scene UK companies have continually to develop to maintain reliable products able to compete with overseas competition. The potential in both the home and overseas sectors is very large and the prospect of achieving success in both these areas is within sight.

As wind energy becomes well established the institutional questions such as standards, insurance, public acceptability, local authority rates and tax incentives will become increasingly important. It is apparent that these matters are now being considered and action taken, as shown for example by the work of the Department of Trade and Industry in co-ordinating discussion on standards, and the recent announcement by the Ministry of Agriculture, Fisheries and Foods concerning grants for agricultural wind turbines (23). However, it is essential that progress is maintained in these areas so that the UK wind energy industry can flourish in domestic and overseas markets.

ACKNOWLEDGEMENTS

The authors wish to thank the Department of Energy, the Science and Engineering Research Council and the Central Electricity Generating Board for their permission to publish this paper. It is pointed out that any opinions expressed are those of the authors alone.

The authors would also like to acknowledge the work of all those participating in the projects referred to in the text and to thank especially those who provided useful additional background material for this paper.

REFERENCES

1. Wind Energy for the UK - A status report L A W Bedford, ETSU, BWEA 6, Reading, March 1984.

2. Wind Energy - A promising renewable energy resource for the UK, E G Bevan et al, BWEA 7, Oxford 1985.

3. ETSU R30 - Prospects for the exploitation of the renewable energy technologies in the UK, May 1985.

4. The 20m diameter wind turbine for Orkney, J Armstrong et al, BWEA 3, Cranfield 1981.

5. The Orkney 60m HAWT - A progress report, L A W Bedford et al, BWEA 7, Oxford 1985.

6. Progress with the UK vertical axis wind turbine programme I D Mays, European Wind Energy Conference, Hamburg 1984.

7. D Energy press release No 6, dated 16 January 1986.

8. ETSU R36, R&D needs in wind energy, BWEA/DEn Workshop September 1985.

9. The UK national wind turbine centre, A G Johnson et al, BWEA 7, Oxford 1985.

10. 'List of manufacturers and suppliers in UK', NEL East Kilbride.

11. European Wind Energy Conference, D J Milborrow, J R W Talbot, R H Taylor and D T Swifthook, ECTD Bedford, Hamburgh 1984.

12. Assessment of offshore siting of wind turbine generators - 3rd International Symposium on wind energy systems, D Lindley, P B Simpson, U Hassan, D J Milborrow,4. BHRA Cranfield, Copenhagen 1980.

13. The UK offshore windpower resource, 4th International Symposium on Wind Energy Systems, D J Milborrow, D J Moore, N B Richardson, S C Roberts. BHRA Cranfield, Stockholm 1982.

14. Wake measurements on the Nibe windmills, G J Taylor, D J Milborrow, D N McIntosh, D T Swifthook, BWEA 7, Cambridge 1985.

15. Estimation of low level winds from upper-air data, , M Bennett, P M Hamilton and D J Moore, IEE Proc Vol 130, Pt A, No 9, 517-522 1983.

16. Risks associated with wind turbine blade failures, J F McQueen, J F Ainslie, D J Milborrow, D M Turner, D T Swifthook, IEE A 130,9, 574-586 1983.

17. Installation and operation of a small wind turbine generator. W G Stevenson & J Strong (Electricity Council Distribution Development) Dec 1981.

18. The Fair Isle Wind Power BWEA System, W G Stevenson et al, BWEA 5, Reading 1983.

19. WEG 25m machine - J R Armstrong et al, BWEA 7, Oxford 1985

19. Press Release by James Howden Ltd 16 Jan 1986.

20. The HWP-30 Wind Turbine , P Jameson et al, IEE Proc, Vol 130 Pt A, No 9, Dec 1983.

21. Press Release by James Howden Ltd, Jan 1986.

22. Press Release dated 10 Dec 1985 by VAWT Ltd in collaboration with Davidson & Co Ltd, Belfast.

23. Agriculture Improvement Regulation 1985, No 1266 dated 1 October 1985. ISBN 0 11 057266 1.

Fig 1   View of the completed 37m high concrete tower for the 3MW HAWT, Burgar Hill, Orkney *(Courtesy NSHEB)*

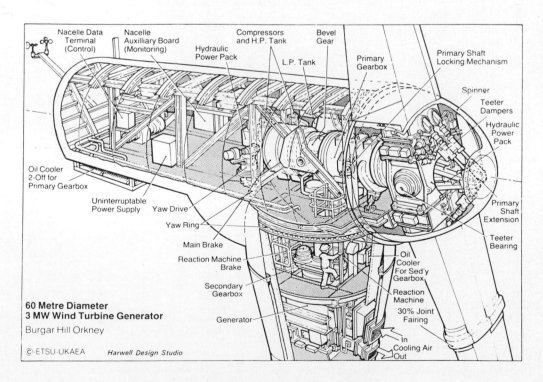

60 Metre Diameter
3 MW Wind Turbine Generator

Burgar Hill Orkney

©-ETSU-UKAEA   *Harwell Design Studio*

Fig 2   Cut-away view of the nacelle and upper tower frustum of the 3MW HAWT

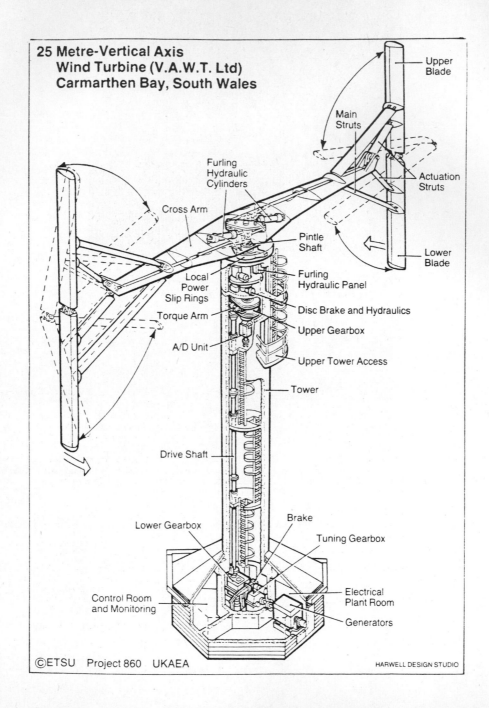

**25 Metre-Vertical Axis
Wind Turbine (V.A.W.T. Ltd)
Carmarthen Bay, South Wales**

Upper Blade

Main Struts

Actuation Struts

Lower Blade

Furling Hydraulic Cylinders

Cross Arm

Pintle Shaft

Local Power Slip Rings

Furling Hydraulic Panel

Torque Arm

Disc Brake and Hydraulics

A/D Unit

Upper Gearbox

Upper Tower Access

Tower

Drive Shaft

Lower Gearbox

Brake

Tuning Gearbox

Control Room and Monitoring

Electrical Plant Room

Generators

©ETSU   Project 860   UKAEA

HARWELL DESIGN STUDIO

Fig 3   Section view of the 25m diameter VAVGWT

Fig 4   Photomontage of the proposed 1MW HAWT at the
         Richborough power station

		ORKNEY 3MW	RICHBOROUGH 1MW
Rotor	– diameter	60 metres	55 meters
	– type	2 blades, teetered, upwind	3 blades, upwind
	– speed	34 rpm	30 rpm
	– control	Variable pitch tips	Variable pitch tips
		Intermediate speed shaft multi-disc brake	High speed shaft disc brake
	– material	Steel spar, fibre glass leading/trailing edges	Wood epoxy composite
Tower	– type	Concrete with steel top frustum	Tubular steel
	– height	45 metres to rotor centre line	45 metres to rotor centreline
	– diameter	3.8 metres	3 metres
	– foundation	Reinforced concrete	Reinforced concrete
Nacelle	– type	Steel frame; Aluminium cladding	Mounded glass reinforced plastic
Generator	– type	Synchronous, 3 MW, 11 kV, 3 phase 50 Hz at 1500 rpm	1 MW, 3.3 kV, 3 phase 50 Hz at 1000 rpm
Switchgear	– type	11 kV, vacuum contactor in base of tower	3.3 kV, vacuum contactor in base of tower
Control	– details	Purpose designed electronic control system plus hard-wire back-up system	Control will be provided by two standard industrial programmable controllers in a similar manner to the existing machine on Burgar Hill, Orkney. In addition a hard-wire back-up system will be provided
Operating details		Cut-in wind speed: 8 m/s Rated wind speed: 17 m/s Cut-out wind speed: 27 m/s Estimated Energy yield at Orkney site: 8.5 GWh (approx) based on 90% availability, 11.1 m/s mean wind speed at hub height	Cut-in wind speed: 5 m/s Rated wind speed: 12 m/s Cut-out wind speed: 25 m/s Estimated Energy yield at Richborough site: 2 GWh (approx) based on 100% avalability and 7.6 m/s mean wind speed at hub height

Fig 5   Comparison of the main features of Orkney and
         Richborough wind turbines

Fig 6   25m diameter HAWT built by Wind Energy Group
near Ilfracombe *(Courtesy Wind Energy Group)*

Fig 7   View of the 25MW wind farm built by James
Howden Ltd, Altamont Pass, California
(Courtesy James Howden Ltd)

The national and international reports of Denmark, Germany, Canada and the European Economic Community

J TWIDELL
University of Strathclyde

SYNOPSIS   This paper reviews the presentations given for the National Programmes Session at the Conference.   No attempt was made to relate the taks with each other, but nevertheless this review covers programmes facing severe economic cut-backs (Canada), technical development (Germany), industrial transformation (Denmark) and cautious sustained support (EEC).

1.   DENMARK

1.1 Local Installations, costs and grants

The spoken paper was presented by Prof. Maribo Pedersen of the Technical University of Denmark.

Denmark is best known for the manufacturing growth and extent of machines in the range of 20 kW to 100 kW. In 1985 the home number of installations increased by 200 to total 1387 at the end of December.   Ex factory costs are about $500 per m^2 of turbine cross section ($750/$m^2$).   A purchaser therefore now pays about $40,000 ($60,000) for a 60 kW capacity machine,   and $60,000 ($90,000) for a 100 kW machine.   It is noteworthy that price per kW capacity does not decrease noticeably for machines greater than 75 kW capacity.   This is despite the very clear development policies of the major companies to produce larger capacity machines to about 150 kW at present.

The well known Danish Government **capital assistance programme** for Danish purchasers is becoming less favourable.   From grants of 30% in the late 1970's, the proportion fell to between 25% and 20% in 1985, and is to be 15% in 1986.   There is an overall limit of $2 million ($3 million) for this Government money in 1986 (i.e. for about 200 home machines).   In addition it is no longer possible for individuals or co-operatives to fund machines at sites beyond their own districts.   This policy effectively limits purchasers to those in rural areas and not in towns.   The growth in numbers of machines also means that there are many examples of local grid power feeder lines becoming saturated with asynchronous aerogenerators.   These latter points are the cause of acrimony between consumer groups,   the Utilities and Government.

A feature in 1985 was the growth of small **windfarms** with, say, five to twenty 75 kW machines.   These were usually established with the strong participation of the manufacturers.

1.2 **Exports** to the United States, especially California, dominated production, Table 1.

Table 1

Year	Number of Machines Exported	Value £ Million	Value $ Million
1983	350	14	21
1984	1600	72	108
1985	3000	160	240
1986	?	?	?

Danish Aerogenerators exported to California

1.3 Research and Development Programme

Government R & D Funding for small, medium and large machines has remained constant at about $1.3 million/y ($2 million/y) for several years. This does not include consumer grants). This support has been used mostly for the Test Station at Risø and for the two 600 kW range machines at Nibe.

The Nibe A machine progresses slowly and has been run for only 18 hours.   A new gearbox has been fitted and blade-stall experiments are being conducted.

The Nibe B machine now has a   3-bladed rotor with wood laminate blades.   It operates automatically and ran during 10 months of 1985 for 2714 hours, producing 729 MWh of power. The machine total was 10,369 hours with 2,520 MWh produced.   The wood laminate blades are judged to be successful, especially in view of the failure of earlier metal blades due to cracks in the central spar.

The main two new national projects are :

i) **Masnedø Wind Farm** This will be of five 750 kW machines developed from the Nibe B design.   One machine is now installed (3 blades, fibre glass on a wood main spar). Machines will be spaced at 5 diameters downwind on a headland facing westward to the sea winds.   The scheme should be completed in 1987 and 7,500 MWh/y output is expected.

ii) **2MW Capacity, 60m rotor diameter machine.**
This is a new design supported by the Government, Utilities and the EEC, for completion in mid-1987. Ex-factory machine cost is £2.6 million ($4 million), and the total project cost is £3.8 million ($6 million). The blades will be of monolythic fibre glass construction and will cost 9% of the total machine price.

1.4 Future Policy

Delegates detected a change in emphasis of the Danish programme towards larger machines. The Danish Government had agreed with the Utilities that the latter should order 100MW of larger machines by April 1986. This was to stimulate a market with the aim of reducing unit price. The Utilities seemed undecided about the scale of "larger" machines, which would be at least larger than 100 kW.

1.5 Conclusion

It is obvious the Danish Wind Turbine industry is moving through noteworthy times under three pressures, (a) the collapse of the American subsidies, (b) the trend to larger capacity machines, and (c) the restrictive changes in the Danish pattern of subsidies and siting. Nevertheless the Danes are still held in awe and most delegates expected Denmark to keep its lead in national wind turbine technology.

2. FEDERAL REPUBLIC OF GERMANY

The German programme was described by Herr Gunther Cramer of the SMA-Regelsysteme.

To the end of 1985, national funding of £58 million (DM 200 million) had supported 70 projects. There was now a move to support smaller grid coupled machines for commercial purposes, rather than megawatt capacity proto-type machines. The Test Station near Stuttgart (average wind speed 5 m/s) was expected to be important for the commercial development of small and medium scale machines.

2.1 Small and medium machines

There are 30 projects with 15 different types of machines. This included the 30 kW Darrieus machine of the Dornier Company and the single-bladed design of Professor Wortman (who had sadly died in 1985). The latest University of Stuttgart single-bladed operational machine has a 4m blade, but a 20m blade design is being developed. A new company would produce commercial 6m single blade machines in 1986.

The Aeroman 2-bladed machines of the MAN company were the best known. These had fan tail yaw and full blade setting control. A 35 kW version had an asynchronous generator for grid connection with 8% generator slip. The 20 kW autonomous machine operated with a speed control to +/- 1%. 300 grid connected machines had been sold in California. Other well known projects were (a) on the Island of Kythnos in Greece (20 x 20 kW machines linked to a diesel system in 1982 and supplying 80% of the island

requirements), and (b) in Eire (2 x 35 kW autonomous machines linked with a battery/invertor system).

With a general trend to increase the size of small machines, the latest development plan was for a 3-bladed, 25m diameter rotor producing 100 kW in an 11 m/s wind. Two generators would be included (for low and high wind speed) and there would be special emphasis for linking into weak rural grids with accurate speed control.

2.2 Large Machines

The large 100m diameter Grovian machine has been tested since June 1983. Much had been learnt about aerodynamics and power control. However concern about faults under high stress conditions had led to a policy to dissemble the machine by the end of 1986. There had been problems with the teetered blade design, and future machines would be expected to have fixed blades.

A single bladed machine of 24m diameter producing 370 kW through an intermediate DC link had been operating since 1983. In July 1985 the blade had been damaged due to blade-control failure. The latest proposal was for three single-bladed machines each of 1MW.

2.3 Remote installations

There was now considerable interest in wind/diesel and weak grid systems. These included one system of two 1.2 MW, 3-bladed, 60m diameter MAN machines for island diesel generation. There would be heat recovery from the diesel engines, and the aerogenerators might be sited offshore in 5m of water. The aerogenerators would be decoupled from the grid using a rectifier, invertor and phase shift link.

As well as electricity generation, there was an interest in water pumping to 50m lift.

3. CANADA

Mr. Jack Templin of the National Research Council of Canada made the presentation. 1985 had seemed "an unmitigated disaster" for the national programme when the new government initially cancelled the R & D funding. However, £500,000 (Can. $1 million) had been reinstated, and commercial wind turbines were now included for "class 34" tax exemption allowing ⅓ of the capital value to be written off against tax in each of 3 years. Almost without exception the national programme involved Darieus rotor machines with which the Canadian programme had gained much experience.

3.1 Large machines

The Hydro Quebec Utility had combined a 50% interest with government and manufacturers in a 4 MW Darrieus machine. Standing 100m high, it would have a 61m diameter 2-bladed rotor to be delivered in June 1986. The ground construction was complete with a 12m diameter synchronous generator directly driven with no gearbox. The generator was a slowed-down conventional hydro generator run at 1/10 speed and producing power at variable speed around 6 Hz. The blade construction had been difficult and consisted of bolted straight sections with GRP leading edges. Blades would be bolted together on site.

3.2 Medium machines

Two 500 kW Darrieus machines had been made. That on the Atlantic Wind Test Site failed after 3000 hours running due to blade fatigue. It appeared that the extruded aluminium had corroded. Retrospective investigation of the monitored strain sensors showed that the absolute strain had steadily increased before failure, but the amplitude of the strain cycle had not increased before failure as had been expected.

A 'low cost' 125 kW machine had been built. The Darrieus rotor was held in a large exterior framework so there was no downward compression on the top bearing (as with conventional guyed supports). The machine had run for 100 hours to date. Mechanical control was used to lessen generated ripple at the rotor frequency. The overall cost was less than Can. $1000 per rated capacity (i.e. £500, US$ 750 per kW).

3.3 Small machines

Four 50 kW DAF-Indal machines now had operation times of greater than 50,000h. The general design was now used for commercial manufacture, however air brakes had been removed. The only braking was now with the mechanical pressure disc brake that comes into operation by dropping the rotor onto a horizontal pad at the base.

4. THE EUROPEAN ECONOMIC COMMUNITY R & D PROGRAMMES

Dr. Palz spoke mainly about the R & D programmes of his Directorate General XII. Other wind related projects were supported as Demonstrations by Directorate General XVII.

4.1 Introduction

The majority of EEC R & D interest had so far related to energy because of Europe's dependence on imported oil. However wind turbine commercial activity had recently been dominated by the sales potential in Carlifornia. Now that U.S. tax credits were being eliminated, European companies had to face considerable marketing difficulties. This was aggravated by the recent fall in the international price of oil.

However cheaper oil had its compensations. It was now possible to plan R & D for longer lead times. Such planning should produce reliable and efficient renewable energy technology for its eventual use when oil price rises again. A further benefit could be an increase in investment capital on the stock markets. Nevertheless it must be realised that EEC funding only provided 30% to 50% of a limited number of projects, and so national resources would be essential.

4.2 The EEC Programmes

The 10 years from 1975 had seen an exploratory phase. For wind interests this largely related to wind meteorological studies. For instance the Wind Assessment Review for Europe was now published. Funding for wind in this first period had been 1 million E.C.U. (£600,000).

The second R & D programme for 1985-88 had been agreed in principle by the Council of Ministers in 1985, and would be administered by Directorate General XII from Brussels. The funding amounted to 18 million E.C.U. (£12 million), and would be divided between :

(a) Assessment of resources
(b) Aerogenerator related experiments
(c) Prototype production

Under (a) it had been agreed to fund a European Wind Atlas to be a data base for power production in Europe, e.g. to assess grid penetration from modelling. The Atlas would relate to the Danish Wind Atlas produced by the Risø Laboratory.

Under (b) research activities included : gust measurement, wakes within assays, materials and aerodynamics.

Programmes involving machines included :

i) the performance of identical machines on a European-wide set of sites,
ii) Megawatt scale machines; of which one had been funded in the UK (1MW by James Howden & Co. Ltd. with the CEGB), one in Denmark (2MW machine), and one in Span (1MW, 60m diameter).

The general EEC conclusion was that best cost effectiveness was with machines in the medium range, and so there was more interest at the 300 kW capacity scale. Unfortunately the programme had only a little money remaining for funding in this range.

4.3 Conclusion

Delegates would be of the opinion that the EEC Wind Energy Programmes (both R & D and Demonstration) were playing a crucial part in developing viable wind power systems. Viewed generally, the EEC projects provided great stimulus and encouragement for both academic researchers and manufacturing industry.

The Irish wind energy programme 1980 – 86

T O'FLAHERTY
Kinsealy Research Centre, Dublin, Ireland

SYNOPSIS   The Irish wind energy programme has involved the installation since 1980 of 13 wind turbines of a range of sizes and for a variety of applications.   The technical features of the individual units and their performance to date are described.

1 INTRODUCTION

The Irish wind energy programme was initiated by the then Minister for Energy in 1980 when he announced plans for investment in several demonstration projects. The aim of these was to investigate the potential contribution wind energy could make to Irish energy needs, and at the same time give Irish engineers first-hand experience of the installation and operation of modern wind energy conversion systems.

The planning of the eventual programme was clearly aimed at acquiring the widest possible range of information and experience, through including a large diversity of applications, site locations and wind turbine sizes and types. In all, twelve wind turbines were installed in the initial programme, and they can be conveniently divided into three groups:

(a) Dept. of Energy projects at Inis Oirr, Ballyferriter, Sligo and Malin Head.

(b) Machines with agricultural applications installed at research stations of the Agricultural   Institute   at   Creagh, Ballinamore, Moorepark and Kinsealy.

(c) Machines owned and operated by the Electricity   Supply   Board   (ESB),   at Poulaphuca, Miltown Malbay, Bellacorick and Portlaoise.

The map in Fig. 1 shows the geographical location   of   the   various   sites.   The specifications of the wind turbines and their performance records up to March 1986 are outlined   below.   A new unit installed at Kinsealy Research Centre in 1985 under the EEC programme of Energy Demonstration Projects is also included. A composite project under the same EEC programme has provided for the continuing maintenance and monitoring from 1983 to 1986 of the five wind turbines at Inis Oirr, Ballyferriter, Sligo, Ballinamore and Moorepark.

2 DEPT. OF ENERGY PROJECTS

2.1 Inis Oirr

The largest machine installed within the Dept. of Energy programme was the 55 kW Windmatic on the island of Inis Oirr off the west coast. The island had a diesel-powered electricity supply with a generation cost of about IR£0.2/kWh. The purpose of the wind turbine was to interface with this diesel system and utilise the high mean windspeeds   to   reduce   the average cost of electricity on the island. Annual electricity consumption on the island is about 160,000 kWh.

The 3-blade fixed pitch 14m diameter wind turbine was commissioned in April 1982. It is mounted on a 15m lattice steel tower on a site 70m above sea level. From an early stage difficulties were experienced with operation of the control interface with the diesel system, with many cut-outs due to relay faults. Also in February 1983 the alternator burnt out due to salt ingress and its replacement involved an outage of nearly six months.

The extreme difficulties associated with the remoteness and inaccessibility of the location continued   to   cause   long   delays following virtually all breakdowns, resulting in low machine availability. In addition, the control strategy which was adopted placed severe constraints on the circumstances in which the wind turbine output was accepted for connection to the load, and thus resulted in a considerable proportion of the output being dumped during the times when the machine was operational.

From about April to October 1984 the machine ran satisfactorily in this mode, but in October it was put out of operation by a voltage regulator fault. Correction of this problem took until March 1985. Then, however, due to the proposed installation of a new diesel generator to meet increased island load it was not possible to re-connect the wind turbine to the load, pending the implementation of a new and re-designed control system.   Hence since then, although the wind turbine has been operating satisfactorily,   all   of   the   energy   it   has produced had to be dumped.

Installation of the new control scheme is now almost complete. This will be based on the principle that if the wind turbine output is greater than the island load by a margin of 10kW for a period of more than 10 minutes, then the load is connected to the wind turbine and the diesel set is switched off. Otherwise the wind turbine output is dumped. Possible uses for the dumped energy, such as the heating of greenhouses, are being investigated.

2.2 Ballyferriter

The first wind turbine on this site, which is an outstanding wind site on the south-west coast just a few miles from the most westerly point in Europe, was installed in mid-1982. This was a 10m diameter, 3 blade, machine rated at 30 kW and supplied by Stork B.V. of The Netherlands. It was the first privately-operated grid-connected wind turbine to be installed in Ireland. Owned by a local cooperative, its purpose was to supply electricity to a 0.8 hectare glasshouse enterprise run by the cooperative. The use of a high-intensity lighting installation to accelerate crop growth, as well as 24-hour operation of boiler fans, hot-water circulating pumps, etc., made the load particularly suitable for supply by a wind turbine.

Unfortunately the performance of the wind turbine proved unsatisfactory from the outset, and culminated in a major breakdown in October 1983. Following this the suppliers agreed to provide compensation which would permit replacement of the machine with one from another supplier. The unit chosen was that of Lagerwey, also of the Netherlands. It has a 10.6m diameter rotor with three glass-fibre blades and was also originally rated at 30 kW at a windspeed of 12 m/s. It was installed in January 1985 on the 18m tubular steel tower which had been used for the original turbine.

Apart from some minor outages in early weeks, the new turbine operated almost continuously from January until late October 1985. In this time it delivered some 36,000 kWh, which represented a substantial saving to the cooperative. A number of problems then developed, the most serious being generator failure due to salt spray build-up; others were a gear-box oil leak, faulty proximity switches on the wind vane sensor, and a faulty yaw motor. All necessary repairs are being carried out under guarantee, with a down-rating of the generator to 25 kW. The unit is due to be returned to operation very shortly.

2.3 Sligo

A Trimblemill wind turbine was installed on the campus of the Regional Technical College at Sligo in May 1982. The output was to provide heating and hot water for a section of the college, and staff of the college's Engineering Department undertook supervision of the installation (1).

The UK-manufactured Trimblemill has the unique feature of incorporating two 6m diameter contra-rotating rotors, one having five blades and the other three. The generator is a direct-coupled 42-pole single-phase alternator, in which the "stator" and "rotor" are driven in opposite directions by the two rotors of the turbine. The blades comprise aluminium spars with a former fixed to the end and a terylene sock pulled over the spar to form the aerofoil. The turbine is mounted on a 9.3m tower and the rated output is 12.5 kW at 23 m/s windspeed. A load controller developed at University College Dublin (2) and designed to maximise energy transfer to the load from the Trimblemill, has been successfully incorporated in the installation at Sligo.

Apart from one emergency in early 1984, in which a blade was shed and travelled about 70m, the unit has operated reasonably well since installation. Currently it is not producing an output due, it is believed, to worn brushes on the generator, but it is expected that this problem will be corrected soon. Due to lack of instrumentation accurate output data are not available.

2.4 Malin Head

Near Malin Head, the most northerly point in Ireland, a 1 kW Aerowatt wind turbine was installed in September 1984 at an elevation of 240m to provide battery charging for the power supply to a remote radio transmitter/receiver station used for marine communications. Installation and supervision of the unit is the responsibility of the Dept. of Communications.

The installation has worked well for most of its life although at present it is out of operation due to an electronic fault which caused over-charging of the batteries.

3 AGRICULTURAL APPLICATIONS

3.1 Ballinamore

A wind turbine was commissioned at this small agricultural research station in Co. Leitrim in April 1982. This machine, called the Phoenix, was a product of Alternative Energy Ltd., at that time the only wind turbine manufacturer in Ireland but now no longer in existence.

The Phoenix has an 8.2 m rotor with two variable-pitch blades, driving a 220 V 10 kW d.c. generator and mounted on an 11m tubular steel guyed tower. The output is connected to heating elements to supply space heating in a laboratory building and also hot water for a milking parlour (3).

Apart from a bearing failure at the end of 1982, the unit has operated with reasonable reliability for most of its life so far, though a significant amount of attention to minor problems has been necessary. Due to lack of data logging equipment no quantified information on performance is yet available, but work on the installation of a logging system is currently under way within the terms of an EEC Demonstration Project. In a storm in January 1986 the tail-vane was damaged and the unit is currently out of operation while this is being repaired. It is expected, however, that it will soon be back in operation.

3.2 Fermoy

Of special interest is the installation at Moorepark Research Centre, Fermoy, Co. Cork, which is the principal national centre for dairy research. Here mechanical energy delivered by a wind turbine is converted directly into heat (3).

The wind turbine, manufactured by Jacobs (U.S.A.), has a 3-blade 7m diameter rotor rated to deliver 10 kW at a windspeed of 11 m/s, with speed control by a blade-actuated governor system. The fluid heating system incorporating this wind turbine was developed and patented by Messrs. Befab Safeland Ltd. of Shannon, who manufacture a rotary hydraulic energy absorber - commonly referred to as a "water twister" - for use in aircraft arresting gear to assist landings on short runways.

In the unit at Moorepark a vertical drive shaft from the wind turbine drives the water twister through a 3.2:1 gear-box at a rated speed of 630 rpm. The twister contains a glycol/water mixture. Heat generated in this primary working fluid is transmitted via a heat exchanger to secondary hot water for dairy use.

Commissioning took place in July 1982 but problems arose in the operation of both the wind turbine and the water twister. Fluid temperatures in excess of 140 C caused failure of the neoprene seal on the underside of the twister. After correction of this problem a period of satisfactory operation followed. Then, however, Messrs. Jacobs advised that metal fatigue in the governing mechanism of their turbines necessitated replacement of this unit, and new blades were also to be fitted.

This was done, but before recommissioning Befab realised that major re-design and modification of the water twister was necessary to ensure reliable long-term operation under the duty involved in a wind energy conversion application. Continuous running under fluctuating load caused problems not encountered in aircraft-arresting applications, resulting in mechanical damage to the twister. A separate problem was poor starting performance due to high inertia.

Following re-design and modification of the entire installation it has recently been restored to operation, and work on completion of the data logging system is continuing.

3.3 Creagh

A land drainage project at a research farm at Creagh, Co. Mayo, has been one of the successes of the programme. A Dutch polder wind pump was installed in late 1981 to drain some 12 hectares of high water table peatland (4). The unit, supplied by Bosman B.V. of The Netherlands, comprises a 3.2m, 4-blade wind turbine mounted on a 7m lattice steel tower and driving a high-volume, low-head pump through a 9m vertical drive shaft. The unit is rated at approximately 0.75 kW and can deliver 1.5 m3/sec. against 1m head at 7 m/s windspeed. It is well adapted to the site, where the outfall drain is too high, by about 1 metre, to drain the land by gravity.

The reliability of this installation has been excellent, with only a small number of very minor outages in over four years. Its effectiveness as a method of drainage has also been fully up to expectation. It has enabled low-lying cut-over bog to be turned into good grassland, and could be applied to many low-lying areas where gravity outflow is difficult or too costly.

3.4 Kinsealy

(a) Multi-blade stand-alone wind turbine

Prior to the formal announcement of an Irish Wind Energy Programme, a 6m multi-blade wind turbine from SJ Windpower of Denmark was installed at Kinsealy Research Centre near Dublin, the main national centre for horticultural research. Rated at 10 kW, this stand-alone unit supplied electricity for heating in a greenhouse. Apart from some lengthy outages resulting from component failures (5), this unit operated until January 1984, producing up to that time 17,999 kWh. In 1983, when availability was virtually 100%, it produced 6,780 kWh, representing a load factor of 7.7%. Mean site windspeed for the year was 3.6 m/s.

In January 1984, the machine suffered a major failure in a storm with gust speeds of 40 m/s. This originated from structural failure of the gear-box housing and resulted in total destruction of the rotor and some tower damage. However, by this time an improved model of the machine was being marketed by a re-constituted company, GSS Power Mills, of Frederikshavn. This new unit, incorporating improvements in gearbox design, rotor bracing, tower reinforcement and electronic control, was installed on a repaired tower, and using the tail-vane of the original machine, in December 1984.

Since then it has operated virtually without interruption. Energy production to March 13, 1986 at 5,666 kWh is, however, slightly disappointing.

(b) Grid-connected wind turbine

On July 4, 1985, a new wind turbine was commissioned at Kinsealy Research Centre under a Wind Energy Demonstration Project of the Commission of the European Communities. Supplied by Scandinavian Wind Systems of Odder, Denmark, it has a three-blade fixed pitch 11m diameter down-wind rotor, driving a grid-connected asynchronous generator rated to deliver 18.5 kW at a windspeed of 13 m/s. The 18m tubular steel tower was manufactured in Ireland to SWS specifications.

The aim of the project is to demonstrate the potential of wind energy, in conjunction with a heat pump, for supplying the base load heat requirements of a greenhouse. An air-to-water heat pump is supplied by the wind turbine output, and is rated to deliver 11kW in water at 45 C, at an ambient temperature of 0 C and a compressor power consumption of 3.75 kW. The heated water is supplied to a 600m2 polythene greenhouse which is double covered to reduce heat loss. Imbalance between the power requirement of the heat pump and the wind turbine output at any time is catered for by power interchange with the grid.

During the first six months of operation the availability of the wind turbine was 70%. The causes of outages were a gear-box oil leak, the need for replacement of the brake lining and a number of overnight outages due to an incorrectly adjusted frequency relay. The energy production in this period was 6,137 kWh, representing a load factor of 7.7%. However, since January 9 availability has been virtually 100%, and energy production between then and March 13 was 6,828 kWh. This represents a load factor for this period of 24.4%, and brings the overall load factor since commissioning to 12.0%.

4 ESB WIND TURBINES

In 1981 - ´82 four wind turbines of different types were installed by the Electricity Supply Board (ESB) at sites in different parts of the country but all adjacent to existing ESB facilities (Table 1). Only three of the machines were commissioned. The performance of the machines up to early 1984 has been previously reported (6). Here this report is summarised and subsequent developments added.

From commissioning to the end of 1985 the overall availability of the three machines was 43.4%, and the % time they were connected to the network was 15.4%. The total energy delivered to the network was 201,375kWh.

In 1985 availability (of two machines) was 57.6%, time connected 21.8% and energy production 57,808 kWh.

4.1 Windmatic wind turbine at Poulaphuca, Co. Wicklow

This wind turbine was first connected to the network in October, 1981. It is a horizontal axis, three-blade fixed-pitch, upwind 14m diameter turbine mounted on an 18m lattice steel tower, and driving a 55kW induction generator through a step-up gearbox.

The first year´s running experience was limited due to a long series of minor teething troubles and a number of more serious faults. The faults included problems with electromechanical controls, fantail replacement, setting of vibration trip, spoiler operation, brake operation, minor cracks in GRP skin on blades. The more serious faults were failure of a gearbox seal on the mainshaft and shearing of holding down bolts for the yawing gearbox.

From December 1982 to September 1983 the performance of the machine was much improved. There was one major outage during this time in January 1983, to replace a spoiler wire. Besides this two-week outage there were a large number of shorter outages due to trips by the protection systems. In the early part of the year there were a large number of trips due to excess vibration. The excess vibration sensor is a fairly crude device and numerous adjustments were required before a setting which avoided unnecessary trips and still provided a safe level of protection was achieved.

The machine is connected to the end of a long 10 kV line and its controls are sensitive to unbalance, locking it out until manually reset. An auto reset device was ordered from the manufacturer and was fitted in March 1984. Other trips have been due to faulty operation of relays and sensors. The machine availability for the period Dec. 1982 - Sept. 1983 was 71.5%.

However the load factor was only 6.2% for that period and the running time 23.3%.

In September 1983 the windmill suffered a complete loss of its control and was found overspeeding with the spoilers out. The main contactor had not released on shutdown due to faulty relay operation, causing the machine to motor against the brakes, wearing down the brake pads and causing the overspeed. Modifications were carried out to prevent recurrence. Following much experimentation the machine was back in service briefly in January 1984 when another overspeed occurred. This time the speed sensor bracket was destroyed and part of it had fractured the brake block leaving the brakes inoperative. Repairs were again carried out and the machine restored to operation.

However, problems with the control logic were still not fully resolved. In addition, in 1985 mechanical faults in the spoiler control wire system were a serious problem, resulting in availability for the year of only 13% and a correspondingly poor energy yield (Table 2).

All spoiler control wires have been replaced and are now giving trouble-free operation. A number of minor electrical faults continue to occur for spurious reasons, requiring manual resetting, but improved performance is hoped for in 1986.

4.2 DAF Indal wind turbine at Miltown Malbay, Co. Clare

This wind turbine manufactured in Canada was erected and first connected to the network in December, 1981. The machine is a vertical axis (Darrieus), two-blade, fixed-pitch wind turbine mounted on a 9m lattice steel tower. The turbine is supported at its top bearing by three guywires. The dimensions of the turbine are 17.6 m high by 11.4 meter equatorial diameter and it drives a 50 kW induction generator through a step up gearbox.

Initially the running performance of the machine was limited as there was a series of problems:

a. with faulty operation of spoilers
b. spurious malfunctioning of the electronic and electromechanical controls
c. water in the lubricating oil

For the first five months the machine was not run unattended.

For the next 6 months the machine performed quite well with an availability of 62.3% and a load factor of 9.1%.

A major fault then occurred with a fatigue failure at the stub shaft flange bolts. The machine was out of service for three months.

From March 1983 to December 1983 the machine performed quite well. The availability for this period was 74.3%, the running time 21.7%, and the load factor 6%. The spoiler problems persisted until it was decided to remove them completely in November 1983.

In December the high speed shaft coupling failed. The braking system operated perfectly to prevent an overspeed. The machine was completely overhauled and the rotor checked for fatigue cracks. As everything was in order, the machine was returned to service in February 1984.

Since then it has operated without interruption. In 1985 its energy production was 55,144 kWh, representing a load factor of 12%.

4.3 Aerowatt wind turbine at Bellacorick, Co. Mayo

This French made wind turbine was erected and connected to the network for the first time in May 1982. The machine was a horizontal axis, two-blade variable pitch, upwind, 18m diameter wind turbine mounted on a 30m guyed steel tube tower, and driving a 120 kW induction generator. The tower and nacelle could be lowered to the ground for access and maintenance.

The experience with this machine was very limited. Firstly there were long delays in the erection of the machine and its eventual synchronisation with the network. The delays were caused by problems with the hoisting gear for lifting and lowering the machine into place, and by major problems with the electronic controls. The machine operated briefly following its first synchronisation but from shortly afterwards was plagued by a long series of problems.

As well as the electronics and hoisting gear, these included disc brake failures, jamming of the blade pitch change bearing and a cable fault. Failure of a blade root casing on a similar machine in Bordeaux led to the Bellacorick machine being shut down pending examination in early 1984, and finally to a decision that the machine should be replaced.

The replacement machine is to be a Ratier-Figeac 100 kW two-blade unit, of which two prototypes are currently being tested in France. Commissioning at Bellacorick is scheduled for August 1986. By virtue of financial support from the French government and the EEC the replacement wind turbine is to be supplied at no cost to the ESB.

4.4 Alternative Energy Ltd. wind turbine at Portlaoise

This is a two-blade, 7.6m diameter turbine identical with the unit at Ballinamore. It drives a 10kW d.c. generator, and a synchronous inverter permits connection to the 380 V, 3 phase, network at the site, which is at the ESB's Line School. Before the machine was commissioned the manufacturer went into liquidation and the unit has since being lying idle. A study is being made of the viability or otherwise of putting it into operation.

5 PROPOSED CAPE CLEAR PROJECT

An announcement was made in December 1985 of a planned wind/diesel project on Cape Clear Island off the south coast. The project involves installing two wind turbines with a combined rating of 60 kW to operate in conjunction with the existing diesel system and battery storage. The turbines are to be supplied by the West German firm SMA, and the project will have financial support from the German Government, the Irish government and the EEC. Commissioning is scheduled for early 1987.

6 CONCLUSIONS

In any attempt at an overall appraisal of the progress to date of the Irish wind energy programme it would have to be ackenowledged that the performance of the wind turbines, and particularly their level of availability has fallen short of initial expectations. This can probably be attributed to two main causes. Firstly, many of the machines when supplied were rather more at the prototype than the fully commercial stage of development; secondly, the consequences of remoteness of machine siting, and of extreme distance from the supplier's base, for the difficulty of diagnosing and correcting malfunctions were more serious than was initially appreciated.

Another factor which has reduced the benefits of the programme has been the delay and difficulty experienced in implementing data logging systems on several of the sites. With EEC aid this problem is gradually being overcome.

In spite of these various shortcomings, it can still be said that a great deal of knowledge and experience of wind energy has been accumulated during the past five or six years, and that this provides an invaluable base on which further progress can be built. There are reasonable grounds for expecting a number of the machines to work well for several more years. If these can be supplemented by the judicious introduction of some new units, incorporating the latest technical improvements, over the next few years, a base can still be laid for wind energy to become a significant component of Irish energy supply in the years ahead.

7 ACKNOWLEDGMENTS

The author wishes to thank the Irish Dept. of Energy for their permission to present the paper and for their financial support. He is also grateful to the ESB for making available the information on wind turbines being operated by them.

8 REFERENCES

1 Doran, M. and Valizadeh, H. Trimblemill at Sligo Regional Technical College. Proc. SESI/EWEA Seminar on Wind Energy, Dublin,1985.

2 de Paor, A.M., O'Malley, M.J. and Ollivier, D. Parameter identification and passive control of a wind driven permanent magnet alternator system for heating water. Wind Engineering, 7, 1983, 193 - 206.

3 Shouldice, C. Dairying and land drainage applications of windpower. Proc. SESI/UCG Windpower Seminar, Galway, 1983.

4 O Fionnachta, P. Wind energy in land drainage. Proc. SESI/EWEA Seminar on Wind Energy, Dublin, 1985.

5 O'Flaherty, T. Three years performance record of a small WECS in the east of Ireland. Proc. 5th BWEA Wind Energy Conf., Reading,

6 Hally, M.J. A utility wind energy programme - findings and conclusions. Proc. IEE 4th Int. Conf. on Energy Options, London, 1984.

Table 1 Details of ESB wind turbines

Site	Make	Type	Diameter m	Rating kW	Date commissioned
Poulaphuca	Windmatic	HA 3-bl	14	55	Oct 81
Miltown Malbay	Daf Indal	Darrieus	11.4	50	Dec 81
Bellacorick	Aerowatt	HA 2-bl	18	120	May 82
Portlaoise	Alt. Energy	HA 2-bl	7.6	10	--

Table 2 ESB wind turbines performance data

	Results to	Availability (%)	kWh generated	Load factor (%)
		from commissioning		
Windmatic	31.12.85	27.1	45,023	2.2
Daf Indal	5. 1.86	67.5	135,168	7.6
Aerowatt	13. 4.84	27.6	21,184	1.0
		1985 only		
Windmatic		13.0	2,664	0.55
Daf Indal		100	55,144	11.9

1 Inis Oirr
2 Ballyferriter
3 Sligo
4 Malin Head
5 Ballinamore
6 Fermoy
7 Creagh
8 Kinsealy
9 Poulaphuca
10 Miltown Malbay
11 Bellacorick
12 Portlaoise
13 Cape Clear

Fig 1 Location of wind turbine sites

The construction of a 26 MW wind park in the Altamont Pass region of California

D L SHEARER
Howden Windparks Inc
A BROWN
James Howden & Company Ltd

SYNOPSIS
The recent success of the Californian Windfarms has demonstrated the feasibility of generating large quantities of electricity from the wind. It has brought a growing demand from the developers and electrical utilities for larger machines to improve land utilisation and simplify installaton and maintenance.

The James Howden Group have completed a significant wind park development in the Altamont Pass area of Northern California using 75 of the Howden 31 metre diameter, 330 kW wind turbine generators: they have also installed a 45 metre diameter, 750 kW machine and 10 - 15 metre diameter 60 kW machines.

The 330 kW machines are based on the Howden wind turbine generator on Orkney, but have been substantially modified for the different methods of installation and operation required in a wind park: the rotor size has been increased to 31 metres from the 22 metre diameter of Orkney, and the generator rating raised to 330 kW.

This paper describes these changes, the design of the 750 kW turbine and also the various stages in designing, constructing and commissioning a windfarm of this size.

INTRODUCTION

In the summer of 1983, James Howden & Company Limited erected a 22m diameter 300 kW wind turbine in the Orkney islands. The turbine had been specifically designed for island and isolated community markets and incorporated numerous features to ease installation and operation in this application. This installation has now generated for over 8,000 hours and produced over 1,000,000 kW hours.

Early in 1984, it became apparent that the major market for wind turbines in the next few years would be in California for windfarm application. After detailed studies of the market, it was decided that the successful Orkney design should be modified to more closely match the different operating environment found in California.

1 DESIGN OF HWP 330/31

1.1 Rotor Diameter

The Orkney turbine was designed for sites with high mean windspeeds. However the windspeeds encountered in California are generally lower, often in the region 6.5 to 7.5 m/s. The use of a 22m rotor with a 300 kW rating would not be economic in these sites, it was therefore essential to increase the rotor diameter to improve the energy capture. Our knowledge of designing and manufacturing wood/epoxy blades had also improved and it was decided that a rotor diameter of 31m would be feasable. This gave a rated output of 330 kW at 13 m/s. The blade construction used was basically similar to that used in Orkney with a laminated Khaya/epoxy leading edge and a foam sandwich trailing edge. The total rotor weight was 5.4 tonnes.

1.2 Drive train

The drive train of the 31m diameter turbine was similar to the Orkney design, however some changes were required due to the larger rotor and higher torque. The shaft diameter was increased to 240 mm as was the bearing size, split roller bearings being retained. The gearbox was produced by the same manufacturer but was of a larger size. The change in generator type meant that a fluid coupling was not required.

1.3 Electrical System

It was a requirement of the Orkney installation that a synchronous generator be used to minimise any disturbances on the local system. With turbines installed on a windfarm which is connected directly to a utility main high voltage line it was possible to use an induction generator which reduces the capital cost of the installation. Solid state protection relays were used in place of the standard utility induction disc type. All the electrical equipment was U.L. approved.

1.4 Tower

The hub height of the rotor was increased from 22m to 25m to increase the blade ground clearance and lower the natural frequency below blade passing frequency of the slower rotor.

1.5 Erection and Maintenance Procedure

When building a turbine on a remote site one of the major problems to be overcome was that of transportation and erection.

The turbine must be transported in small sections and erected without the use of large cranes. When building a large windfarm on a mainland site however, transportation was much simpler, with large cranes being readily available. It was also important to speed the erection and commissioning process by as much factory assembly and testing as possible. It was therefore decided to fully assemble and cable the bedplates at a factory near the site, which could then be tested and shipped as a complete assembly. The towers were assembled vertically and the nacelle assembly lifted into position.

2   WINDFARM CONSTRUCTION

In October 1984, Howden Wind Parks Inc., of San Rafael, California, was formed as a wholly owned subsidiary of the Howden Group PLC. The company was formed to develop windfarms within the United States of America using Howden Wind Turbine Generators.

By the end of November 1984, the subsidiary had set up its management and engineering team and was preparing to finalise a site acquisition for its first wind park development - Howden Wind Park I- in the Altamont Pass area of California.

The first development was on a 640 acre site consisting of 75 - 330 kW, 1 - 750 kW, and 10 - 60 kW wind turbines, with an operations and maintenance building and a 60 MW 21/230 kV substation inter-connecting to the grid of the local power utility - Pacific Gas & Electric.

After extensive negotiations in the latter part of 1984 and early 1985, a power purchase agreement with the local utility was obtained and all necessary planning permission, construction and grading permits approved by the respective local government offices. Ground-breaking for the project commenced in April 1985, with turbine construction and connection to the grid completed on schedule for sale to an institutional investor on December 20th, 1985.

The main pre-requisite for a wind park is an adequate wind resource. Final siting and optimisation of turbine spacing was dependant on:-

a) Long term wind data on site,
b) Extensive subsequent co-relation of wind data from central wind station to satelite stations,
c) Extensive local meteorological Tala Kite surveys to determine specific conditions - wind speed and direction - of potential sites.

This detailed information was then located on a three dimensional model of the site and final siting determined and subsequently confirmed by staking on site.

The site infrastructure necessitated over 11 miles of access roads, the majority of these engineered to allow for two-way heavy vehicular site traffic (ready mix concrete deliveries). The minor roads and the pads to each turbine required civil engineering commensurate with the machine size and lay-down requirements for construction. Major design consideration was made for soil conservation through water run-off control and high quality hydro-seeding after construction. Turbine foundation design was preceeded by a detailed soil investigation with borings and soil samples taken at each turbine site. These tests determined the reinforced concrete foundation design for the turbines. See Table 1 for details.

Power collection design parameters were set to have as many collection lines as possible underground. The final design incorporated two or three 330 kW turbines per step-up transformer - 480V to 21 kV - depending on siting and numbers, with underground collecting lines for these transformers surfacing at a central single overhead pole line collecting system from site to substation.

To facilitate erection, operation and maintenance of the wind park, a service building was constructed on site. This 2,600 square feet facility had sufficient space for local repairs, maintenance and spare part storage, management and service rooms and a central data collecting/turbine monitoring system.

Delivery of all turbine parts was scheduled for marshalling in two locations:

a) The Bedplate Assembly and test facility at Stockton - 20 miles from site
b) Direct to site for parts such as tower fabrications and blades.

The Stockton facility was capable of assembling and testing seven bedplates per week (the peak erection schedule) and manufacturing all wiring harnesses. On arrival at site, the nacelle covers were fitted to the bedplates in preparation for erection. Tower erection commenced with the setting of the control panels on the foundation, lowering base cone and erecting three tower sections. The fully tested bedplate was then located at the base of the tower and two of the three blades attached. The complete bedplate assembly was then lifted onto the top of the tower using a 120 ton crane and the third blade attached. Commissioning of the turbine commenced with the interconnect to the transformer and pre-commissioning checks.

Full commissioning was dependant on the wind resource being available.

The monitoring of the turbines and the collection of operational and wind data was effected using a radio transmitter to each of either two or three groups of turbines. The data was collected from the turbine or anemometer mast, fed into an interface card and the signals injected into the 480 volt power line. This data was subsequently retrieved at each of the 480V/21KV transformers and relayed to the operations and maintenance building using high frequency radio transmitters. With a large extensive site, this system replaced the conventional hard wired circuitry.

3 750 kW TURBINE DESIGN

As experience of designing and constructing wind turbines was gained it became apparent that the philosophy of using standard components could be extended to larger sizes. It was considered that this would offer potential economic advantage and a series of studies pointed towards a rating of 750 kW and a rotor diameter of 45m as a feasible next step. A brief system specification is given in Table 2.

3.1 Rotor

The three-bladed upwind rotor has a diameter of 45m and rotates at 30 r.p.m. 4 metre long blade tips are provided for speed control during start-up and shut-down and for power limiting. The tips are fail-safe, being held hydraulically in the run position against the action of a spring. The construction of the wood/epoxy blades is similar to that already developed for smaller machines. The cast S.G. iron hub is attached to the low speed shaft via a bolted flange connection.

3.2 Low Speed Shaft and Bearings

The low speed shaft is machined from a steel forging and is complete with a flange at the front for the hub attachment and a central passage to supply hydraulic oil to the blade tips. The shaft is supported in two split roller bearings to facilitate maintenance without disturbing the rotor gearbox.

3.3 Gearbox

The gearbox is mounted directly onto the low speed shaft with a shrink disc and has a planetary first stage and a parallel second stage. Oil cooling is by circulating pump and air blast radiator. Hydraulic oil for the blade tips is fed to the low speed shaft via a rotating hydraulic union.

3.4 High Speed Shaft

The high speed shaft is formed from a spacer-type gear coupling and provides a mounting for the new brake disc. The brakes are fail-safe and are capable of stopping from overspeed without the use of the blade tips.

3.5 Yaw System

The nacelle is supported on a crossed roller type slewing ring. The yaw system is powered by two geared hydraulic motors Three fail-safe yaw brakes are provided to lock the nacelle when not yawing.

3.6 Electrical System

The three phase induction generator produces power at 4.1 kV, which is fed to ground level via twistable cables. The main switchgear and instrumentation cabinets are situated in the base of the tower. Control of the turbine is by two industrial programmable controllers, one in the nacelle and one in the tower base. The controllers are linked by fibre-optic cable.

3.7 Tower

The tower is a tubular steel structure 2.5m in diameter mounted on a conical base. The tower is assembled via internal flanges. Internal ladders and platforms are provided for access.

The first HWP 750/45 was erected on the Altamont Pass windfarm in December 1985.

Two cranes were used, however it is important to note that they were standard construction equipment and that, apart from the spreader beam, no special equipment was required. The turbine is currently undergoing an extensive test programme to fully verify the operating parameters.

Table 1: Turbine Foundations

kW	SLAB SIZE	NO. OF PILES
750	32'sq x 5' deep	12 - 2'dia x 15'deep
330	20'sq x 5' deep	8 - 2'dia x 10'deep
60	12'sq x 4' deep	4 - 2'dia x 10'deep

Table 2

HWP 750/45 Altamont Pass
System Specification

ROTOR

Diameter	45m
Blade number	3
Position	Upwind
Material	Wood/epoxy composite
Control	Variable pitch tips

DRIVE TRAIN

Low speed shaft bearings	Split roller
Brakes	Spring applied - hydraulic release, on high speed shaft
Gearbox	2 stage planetary and parallel shaft

YAW SYSTEM

Bearing	Crossed roller slewing ring
Drive	Twin hydraulic motor
Brakes	Hydraulic fail safe

ELECTRICAL SYSTEM

Generator	4.1 kV, 3 phase, induction
Control	Twin industrial programmable controllers with hard wired back-up

TOWER

Material	Tubular Steel
Diameter	2.5 metres
Hub Height	35 metres
Access	Internal Ladder

Analysis of the WEG MS1 wind turbine generator performance using continuously logged data

R S HAINES, C WHITE, R McDONALD and D C QUARTON
Taylor Woodrow Construction Ltd

SYNOPSIS The scientific uses of regularly logged summary data are contrasted with investigations using detailed but short-term data blocks. Computer programs to facilitate the former approach are described. This is illustrated by the comparison of rotor power and generating losses in fixed and variable speed running, the investigation of blade resonance in variable speed running, the determination of the swirl angle behind the rotor, and the identification of transmission damping factors.

1. INTRODUCTION

Wind turbine generators (WTG's) present particular difficulties for experimental data collection. A forcing function with random or unpredictable variations over time scales from the momentary to the historic is combined with a cyclicly varying structure and significant non-linearities. Thus data collected over too short a period is likely to be unrepresentative, but too long a period is likely to cover too wide a range of operational conditions so that interpretation is obscured.

Perhaps the best bet is a mixture of long-term data logging and close-up studies of fast-moving events. Both facilities are available in the performance monitoring system installed by Wind Energy Group (WEG) at Burgar Hill, Orkney. Results from this system presented previously to the BWEA by Garrad et al (3) and Warren et al (4) have concentrated on the use of the facilities for short run studies. To provide a more complete picture, this paper will describe the context and use of the long-term logging facilities, and then present a number of applications of this data.

2. THE MONITORING SYSTEM

2.1 Background

The Monitoring system was commissioned in early 1984 and was used until 1986 to monitor the performance of the "MS1" WTG. The MS1 is an experimental horizontal axis machine designed by WEG. Its features have been described by Armstrong et al (1). Those of interest here include a 20m rotor, a flexibly mounted gearbox, and 250 kW rated output with the option of running in constant or variable speed mode.

The Monitoring System, itself comprises a Pulse Code Modulation based data acquisition system, a Data General Nova 4/X computer configured as a fast preprocessor, and a DG Eclipse S140 computer with peripherals for general data handling purposes. The specification has been described by Hassan et al (2). Over 100 channels of data were sampled for the MS1, at frequencies from 31 to 500Hz.

2.2 Facilities

The main forms in which data is recorded are indicated in Figure 1. Raw PCM data is recorded for back-up purposes only. "Campaign" data is the facility used for close-up studies, and typically constitutes a block of data at full sampling frequency from selected channels for a period around an event of interest. Long-term logging, on the other hand, occurs in four different ways, for each of which daily files are created.

Continuous cycle classification by the "rainflow" algorithm is applied to up to 20 strain signals for fatigue damage estimation. A very condensed outline of the machine and environmental state is provided by a "Status Vector" which is recorded whenever it changes. The "Daily Report" provides operational data in an accessible form with the aid of histograms, tables and text. Finally, "Periodic Summary Files" provide a database of summary information from a wide range of signals.

The campaign process is considered to be the basic tool for experimental data capture, and was used to obtain the previous results referred to (3, 4). The focus of this paper, however, is the use of Periodic Summary files.

2.3 The Periodic Summary Facility

A daily Periodic Summary File is produced for each of three summary periods, namely one minute, ten minutes and one hour. The block of data generated at the end of each summary period comprises up to a hundred elements calculated in various ways from the data input. (See Figure 1).

Unlike the Campaign and On-line Rainflow processes which have access to data at full sampling rate, the Periodic Summary

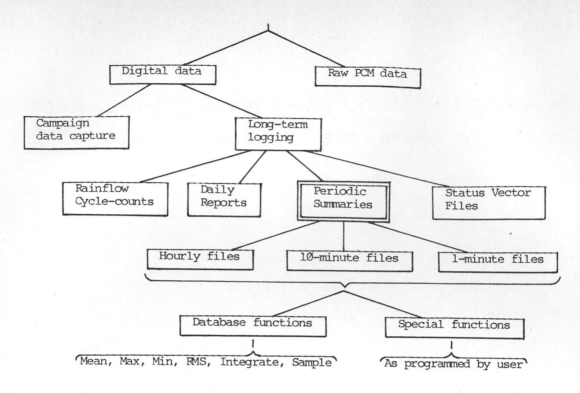

Fig 1   Types of data captured by the monitoring system

calculations are confined to a set of data extracted by the pre-processor and passed on in once-per-second packages. Such a package includes a single sample from each channel, and may include flags to indicate whether specified channels have exceeded pre-set threshold levels: the "Alarm" facility. Another useful element in the package gives the maximum and minimum value within the second of each signal handled by the On-line Rainflow process.

The calculations used to summarise this data fall into two categories. "Database functions" are preprogrammed functions selected from a menu and applied to the once-per-second samples, for channels that have been specified by the user in a conversational mode prior to the start of a period of monitoring. The results are automatically converted to physical units where applicable. "Special functions" must be programmed by the user in two parts, a part called once per second whch may access any element in the once per second package, and a part called at the end of the summary period. At the price of reduced convenience, the use of special functions adds greatly to the versatility of the Periodic Summary facility.

The commonly used database functions are mean, maximum, minimum and root mean square. The "integrate" function is typically applied to binary state signals, for example the contact breaker status is integrated hourly to give connect time. The "sample" function would be appropriate to slow moving signals such as temperature.

Two groups of special functions may be cited by way of example. One is applied to directional variables such as nacelle orientation or wind direction. To avoid misleading results if the received value jumps

by a revolution, a flag is set according to the range of the first value in a sampling period, and subsequent values within the period are adjusted to a consistent range before the mean or other summary term is determined.

The extreme strain values incorporated in the once-per-second package are utilised by four special functions. As has been stated, these values are the maximum and minimum within each second of the strain signals analysed by the on-line rainflow process. Since the rotor rotates nearly $1\frac{1}{2}$ times per second, the values will approximately correspond to the extreme values of one rotation. For each pair of values, the range (difference) is found and its mean and standard deviation over the summary period determined. The other two special functions are determined similarly from the mid-value of the two extremes. Thus a measure of the level and variability of both the cyclic and steady parts of the strain signal is separately provided.

3.   PROCESSING SOFTWARE

Each Periodic Summary file is in two parts, a header part defining the values that make up each data block, and a data part comprising one block of values for each summary period. For offsite use, selected files have been converted to ASCII text at site, on tapes which have subsequently been read by the Prime-based computer system at Southall. Recently, however, a small program has been implemented to convert binary values from Data General format to Prime format, so that now the files can be accepted by the Prime system in their original, much more efficient form. The next stage is to create a file in ASCII format comprising up to twelve columns of data extracted from the data blocks, one variable in each column. This is achieved by a program called "EXTRACT", which also provides the option of several commonly used combinations of variable: for example, linear combinations of blade strain signals to give in and out of plane bending moments. Output from EXTRACT may be listed out for preliminary review, or may be read by further programs, of which the most versatile is called MANIPULATE.

MANIPULATE is an interactive program with a wide range of simple functions developed to aid the evaluation of EXTRACTed data. The twelve columns of the workspace may be transformed, rather as in 'spreadsheet' programs, by applying simple one or two argument functions; (e.g. one column may be added to another). Rows may be selected or excluded according to chosen criteria (for example if the wind direction falls in a certain range); the excluded rows may be subsequently restored. Selected columns may be summarised (by finding the overall mean or the distribution in bins, for example). The relationship between two or more variables may be explored by multiple linear regression analysis, by averaging in bins of a chosen variable, or by constructing a two dimensional distribution table.

An example of the last tool is shown in Figure 2. A two dimensional array of bins is constructed according to the limits indicated in the automatic caption below, the columns being identified by mnemonics. The number of occurrences in each bin is shown by an integer in the range 1 to 999 inclusive. A dot represents no occurrences, and asterisks represent more than 999 occurrences. The extreme bins in each direction hold all occurrences falling outside the upper or lower limits specified. The limits may be specified by the user or chosen by the program. The effect is of a crude 'X-Y' plot, over which it has the advantage of being constructed at a VDU terminal in literally two or three seconds of elapsed time. The consequence is that a large number of possible relationships and/or data

selection criteria may be be tested and immediately evaluated in turn. The interpretation of the example shown is considered in Section 4.4.

Once presented on the screen, output from MANIPULATE may be duplicated to a 'hard copy' file, or prepared for a true plotting program.

4. APPLICATIONS

4.1 Generating losses

The first application to be considered is the determination of the overall energy loss between the rotor shaft and the electrical output, including both mechanical and electrical losses.

The mean power at the rotor shaft was estimated for these purposes from 10-minute summary data from three transducers, two giving the reaction load at the gearbox mountings and the third giving rotor speed. In determining the rotor torque from the gearbox reaction torque, a gearbox efficiency must be assumed, but because of the high velocity ratio a small error in this assumption has little effect on the total loss figure.

With the aid of MANIPULATE, the total losses were calculated and binned against the output power. The results are shown in Figure 3. Also shown are the predicted total losses according to component manufacturers' data, and a more conservative relationship assumed by WEG for certain design purposes. The actual losses are seen to follow a similar trend, and to represent a modest improvement on the design assumption.

A few points in this figure, highlighted by solid symbols, are seen to diverge from the general trend. The linear regression facility of MANIPULATE was used to estimate the departure from the local trend line and the discrepancy is plotted against time in Figure 4. This reveals the cause to be additional losses after starting from cold.

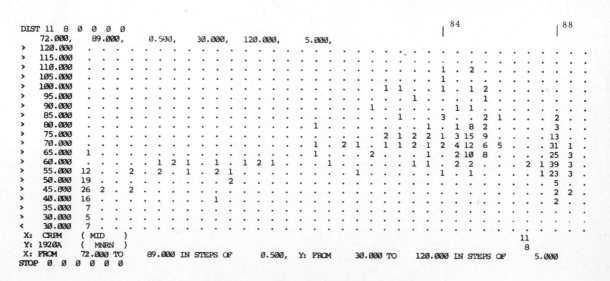

Fig 2 MANIPULATE distribution table: Mean cyclic amplitude of blade strain as function of rotor speed

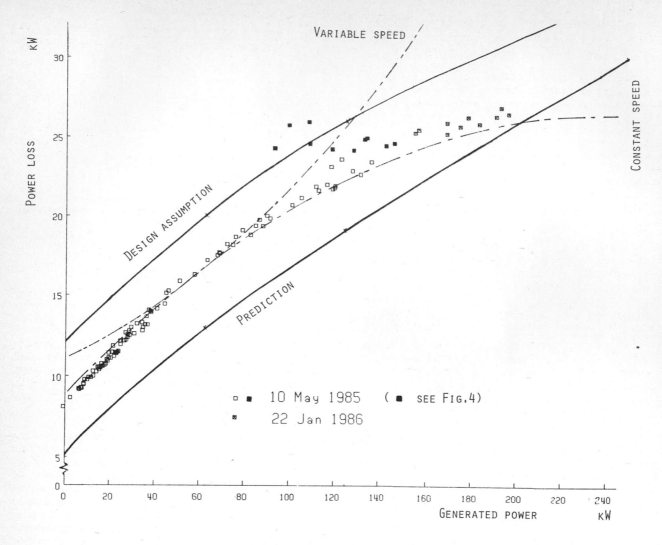

Fig 3   Power loss as function of power at generator

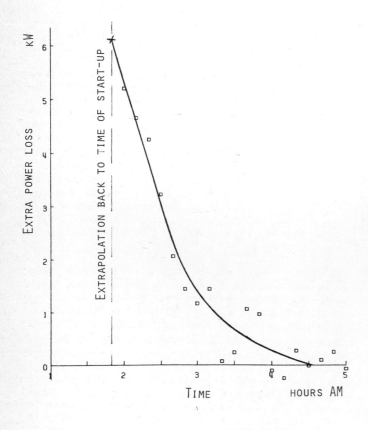

Fig 4   Decrease of power loss as WTG warms up

As an alternative to constant speed (synchronous) running, the MS1 can run in a variable speed mode, using a thyristor-based Power Conditioning Unit (PCU) to interface with the grid. The losses in this case were determined in a similar manner, and the quadratic relationship found by linear regression on the output power and its square is shown by the upper chain line in Figure 3. This may be compared with constant speed mode data from the same few days which is reduced to the lower chain line. As may be expected, the operation of the PCU increases losses at higher powers.

4.2 Power curves

Determination of the power/wind speed relationship is probably the most familiar application of continuously logged data in the field of wind power. In the case of the MS1 it is of interest to determine the additional power captured by the rotor when running in variable speed mode.

Because the amount of time in variable speed mode has been very limited for operational reasons, it is particularly necessary to exclude any bad data. It was decided for this purpose to exclude data points if the tips were active at any time during the summary period, if the average wind direction deviated by more than 10° from the nacelle axis, or if the wind direction was such that wake effects could distort the results.

This left insufficient points to give satisfactory definition using ten minute mean values, so one-minute files were used. This also has the advantage of reducing the distortion of non-linearities, because the likely variation of the windspeed within the summary period is reduced. One-minute mean data is not available for the gearbox reaction signals, so one-minute means of output power were used and corrected using the power loss relationships indicated in Figure 3 (chain lines).

Results from four periods of running in variable speed mode were used, comprising about five hours in total before screening. The calculated rotor power was averaged in bins of hub-height wind-speed to give the broken curve in Figure 5. The integers in round brackets against the points indicate the number of values in each bin.

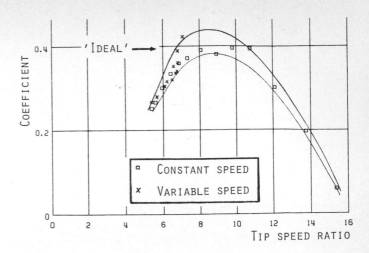

Fig 6   Coefficient of performance

temperature changes or long-term signal drift. Using the same data selection criteria and averaging method as before, the full line shown in Figure 5 was obtained. The integers in square brackets indicate the number of values in each bin.

It is seen that variable speed running increases the power at the rotor for wind-speeds below 8m/s, but reduces it in higher winds. To understand this behaviour it is helpful to transform the measured constant speed performance curves to the dimensionless form of Figure 6, where the abscissa is the coefficient of performance : square symbols for constant speed mode, crosses for variable speed. This shows that optimum performance is obtained with a tip-speed ratio in the range 7 to 11. If this ratio were maintained, in wind-speeds below 11.5 m/s, the "ideal" power:wind speed relation shown by the chain line in Figure 5 would be achieved.

In practice, the control law for variable speed running is set to give a rather lower tip-speed ratio, partly because lower shaft speeds lead to slightly lower generating losses, and partly because there is a trade off with the quality of power control. The effect is shown in Figure 7, where the actual variation of tip speed ratio with windspeed is shown by a broken line, based on the same summary data sets as the previous figures. Because the ratio is generally below 7.0, a power output rather below the "ideal" curve would be expected, and in general this is borne out in Figure 5.

In winds above 11.5 m/s the rotor speed is the limiting factor. To maintain the design speed limit without the benefit of direct connection to a stiff grid it is necessary to control the speed about an average of 85 rpm in contrast to the synchronous speed of 88 rpm. Because the small reduction in tip speed ratio takes effect on a sensitive part of the characteristic (see Figure 6), the power in variable speed mode would be expected to fall below that in constant speed mode for winds up to 16 m/s, consistent with the results in Figure 5.

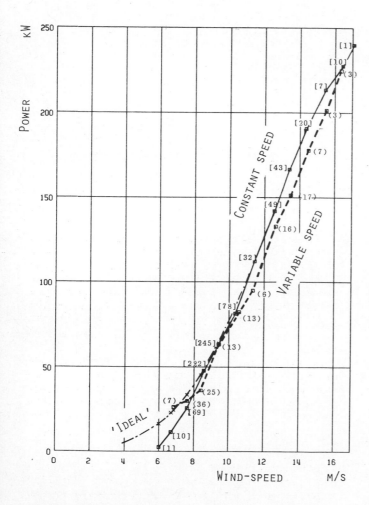

Fig 5   Rotor power versus wind-speed at constant and variable speed

The constant speed mode data used was taken from periods immediately preceding or following the periods of variable speed mode running, comprising about 24 hours before screening. By this means it was hoped to eliminate errors that might arise from

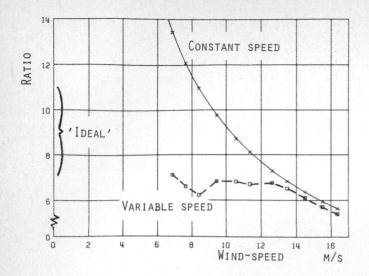

Fig 7   Tip speed ratio versus windspeed

For the purposes of practical comparison the greater generating losses in variable speed mode must also be taken into account. Figure 8 shows the measured performance curves for electrical power generated. The chain line is derived from the chain line in Figure 5 with allowance for the generating losses. Even if this line were followed at the lower wind speeds it is clear that for a site with an annual mean wind speed of 9.6 m/s hub height, the energy lost in variable speed mode in high winds easily outweighs the extra energy captured in low winds.

4.3   Gearbox mounting damping factor

The WEG MS1 has a gearbox which swings about the shaft centreline and is restrained by two springs with viscous dampers in parallel (see (Ref. 3). Each damper acts through a load cell and the swinging motion is recorded by a velocity transducer. As the relationship between the velocity and damping force is a dynamic one, the use of summary data to monitor damper performance may seem, at first sight, surprising.

On reflection, however, it will be recognised that the maximum and minimum values in a summary file are in fact "spot" values, and in a correctly functioning damper the extreme load values should correspond to the same sample as extreme velocity values. Thus provided that the relationship between the extremes is a consistent one, it can be taken as representing the damper characteristic.

Figure 9 shows the relationship between the minimum velocity and the maximum load, and also between the maximum velocity and the minimum load, compared with the nominal design

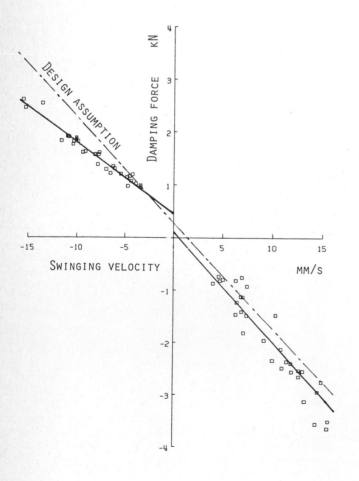

Fig 8   Electrical power versus wind-speed at constant and variable rotor speed

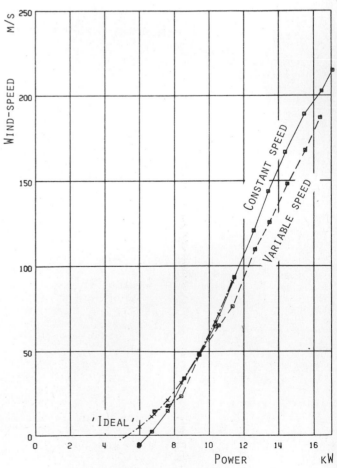

Fig 9   Gearbox damping law

relationship, shown by a chain line. The vertical axis shows the sum of the loads at the two dampers, a preliminary check having confirmed that these were very similar. Although there is some scatter, a clear relationship is apparent. There is seen to be a degree of imbalance between the two restrictors which respectively control forward and reverse swings, but each gives a damping effect acceptably close to the target value.

The offset between the two halves of the characteristic gives an indication of the level of sliding friction in the dampers. The zero on the load scale has no significance as it is known to be subject to thermal drift.

4.4 Blade resonance

The first flatwise natural frequency of the MS1 blades has been measured at 5.6 Hz under running conditions. In variable-speed mode this would be expected to give rise to a resonance excited by the fourth harmonic of the rotational speed centred on a speed of 84rpm. The question arises whether this has a significant effect on blade strains.

This is another application where the use of summary data is not an obvious choice. However, the special functions described in Section 2 do offer potentially relevant information. In Figure 2 the special function used is the one-minute mean of the one-second range of a strain signal 'R-SG19&20A', which gives mainly flatwise bending strain at the 33% radius on Blade A. The table gives the distribution of this parameter against the shaft rpm during a period of mixed constant-speed and variable-speed running, data being excluded where the mean power exceeded 200 kW.

The increase in flatwise bending amplitude as the speed increases to 84 rpm is clearly visible, although there is also considerable scatter attributable to turbulent effects etc. A reduction of amplitude above 84 rpm can also be seen. Constant-speed mode running corresponds to entries in the five right hand columns.

It is concluded that the resonant peak gives appreciably larger bending cycles at 84 rpm compared with 88 rpm, although the increase is not dramatic and within a short time period such as a single minute may be swamped by other effects.

4.5 Aerodynamic induction factor

The wind yaw sensors mounted behind the rotor give a signal which in general diverges from the yaw as calculated from the free-stream wind direction sensor and the nacelle orientation sensor. This divergence is partly due to imperfect correlation between the windspeed at the mast and at the WTG, and partly due to deflection of the stream on passage through the rotor disk – the "swirl angle" which appears as a significant term in rotor aerodynamics. In averaged data the first cause should cancel out so that the mean yaw sensor "error" can be taken as a measure of the swirl angle.

Elementary in principle, the satisfactory determination of the swirl angle in this manner poses a number of practical difficulties, not least the poor signal-to-noise ratio arising from the subtraction of wind-vane signals. In the present case an additional problem is introduced in that the signal from the nacelle-mounted sensor is received via the WTG control system, which applies a moving average to the signal. Thus this signal tends to reflect conditions some minutes earlier, whereas the other signals are current.

A number of techniques for counteracting this distortion of the data have been tested with the aid of MANIPULATE. One simple technique is to include only data where the recorded mean yaw signal for the current minute is little changed from that in the previous minute. When this is the case, the yaw during the current minute must be close to the current moving average value.

Using this technique to select data from a group of one-minute summary files, the swirl angle plotted against wind speed is shown in Figure 10. The mean values in bins of wind-speed (broad line) and the relationship predicted by a steady-state aerodynamic blade element theory program (broken line) are also shown. It is seen that there is a scatter band a few degrees wide and widening at the lower wind speed end. This is the expected effect of

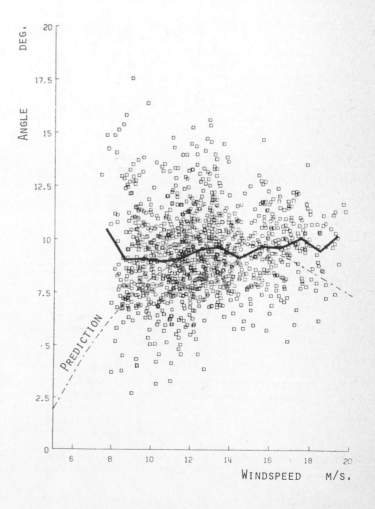

Fig 10   Swirl angle at nacelle wind vane

the imperfect mast: WTG correlation. Considering the binned means, it is seen that the swirl angle at intermediate wind speeds is well predicted, but the expected variation with wind speed is not confirmed.

4.6 Other applications

Some further examples will be noted briefly. The special functions described for rotor strain, considered in Section 4.4, are currently being used in the evaluation of the effect on cyclic blade strains of immersion in the wake of another WTG. Simple means of blade strain over ten minute periods have been used for the validation of blade aerodynamic performance, as described by Garrad et al (5).

The latter reference also describes the evaluation of axial induction factor. Finally, a set of special functions that classify variations in the grid voltage have been used to assess the effect of WTG running on the network quality.

5. CONCLUDING REMARKS

The examples given here illustrate the versatility of continuously logged summary data, and also the special measures that are often necessary to extract good quality information form the recorded data. The approach is a convenient one, in that once the channels and summary functions are defined attention can be concentrated on the operation of the WTG. The actual periods of data to be used can be chosen at a later date, with the benefit of hindsight. Data collected under an assortment of different operating conditions can efficiently be pooled so that secondary effects such as machine temperature can be eliminated by averaging, or isolated as in the power loss study.

On the other hand, there are certainly many applications for which summary data is not enough, most obviously those involving analysis in the frequency domain, or the monitoring of a sequence of events such as a shut-down. Moreover, continuous summary data from all channels would be extremely unwieldy; data from channels not chosen for inclusion must perforce be collected on selected occasions, by the Campaign process.

Turning to the applications presented above, the conclusions may be summarised as follows. For the site and machinery concerned, the increased energy capture by the rotor obtained by running at variable speed in lower winds is outweighed by the increased energy losses. Variable speed running also increases the cyclic rotor strains in strong winds because of a minor resonant magnification of the strains at a speed below the synchronous speed. In constant speed running generating losses and transmission damper behaviour have been identified and acceptable agreement with design assumptions confirmed. Predictions of the rotor swirl angle in medium winds have been supported, but the predicted variation with wind speed has not been found.

ACKNOWLEDGEMENTS

The authors are pleased to acknowledge the support of the UK Department of Energy and the directors of WEG for their financial support and their permission to publish this paper. (WEG is the Wind Energy Group, a consorium of Taylor Woodrow Construction Ltd., British Aerospace plc., and GEC Energy Systems Ltd.).

REFERENCES

(1) ARMSTRONG, J.R.C., KETLEY, G.R. & COOPER, B.J. The 20m diameter wind turbine for Orkney. 3rd British Wind Energy Association Conference, Cranfield, UK, April 1981, pp 54-62.

(2) HASSAN, U., HENSON, R.C. & PARRY, E.T. Design of a microcomputer based data acquisition and processing system for the Burgar Hill, Orkney, wind turbines. 4th International Symposium on Wind Energy Systems, Stockholm, Sweden, September, 1982, 2, pp 137-152.

(3) GARRAD, A.D., HASSAN, U. & LINDLEY, D. Monitoring results and design validation of the Orkney 20m wind turbine generator. 6th British Wind Energy Association Conference, Reading, UK, March 1984, pp 76-84.

(4) WARREN, J.G., QUARTON, D.C. & LINDLEY, D. An evaluation of the measured dynamic response of the Orkney 20m Diameter WTG. 7th British Wind Energy Association Conference, Oxford, UK, March 1985, pp 259-268.

(5) GARRAD, A.D., QUARTON, D.C. & LINDLEY, D. Performance and operational data from the Orkney 20m diameter WTG. European Wind Energy Conference, Hamburg, W. Germany, October 1984, pp 170-176.

Estimation of the fuel savings by wind energy intergration into small diesel power grids

T D OEI

SYNOPSIS   Fuel consumption studies were performed by a computer model for small diesel stations for various degrees of wind energy penetration. For given load duration curve of the station and power duration distribution of the wind plant, fuel consumption has been calculated with a diesel station of various power mix. Further it is assumed that at least one diesel unit remains connected to the bus to serve the load. Spinning reserve at various levels of the demand is defined in relation to the power mix of the station. Required connected capacity to the bus at various load levels of the station takes account of this spinning reserve and the power variation of the wind plant. By further consideration of the Willans equation for the diesel fuel comsumption, properties of this equation show methods for improved fuel displacement. At low penetration of the wind energy, only an amount of fuel equal to the marginal fuel consumption of the unit on load frequency control is displaced. With increasing wind energy penetration, more than this amount may be gained by reducing the no-load fuel consumption, provided the power mix of the station is suitably chosen. To further increase the wind energy penetration, in particular for stations with higher peak to base load ratios, an amount of excess wind energy may be used to drive the diesel engines to overcome the no-load resistance of the engine. By externally supplying this power, we directly influence the throttle action of the machine while power output is maintained.

1   INTRODUCTION

Many are the questions that still need an answer for the successful integration of wind energy into small isolated diesel power stations. Among these is the question of the effectiveness of wind energy as a fuel saver in such a system [1,2,3,4,5,6]. To investigate the efficiency of such plants and their operational boundaries, estimates of power demand and total energy requirements are needed. Unfortunately such data are not easily obtained.

In literature, fuel saving or displacement figures are given for some test plants [1,2,3,4] but no uniform way of assessing the amount of fuel saved has been given. This paper is an attempt at estimating the fuel consumption of diesel stations, with and without the wind contributions, taking into account the load curve as well as the power mix of the diesel station. Throughout, the model assumes at least one diesel unit on load and with minimum load setting for this unit. For the calculation with the wind energy, the power duration curve of the wind plant is required.

As longer time scales are considered, ideal control of the system is assumed, while diesel starts and stops fall outside the scope of this study. Simulation models using time scales of the order of seconds or less are more appropriate for such studies. Calculated fuel displacements are therefore optimistic estimates, or upper limits of what may be reached with wind diesel systems.

Fuel economy of small stations, with or without wind, depends on parameter setting such as allowed minimum spinning reserve. Control of spinning reserve to follow the load variation at various levels of the load is determined by the mix of the installed power. In order that we draw conclusions which are pertinent to the system, only small stations of 1 to 2 MW and less, consisting of up to 6 units with maximum unit rating of 600 kW, will be considered in this paper.

2   THE LOAD

Small power stations in remote areas have only residential and commerical loads. Many remote areas with prospects for wind diesel systems, are holiday resorts, so the commercial loads in these areas are highly seasonal. As the population then may rise threefold or more, the peak load of the station will coincide with the high season lasting for a month or two. Off season, the load is mainly residential. As life in small villages is less regular than in suburban areas, day to day load variation for the same clock time between 8 AM and 5 PM may be expected to be large for these small power stations. Loads in the evening hours will show less variations from day to day as they are more determined by habit like watching T.V. and cooking and further illumination to follow the cycle of the day. Street lighting and refrigerators if they are in use, finally form the greater part of the base load occurring in the late night and early morning hours and with greater constancy. This is illustrated by Fig. 1. The curves are composed from measured data of the station in Kythnos, Greece [7], for three winter days.

We may conclude from the above that the peak to base load ratio of these small power stations will be high when heating and airconditioning purposes are excluded (this is true only outside the USA). For the Greek islands this ratio varies between 5:1 (larger islands) and 15:1 (smaller ones) [8].

In order to prevent too low loadings of the diesels which is detrimental to diesel life and efficiency, the installed capacity of the station with high peak to base load ratio is split into two or more units. Depending on the value of this ratio, the ratings of the units should be chosen such as to facilitate the scheduling of the units on load at sufficiently high powers. Estimations of the power swings around identified mean demand levels of the load is then a means to determine values for the spinning reserve at these demand levels. This will be further explained in the following sections.

Diesel stations with low peak to base load ratios, about two and less, need special consideration. For these stations the load can be served by a single machine. By definition the fuel saved by moderate introduction of wind energy only amounts to the marginal fuel consumption (see section 4) of the diesel engine. Massive introduction of wind energy and on/off control of the diesel engine is required to save larger amounts of fuel than marginal.

3 DAILY LOAD CURVE AND POWER MIX

By rearranging the load pattern of Fig. 1 into one with decreasing order of magnitude one obtains the load duration curve of Fig. 2. In this figure the largest day-to-day variation of the load between noon and 5 PM is still visible and at the tail of the curve is the minimum load during the night and early morning. So the curves indicate that the nature of the load variations depends on the level of the load and time of day. The calculated minimum mean load

for these three days and for the last 7 hours of the curve of Fig. 2 is 94 kW with maximum variation of 19 kW around this mean. Should more data become available for analysis, then the frequency by which the mean value is exceeded by more than a certain absolute value may be counted. From the corresponding frequency distribution and reliability requirements, criteria may be drawn with regard to the required connected capacity to serve the load of certain mean value.

Suppose the forecasted variation of 19 kW in the above figures satisfies the reliability criterion for this mean load, then the corresponding required connected capacity is 113 kW. To round off the power figures, two machines of 60 kW each may be chosen. The maximum load measured in Fig. 1 is 370 kW. Suppose again this load is the actual peak load of the system, then the total required installed capacity is 370 kW plus a certain safety margin, say 30 kW. So a total of 400 kW needs to be installed. Two times 60 kW are to serve the base load and thus another 280 kW of capacity is still required. If one unit of 280 kW is chosen, then connected capacity varies step-wise according to 120-280-340-400 kW. If two units of 140 kW are chosen then the connected capacity varies as 120-140-200-260-280-340-400 kW. From Fig. 2 and these capacity combinations, mean demand levels and their variations can then be constructed as shown in Fig. 3 and 4 for the two systems. It may be expected that the system with four units will have the lowest fuel consumption as all units will on the average run at higher power and therefore at higher efficiencies. These load curves and these two fictitious power systems will be used for example calculations in section 8.

4 THE DIESEL GENERATOR

Many investigators in the field of wind energy have measured the fuel consumption of diesel generators as a function of the load [1,2,3,4,9]. All experiments prove the adequacy of the Willans line - originally used to represent the amount of steam required by a steam turbine as function of load [10] - to represent the fuel consumption at various loads of the generator. The Willans line representation of the fuel consumption is given by the equation:

$$F = a(P) + b(P).X, \tag{1}$$

where F = fuel flow in kg/hr and the thermal energy equivalence of the diesel fuel is 11.9 kWh/kg

a(P) = no-load fuel consumption of a machine with rated power P, kg/hr

b(P) = slope of the curve of a machine with rated power P or marginal fuel consumption, kg/kWh

X = load, kW.

Henceforth we will call b(P).X the power rated fuel consumption and b(P) the marginal fuel consumption of the unit. As almost no speed variations are involved in the operation of a diesel generator, the amount of fuel consumed per kWh electricity produced with fluctuating loads around a mean is the same as at constant mean load. By the linear relation between the fuel flow and the load, the specific fuel consumption is thus only determined by the average power at which the machine has been operating during the period considered. For small stations with a few units and limited number of power combinations, a discrete mean load duration curve as shown by Fig. 3 and 4 is very suitable. Fuel consumption calculation is then a matter of a few multiplications and summations only.

Let us further investigate what can be learned from the Willans equation on the sizing of the constituting diesel units. At full power of the diesel generator the ratio between no-load to power rated fuel consumption roughly varies between 1:2 and 1:3. Suppose a fraction η of the power rated fuel consumption is converted into useful power and a fraction (1-η) is the energy lost through the exhaust and the cooling system, then at full power the thermal efficiency of diesel generators varies between (2η/3) and (3η/4). Assuming equal division of the power rated fuel consumption into useful power and the energy in the exhaust, the ratios then correspond with thermal diesel efficiencies between 33.3 and 37.5%, yielding specific fuel consumption of 260 and 230 g/kWh respectively. The no-load fuel flow a(P) of naturally aspirated diesel engines of various powers in fact is almost linearly proportional to the power rating of the machine, while b(P) is nearly independent of P. So, by replacing a diesel generator while operating at power X by another unit which is n times smaller (n>1), but still with power P>X, operation at the same power yields a reduction of the no-load specific fuel consumption by an amount approximately equal to

$$\frac{\Delta F}{X} n\ell \approx \frac{a(P)}{X} \left(1 - \frac{1}{n}\right) \quad \text{[kg/kWh].} \qquad (2)$$

For X is 50% of rated power P, replacement of a unit by another with half its rated capacity will reduce the specific fuel consumption by an amount $a(P)/P$ kg/kWh which in reality is between 70 and 100 g/kWh. This amount rapidly rises when units are allowed to operate at powers below 40% of their rated capacity.

Introduction of wind energy will tend to intensify the load variation of the station with higher installed powers of the wind station. At low installed wind powers and without consequence for the scheduling scheme of the diesel units, the displaced amount of fuel per kWh wind energy is equal to the marginal amount $b(P)$ of the unit on load frequency control [2,6]. As more wind energy is introduced an additional amount, proportional to $a(P)/X$ per unit wind energy, is displaced during the time a unit with rated power P and operating at power X can be switched off or replaced by a smaller unit. To have full advantage of this additional fuel gain, frequent on and off switching of the units is required and unit sizes have to be carefully chosen with regard to the load pattern of the station and the supply pattern of the wind station as well.

As still more wind energy is introduced, then at times of low demand and high supply of wind power, part of the wind energy has to be dissipated. In order to increase the saturation level by which wind energy can be absorbed, an amount of excess wind energy may be used to drive the diesel engine to overcome the no-load resistance. By externally supplying this power we directly influence the throttle action of the machine while power output is maintained, see also [3,11]. For each kWh of wind energy used this way, about $90/\eta$ g of fuel is saved.

5 THE WIND TURBINE

For the estimation of the fuel displacement, two kinds of calculations need to be performed. One is the reference calculation without the wind, by which the influence of the scheduling scheme and the size of the constituting units on the fuel consumption can be studied.

For the calculation with the wind contribution, a power duration curve of the wind station is required. The construction of such a curve requires the power characteristics of the turbine and the wind speed distribution at hub height, at the turbine site. An example of the respective histograms of the turbine power characteristics and the wind speed distribution is shown by Fig. 5. Fig. 6 shows the wind power duration curve obtained from the two curves of Fig. 5. The dotted line of Fig. 6 shows the same wind duration curve but with lesser resolution. For regions with reliable wind data and together with detailed turbine specifications, such curves can be calculated with the code WEPP [12].

We further assume that for each hour of the day the same probability exists for a certain wind speed to occur and when more than one turbine is installed, the effect of dispersed sitings of the turbines is taken into account at the determination of the wind power duration distribution of the turbine cluster.

In Fig. 6 the ratio $b_j = T_j/T_{tot}$ is thus the probability for wind power W_j to occur. We further assume in our model that a minimum of one diesel unit is on line and that the load meets the minimum load requirements P_{min} of the unit. This means that during low loads and high supply of wind power an amount of wind energy has to be dissipated in order that the diesel load does not drop below its minimum allowable value P_{min}. In other words, during demand D_i with probability p_i and wind supply W_j with probability b_j the diesel station has to supply:

$$X_{ij} = D_i - W_j \quad \text{[kW]} \qquad (3)$$

$$\text{during } T_{ij} = p_i \cdot b_j \cdot T \quad \text{[hr],} \qquad (4)$$

with the constraint $X_{ij} \geq P_{min}$.
For $D_i - W_j < P_{min}$ then an amount of wind energy is dissipated with power

$$W^W_{ij} = W_j - (D_i - P_{min}) \quad \text{[kW]} \qquad (5)$$

during

$$T_{ij} = p_i \cdot b_j \cdot T \text{ [hr]; otherwise } W^W_{ij} = 0. \quad (6)$$

The total energy production of the diesel units is calculated from

$$E^d_{tot} = \sum_i \sum_j X_{ij} \cdot T_{ij} \quad \text{[kWh]} . \qquad (7)$$

The amount of dissipated wind energy is

$$E^W = \sum_i \sum_j W^W_{ij} \cdot T_{ij} \quad \text{[kWh]} . \qquad (8)$$

6 DIESEL SCHEDULING

The scheduling of the diesel units is now done stepwise and for the cases without wind, possible scheduling schemes are shown by fig. 3 and 4. For each step i of the histogram, the connected capacity should thus satisfy the relation

$$\sum_k P(k) \geq D_i + \Delta D_i \quad \text{[kW],} \qquad (9)$$

with $P(k)$ = rated capacity of diesel unit k. The summation of eq. 9 extends over the K units on load during period i (e.g. in the example of Fig. 3, K varies between one and three). Suppose $X(k)_i$ is the average power diesel unit k produces at step i, then the fuel consumption of all units over the period considered is

$$F_{tot} = \sum_k \sum_i [a(k)+b(k) \cdot X(k)_i] \cdot T_i \text{ [kg]} . \quad (10)$$

where $a(k)$ and $b(k)$ are abbreviations for $a(P(k))$ and $b(P(k))$ respectively.

For the determination of the connected capacity in the situation with wind, estimations of wind power variations at each wind speed interval have to be made. At lower mean values of the wind speed, high relative turbine output variations may be expected as turbine output varies with the third power of the wind speed. In the wind speed class between 4 and 5 m/s, the relative power variation $\Delta W/W$ is around 100% while mean power output is about 10% of rated value. At wind speeds greater than rated speed, less power variation may be expected as turbine control, either active or passive, will damp the power fluctuations due to gusts. We may estimate power fluctuations $\Delta W/W$ of 25% in this region of the wind speed, to match with the overload capacity of the generator. In this case 75% of rated capacity is firm capacity during wind speeds above rated and may replace the same amount of connected diesel capacity. For most wind turbine locations the variance of the mean wind speed v is proportional to the square of the mean speed. In this case, the relative variation $\Delta W_j/W_j$ then varies linearly with W_j. For wind speeds between 4.5 m/s and v_{rated} then $\Delta W_j/W_j$ is found by linear interpolation.

Now at demand D_i with variation ΔD_i and wind power W_j with variation ΔW_j the required minimum connected diesel capacity will be:

$$\sum_k P(k) \geqq D_i - W_j + (\Delta D_i + \Delta W_j) \quad [kW]. \quad (11)$$

At low wind speeds with $\Delta W_j \approx W_j$ the required connected capacity is thus the same as for the case without wind.

Fuel consumption over the period considered is now calculated according to

$$F_{tot} = \sum_k \sum_j \sum_i [a(k) + b(k).X(k)_{ij}].T_{ij} \quad [kg]. \quad (12)$$

Here, $X(k)_{ij}$ is the average power diesel unit k produces during demand D_i and wind power W_j. When excess wind energy is used to drive the diesel engines for which an amount of power W_{ij}^w as given by eq. (5) is available, the re-reduction of the fuel consumption of diesel unit k at step ij is:

$$\Delta F(k)_{ij} = a(k) . T_{ij} \quad [kg] , \quad (13)$$

for

$$W_{ij}^w \geqq c.\eta.a(k) \quad [kW] , \quad (14)$$

that is, when available excess wind power exceeds the no load power of the unit. When less wind power is available the equation becomes:

$$\Delta F(k)_{ij} = \frac{W_{ij}^w}{c.\eta} . T_{ij} \quad [kg] . \quad (15)$$

with

c is the thermal energy equivalence of 11.9 [kWh/kg].

7 COMPUTER MODEL

The programme WIN(d) DIE(sel) S(ystems) is written in BASIC for an Apple IIe computer with 64 K memory. An Epson printer is required since formatted data output makes use of the printer's ROM.
The precaution against memory overflow due to the large number of variables used, led to the following restrictions. The number of steps used to represent the mean load duration curve and the wind power distribution is limited to ten for each with the constraint that their product does not exceed sixty steps. Maximum diesel unit input is also limited to ten.

Four parts distinguishing the programme structure are: input, calculation of eq. (3) and (4) of section 5, diesel scheduling input

and final output. Programme input is divided into diesel specification input, mean load duration and wind power distribution input. For the determination of the power variation of the wind station at various powers, four input data are required i.e. the maximum and minimum relative values of the variation and the powers for which these values are valid. A maximum of 100% at 10% of rated power and minimum relative variation of 25% at rated power of the wind station seem to be a good guess for this input. Scheduling input for the diesel units accounts for number of diesel units K, used at each step of the load duration curve and so many diesel unit numbers k, distinguishing the units and scheduled power of each unit. Reduction of the number of power steps used in both duration curves, makes this input less tedious. Graphic output is for all relevant duration curves. Printer output in table form for each power step is for various quantities i.e. required connected capacities, truely connected capacities, average diesel efficiency, fuel consumption, energy produced etc. Individual diesel production, efficiencies, operation time, mean load, fuel consumption, fuel saved by driving are also printed in table form. Finally, integral quantities of interest are printed.

8 APPLICATION

Calculations were performed for two fictitious diesel stations with three (system A) and four (system B) diesel units as of section 3 and corresponding mean load duration curves as shown by Fig. 3 and 4 respectively.
Table 1 lists the input of the specifications of the diesel units of various capacities. Fuel

consumption characteristics are estimations by the author from diagrams shown in [13] with manufacturers data given for stripped engines. No-load fuel efficiency η is 50% for all engines assuming equal distribution of useful power and energy in the exhaust [10].
Table 2 lists the values of the mean load duration histogram of Fig. 3 and table 3 the same as of Fig. 4. The wind power distribution for two resolutions as shown by Fig. 6 is given by table 4 and listed for nominal wind power value of 100 kW. Wind speed distribution of Fig. 5 is calculated for 20 m hub height and Weisbull distribution with shape parameter of 1.45 and mean wind speed of 6.8 m/s measured at 10 m height. For varying nominal powers of the wind station, the same power duration distribution is assumed. Calculations were performed for a period of one year and total energy demand is 1645 MWh. For all calculated cases a minimum diesel load of 30 kW is valid.

9 RESULTS

The reference calculations without wind show negligible difference in fuel consumption between system A (446 ton/yr) and system B (440 ton/yr). So the specific fuel consumption is around 269 g/kWh for both systems. Results of the calculations with various levels of installed wind power is summarized in table 5. The amount of fuel saved by driving the diesel engines is listed in the column under the heading "fuel extra saved". By splitting the 280 kW unit of system A into two equal units an averaged improvement in fuel saving by 8% is obtained as shown by table 5. The fuel displaced per kWh of wind energy absorbed by the grid versus the installed wind power is plotted in Fig. 7. This is shown by curves F_A and F_B. Curves F_A' and F_B' represent specific savings when excess wind energy is used to drive the diesels and amount to 224 g/kWh for system B at 200 kW installed wind power. The fuel savings per unit installed wind power is shown by curves L_A and L_B. Decline of curves L_A and L_B starts at 64 kW wind power, being the difference of the minimum mean load of 94 kW and the minimum diesel load set at 30 kW. Curves L_A' and L_B' again represent the cases for wind driven diesel engines. At installed wind powers exceeding 126 kW not all excess wind energy can be absorbed to drive the two 60 kW diesel engines. At 140 kW wind power, about 8 MWh wind energy has to be dissipated which rapidly rises to 61 MWh at 200 kW. This dissipation of wind energy then causes the curves L_A' and L_B' again to decline.

At these values of installed wind power, on/off control of the last diesel unit may save an extra amount of fuel by 170 g/kWh, but more wind energy which has been used to drive the diesels has then to be dissipated see table 5. For system B and for installed wind power of 200 kW, operation times of the two bigger diesels reduce by 2260 hrs compared to the reference case while increase of the operation time of the smaller units amounts to 1690 hrs. This means that by increasing wind energy penetration the load is more and more taken up by the smaller units. In order that on/off control becomes effective at relative low levels of wind energy penetration, the smaller

units have then to be replaced by a bigger one and more wind power has to be introduced than is assumed in these calculations.

A plot of the fuel consumption versus the degree of wind energy penetration is shown by Fig. 8. For the highest effectiveness of wind energy as a fuel saver curve R_B' shows that at least a penetration of 17% of wind energy is required. Almost 1:1 relation between reduced fuel consumption and wind energy penetration is observed. However, beyond 24% of wind energy penetration, the required installed wind turbine power increases more than linear. The influence of the scheduling scheme of the diesel units has been investigated for system B and with 120 kW of wind power. If no capacity value is attributed to the wind turbines then the scheduling scheme of the diesels has to follow the same scheme as if there were no wind, i.e. the reference case. In that case, an increase of fuel consumption by 10 ton/yr is calculated. Finally, to investigate the influence of the resolution of the wind duration distribution of Fig. 6 on the calculational results, comparison of the results with both curves for system B and with 120 kW wind power, show negligible difference for all calculated quantities. This means that for system B and with connected capacities varying as shown in section 3, the coarser resolution of the wind duration distribution suffices.

10 CONCLUSIONS OF CALCULATIONS

At moderate peak to base load ratios as is assumed in this exercise, reference calculations for system A and B show no further use of

splitting up the 280 kW unit for the diesel station as such. As wind energy is introduced, fuel savings for system B are about 8% higher than for system A irrespective of the degree of penetration. The absolute amount of fuel saved to justify the cost of the extra unit of system B therefore requires a minimum level of wind power contribution. Fuel savings due to spinning reserve limitations amount to about 10 ton of fuel per year. However automatic control of the operation of the station will be indispensable to adjust connected capacity at each combination of demand and wind station output. Diesel driving improves the specific fuel savings significantly as wind energy penetration can increase by about 11% or more while wasted wind energy is obliterated. Many small effects in wind diesel operation add to the effectiveness of wind energy as a fuel saver in small diesel grids. To come to that end, good engineering and control, meeting this special purpose, is indispensable. Fuel savings around 30% can then be obtained with only a slightly higher percentage of wind energy contribution.

ACKNOWLEGDEMENT

Programming support by R.M. Moor is gratefully acknowledged. The author is much indebted to Dr. J.B. Dragt for his comments and helpful discussions.

11 REFERENCES

[1] Wind Turbine Assisted Diesel Generator Systems, DAF Indal Ltd., Meisisauga, Ontario, Canada.

[2] STILLER, P.H., SCOTT, G.W. and SHALTENS, R.K. Measured Effect of Wind Generation on the Fuel Consumption of an Isolated Diesel Power System, IEEE Trans. on Power Apparatus and Systems, Vol. PAS-102, no. 6, June 1983, pp 1788-1792.

[3] FRERIS, L.L., ATTWOOD, R., BLEIJS, J.A.M., INFIELD, D.G., JENKINS, N., LIPMAN, N.H., TSITSOVITS, A. An Autonomous Power System Supplied from Wind and Diesel, Proc. Eur. Wind Energy Conference, Hamburg, 1984, pp 669-673.

[4] BONTE, J.A.N. de. The Dutch Autonomous Wind Diesel System, Ibid., pp 685-689.

[5] LUNDSAGER, P., MADSEN, H.A. The Wind/ Diesel Development Program at Risø National Laboratory, Ibid., pp 663-668.

[6] OEI, T.D. Eenvoudige Berekeningen van de Brandstofbesparingen bij Windturbine - Dieselcombinaties voor Elektriciteitsopwekking, Tweede Nationale Windenergie Conferentie 1983, Noordwijkerhout, Paper E4.

[7] Public Power Corporation, KYTHNOS, Greece, Private communications.

[8] TASSIOU, R., CHADJIVASSILIADIS, J. Wind Energy on the Greek Islands, Proc. Eur. Wind Energy Conference, Hamburg, 1984, pp 41-46.

[9] CASELITZ, P., HACKENBERG, G., KLEINKAUF. Short Time Behaviour of Diesel Generators when Integrated with Wind Energy Converters, Ibid., pp 674-680.

[10] POTTER, Philip J. Power Plant Theory and Design, The Ronald Press Cy, New York, 1959.

[11] INFIELD, D.G. et al., Further Progress with Wind/Diesel Integration, Wind Energy version, Proc. of the 1985 seventh BWEA Wind Energy Conference, London, 1985, pp 193-199.

[12] OEI, T.D., CURVERS, A., VAN DE HEE, H. Energy Production Estimation and Parameter Sensitivity Analysis for WECS, ECN-165, Netherlands Energy Research Foundation, Petten, 1985.

[13] Diesel & Gas Turbine Worldwide Catalog 1981, Diesel Engines, Inc., Milwaukee.

Table 1. Specification of diesel units

rated capacity in kW	min. power in kW	const. loss flow kg/h	power rated fuel kg/kWh	eff. full power %	eff. no-load %
60	15	5.1	0.170	33.	50
140	35	11.2	0.160	35.	50
280	60	21.0	0.165	35.	50

Table 2. Input: mean load duration and variation

step nr	demand in kW	variation kW	duration hrs
1	343	25	730
2	286	49	1825
3	173	93	3650
4	94	19	2555

Table 3. Input: mean load duration and variation

step nr	demand in kW	variation kW	duration hrs
1	343	25	730
2	305	30	1095
3	258	32	730
4	198	68	2190
5	136	35	1460
6	94	19	2555

Table 4. Input: wind power duration and variation for two resolutions

step nr	wind power in kW coarse	wind power in kW fine	variation in kW coarse	variation in kW fine	duration in hrs coarse	duration in hrs fine
1	100	100	25	25	1743	1743
2	86	96.5	31.5	26.9	902	420
3	48	76	32.8	34.2	1840	482
4	15	58.5	14.3	34.8	1489	552
5	0	48	0	32.8	2786	613
6	-	35	-	27.7	-	675
7	-	21	-	19	-	727
8	-	10	-	10	-	762
9	-	0	-	0	-	2786

Table 5. Wind energy produced and fuel saved by a wind/diesel system.

Wind power kW	Wind energy MWh	System A 3 units — Wind energy: to grid MWh	diesel drive MWh	wasted MWh	Fuel: saved ton	extra saved ton	total saved ton	System B 4 units — Wind energy: to grid MWh	diesel drive MWh	wasted MWh	Fuel: saved ton	extra saved ton	total saved ton
20	73	73	-	-	11.8	-	-	73	-	-	13.5	-	-
40	146	146	-	-	25.2	-	-	-	-	-	-	-	-
80	289	280	9.4	-	49.8	1.6	51.4	280	9.4	-	53.5	1.6	54.9
100	363	339	24	-	59.5	4.1	63.6	336	24	-	65.1	4	69.1
120	436	-	-	-	-	-	-	393	43	-	76	7	83
140	509	453	48	8	80	8	88	441	60	8	86	10	96
160	581	494	66	21	87	11	98	485	75	21	93	13	106
200	725	-	-	-	-	-	-	558	106	61	107	18	125

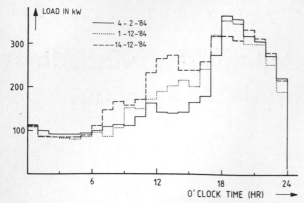

Fig 1   Daily load curves in winter of the diesel station on Kythnos (Greece)

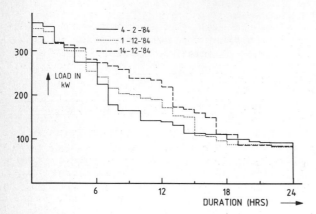

Fig 2   Daily load duration curves in winter of the diesel station on Kythnos (Greece)

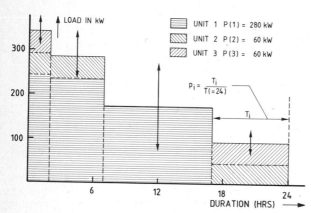

Fig 3   Mean load duration and variation of station with three units

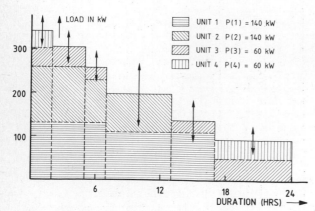

Fig 4   Mean load duration and variation of station with four units

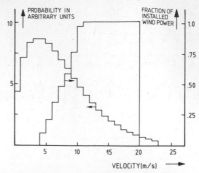

Fig 5   Wind speed distribution and wind station power curve

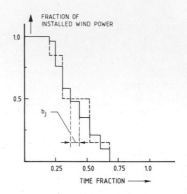

Fig 6   Wind power duration distribution

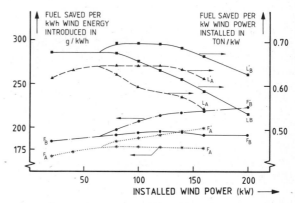

Fig 7   Fuel savings as a function of installed wind power and diesel power mix

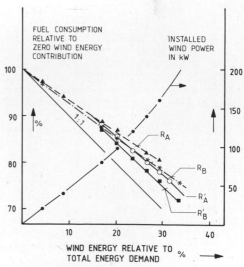

Fig 8   Relative fuel consumption and wind energy relative to total demand

Longterm performance modelling of a wind/diesel system with hydraulic accumulator storage

G SLACK and **P J MUSGROVE**
Department of Engineering, Reading University

SYNOPSIS. The use of hydraulic accumulator storage as a small capacity, high power, energy store to improve operation of a wind/diesel system has been validated by tests on a prototype system at Reading. Although measured data has made possible the formulation of a computer model which accurately represents the performance in the short term, assessment of the long term performance and economics causes difficulty due to the large amount of computer time required. This paper describes a fast and flexible method of combining energy buffer performance parameters derived from fast frequency wind data so as to provide an assessment of the long term performance and economic potential, for a wide variety of system configurations and operating conditions.

1 INTRODUCTION

The use of wind turbine generated electricity to supply small groups of consumers not connected to a large grid system is currently the subject of much interest. A large number of such potential applications exist world-wide with consumers presently being supplied by diesel-electric generating sets. These devices can be expensive to run and maintain, particularly given the demanding load pattern of a small community, with long periods of low load, a high peak/mean demand profile and limited supply system flexibility. These factors, together with the increased costs and poor reliability of fuel supply to remote locations, often lead to very high electricity costs.

Although modern wind turbines have been shown to be capable of producing electrical power at competitive cost, the use of this power to supply an isolated community presents several problems. The speed and hence the power content of the wind is predominantly variable on two distinct time scales. Synoptic variations occur due to macro-meteorological changes causing the passage of weather fronts, with the resulting wind speed variations having periods of the order of a number of hours. By contrast atmospheric turbulence, i.e. small scale irregularities in the overall flow field caused by interaction with the Earth's uneven surface, lead to wind speed variations predominantly with periods of seconds up to a few minutes. Although these variations are reduced with increasing height, and their importance reduced by the spatial averaging effect over the rotor disc, the resulting wind turbine power output variations are a problem when trying to meet a steady load. This is particularly true for a small machine where height and rotor size are limited, and the rotor inertia time constant is small.

The poor efficiency and increased maintenance problems of diesels under prolonged low load operation means that in order to save appreciable amounts of fuel in a wind/diesel system the wind turbine should be capable of meeting the load with the diesel turned off. Due to fluctuations in the wind turbine's power output, however, this can lead to either a damagingly high number of diesel start/stop cycles, or to some compromise in supply continuity. These problems can however be avoided by the provision of some form of energy storage such that any instantaneous excess of wind energy over the demand may be stored for use during a subsequent period of deficit. As described above, a store with capacity of a few minutes is sufficient to buffer variations due to atmospheric turbulence, and is thus able to reduce to an acceptable level the required number of diesel start/stop cycles. This conclusion has been confirmed by a number of computer modelling analyses (1,2).

2 HYDRAULIC ACCUMULATOR STORAGE

One way of providing a short term, high power rating energy store is the hydraulic accumulator. This device is a pressure vessel containing a bladder full of Nitrogen gas (Figure 1). When high pressure hydraulic fluid is pumped in, the bladder is compressed and the gas pressure rises, thus storing energy. When the gas is allowed to push the fluid back out of the accumulator, energy is available for conversion in an hydraulic motor to mechanical power. By using a variable displacement swash plate pump/motor, the same device may be used both to charge and discharge the store, with the input/output torque being instantly and continuously variable by means of electro-hydraulic control of the displacement. The system can thus also be used to control the load on the wind turbine and hence its rotational speed. This allows tight control of the supply frequency. The overall system operation has been more comprehensively described elsewhere (3, 4).

For application in wind/diesel systems, hydraulic accumulators have several advantages; they have high charge/discharge efficiency, low maintenance costs, low standing losses and a long life-time which is not degraded by rapid and repeated charge/discharge cycles. Disadvan-

43

tages include a low energy density and a high capital cost.

A prototype hydraulic accumulator energy buffer system was built and tested using the Reading University wind turbine simulator (5), a computer controlled D.C. motor programmed to respond to a recorded time series of wind speed measurements in the same way as the wind turbine with whose characteristics it is programmed. A computer controlled electrical load is also programmable, and can mimic a changing consumer load up to approximately 5 kW. The prototype storage system consists of an RHL A70 port plate pump/motor, with maximum displacement of $\pm$12.3 cm³/rev. and a Fawcett Hydraulics Ltd. 49.2 litre, 210 Bar rated pressure accumulator. The system is controlled via a Dowty Hydraulics Ltd 4552 Electro-Hydraulic valve energised by a CIL PCI 6300 data interface connected to an Apple micro-computer, which also serves as a data logger for the extensive instrumentation of the rig.

Tests on the prototype allowed the characterisation of the static and dynamic operation of the individual components and of the system as a whole over a wide range of operating conditions. A computer model could then be generated, to give a fast and accurate simulation of the system's operation for any combination of wind turbine size and type with the store, and for any time series of wind speeds and loads.

3 LONG TERM PERFORMANCE

In the design of a wind/diesel installation a large number of variables are involved; the size and type of wind turbine and diesel(s), together with the characteristics of any storage device and the overall system control strategy are of obvious importance. In addition, each isolated community will have a unique pattern of electricity use depending on its location, social structure and economic base; and the effects of changes to the generating plant (e.g. the addition of a wind turbine) and consequent changes to the tariff structure on the magnitude of the demand and on the daily and seasonal demand variations will be uncertain. It may also be desirable to influence the demand profile by means of a load control strategy.

Computer modelling is an important tool in the design process since estimates of the system performance, such as the fuel saving potential of the installation and the likely generating cost, may be obtained. Stochastic models (6) describe a process by means of probabilistic relationships between the states of the system. Probability theory is then used to directly calculate the quantities of interest. Simplifying assumptions are usually required, however, due to the mathematical complexities, and this may degrade the model's accuracy. Time step simulation models offer the most accurate method of representing the performance of a complex system, although careful consideration of the nature of the system modelled is required in order to determine the appropriate time step. Where plant operating characteristics change only slowly, time steps of one hour may be appropriate (7). For an application such as that examined here, however, the use of data averaged over such a period would exclude the effects caused by turbulence, and a much shorter time step (of the order of seconds) is required. This may lead to

the need for excessive amounts of computer time if performance over a long time must be assessed. In addition, time step models simply generate information on the future state of a system, without giving insight into the changes in system behaviour as the input variables are altered. Further data processing and repeated runnings of the model are therefore required, again increasing the expense in terms of both time and computing resources.

The approach adopted here is therefore that of a hybrid model, with the medium term (hourly) performance of a wind/diesel/hydraulic accumulator system being assessed using short time-step data (2 second) for a variety of combinations of wind, load and system configurations. This data is then combined using firstly (in section 6) time series of measured hourly wind data and secondly (in section 7) a probabilistic method together with a Weibull distribution of hourly wind speeds, so as to give information on the long term (yearly) performance of the system.

4 INPUT DATA AND SYSTEM OPERATION

A large amount of 2 second averaged wind data was available, recorded by NSHEB at the site of their 22 kW wind turbine on South Ronaldsay, Orkney in November 1981. Out of this data, a number of hours were chosen to represent mean wind speeds between 5 and 15 m/s, at 1 m/s intervals and within limits of $\pm$ 0.1 m/s of each value. Hours having a turbulence intensity outside the range 10-12% were then rejected, leaving between 3 and 10 individual hours of data representing each mean wind speed.

The wind turbine characteristic used with this data was based on NSHEB's 22 kW Windmatic machine on South Ronaldsay. This is a 3 bladed, fixed pitch machine of 10 m diameter and fixed rotor speed of 67 rev/min, with rated output being achieved at a wind speed of 13 m/s (8).

A number of year long records of hourly averaged wind data were also available for a number of Meteorological Office stations for the years 1970-79, with the yearly mean wind speed and the best fit Weibull parameters for each record already tabulated (9).

The diesel generator fuel consumption data used to calculate fuel saving potential was measured using the 6.3 kW Lister ST2 unit at Reading, having a peak efficiency of approximately 28%. Although both wind turbine and diesel characteristics were measured in terms of electrical output, the performance of both was redefined in terms of mechanical power output using a typical synchronous generator characteristic, in order to interface with the section of the model describing the operation of the hydraulic storage system.

5 METHOD OF ANALYSIS

Detailed operation of the storage system is described elsewhere (4) and is summarised in the flow diagram in Figure 2. Basically, for each hours run using 2 second data an initial system configuration is declared, specifying variables such as accumulator size and initial pressure (state of charge being directly proportional to pressure) and a constant mechanical load on the system. (A random variation of load about this mean could easily be introduced if required, to simulate higher frequency load variations). For each 2 second time step, the appropriate wind speed is read and the relevant wind turbine mechanical power calculated and compared with the load. Depending on whether an excess or deficit of power is indicated, together with other current operating variables such as the store's state of charge and whether or not the diesel is running, the store is either charged or discharged and the operating variables updated. This process is then repeated for each 2 second time step in the hour. At the end of the hour, parameters of interest are recorded, such as the number of diesel starts (N), the diesel running time (D), whether or not the diesel was running at the end of the hour (DR=1 or 0) and the final accumulator pressure (FP).

This process was then repeated for each hours data withing each wind speed interval, or bin, (i.e. 4.9 to 5.1 m/s, 5.9 to 6.1 m/s etc.) and average values of N, D and DR over the whole number of hours in each bin were then calculated, giving the probable values $P'(N)$, $P'(D)$ and $P'(DR)$ for that wind speed and load value and assumed wind turbine size, accumulator configuration and initial pressure. In practice it was found that given the narrow band of mean wind speeds and turbulence intensities, the values for different hours within each bin were very similar.

Runs were then also completed using the same input assumptions and wind data sets, but varying the assumed initial accumulator pressure (i.e. the initial state of charge of the store). Due to the small capacity of the store (equivalent to the wind turbine's rated output for just a minute or two), this parameter was found to have only a small effect on the values of N, D and DR. A check was kept on the final accumulator pressure, FP, i.e. the final state of charge of the store at the end of each hour/initial pressure condition. This parameter is important if the data for one hour is to be combined with the data for subsequent hours; the final pressure for any hour is the initial pressure of the next hour. This final pressure was found to vary widely between the "empty" (pre-charge) and "full" (relief setting) pressures, both for different initial pressures and for the different time series of wind speeds in hours with similar mean values. As mentioned previously, this is to be expected due to the small capacity of the hydraulic store relative to the assumed range of loads.

Consequently, for any given system configuration it can be assumed that the final pressure (FP) at the end of any hour is randomly distributed between the pre-charge and relief pressures i.e. that the probability of any particular value of FP is constant. Since the FP value for any one hour is the IP (initial pressure) value of the next hour, this implies that over a long period the probability of various initial pressures, P(IP), is also constant. Thus $P'(N)$, $P'(D)$ and $P'(DR)$, the probable values of N, D and DR averaged over each set of similar hour long time series, may also be averaged over the range of IP values to give the overall probable values P(N), P(D) and P(DR), irrespective of both time series and initial state of charge. Since the values of $P'(N)$, $P'(D)$ and $P'(DR)$ vary little with IP anyway, any bias in the assumed random distribution of IP will have very little effect.

The value P(DR) gives the probability of the diesel running at the end of the hour in question. As with accumulator pressure, this value also gives the probability of the diesel running at the start of the next hour. Whether or not the diesel is assumed to be running at the start of the hour has a small, but in some cases significant, effect on P(D) and P(DR) particularly if, for the system configuration being examined, the average wind power is approximately equal to the load. For any hour in which the diesel must start at all, the number of diesel starts, P(N), is also altered by whether or not the diesel is running at its beginning. Therefore, for any given hour and system configuration, two sets of probability values P(N), P(D) and P(DR), and $P_n(N)$, $P_n(D)$ and $P_n(DR)$, must be generated corresponding to whether the diesel is running or not running (subscript n) at the start of the hour.

The modelling process was therefore as follows; a certain configuration of wind turbine type and size, storage capacity and mechanical load torque were specified. For each of the sets of hour long files of measured, 2 second-averaged wind data sets, (having mean wind speeds of $X \pm 0.1$ m/s where X is an integer wind speed between 5 and 15 m/s and each with a turbulence intensity of between 10 and 12%), the six probable values required, $P'(N)$, $P'(D)$, $P'(DR)$, $P'_n(N)$, $P'_n(D)$ and $P'_n(DR)$ were computed. These values were then averaged over a range of IP values to give the overall probable values P(N), P(D), P(DR), $P_n(N)$, $P_n(D)$ and $P_n(DR)$. This process was repeated with different mechanical load torques on the system between 5 and 35 Nm, in steps of 5 Nm. An (11x7x6) matrix was thus constructed, expressing the performance of the specified system over a range of windspeeds and mechanical load torques. By a two stage process of linear interpolation, the performance for any given wind speed and load condition within these ranges may therefore be approximated. At wind speeds below 5 m/s, it is assumed that the diesel must run all the time. At wind speeds greater than 15 m/s, performance is assumed to be the same as that for 15 m/s. That is, no furling speed is included. Overall yearly performance is significantly affected by this assumption only at high annual average wind speeds, and this will be discussed where appropriate. The non-zero values of P(D) obtained at high wind speeds indicate the average time required for the accumulator to become fully charged if the diesel is already running at the start of that hour, averaged over the range of IP (initial accumulator pressure) values.

6 YEARLY PERFORMANCE ASSESSMENT USING TIME SERIES OF MEASURED HOURLY AVERAGE WIND SPEEDS

The interpolated hourly performance values calculated as described in the last section were then used together with the year long files of Meteorological Office hourly mean wind speeds described earlier, and with an assumed daily variation in load. For each hour, the appropriate load and average wind speed is read, and the six overall probable values interpolated;

P(D)   = Probable running time if diesel runs at start of the hour,

P(N)   = Probable number of starts if diesel runs at start of the hour,

P(DR) = Probability of diesel running at end of the hour if it is running at the start,

$P_n(D)$ = Probable running time if diesel is not running at start of the hour,

$P_n(N)$ = Probable number of starts if diesel is not running at start of the hour,

$P_n(DR)$ = Probability of diesel running at end of the hour if it is not running at the start.

The actual diesel running times and number of starts for the hour in question are then calculated;

$$P_A(D) = P_A(DR)*P(D) + (1-P_A(DR))*P_n(D)$$

$$P_A(N) = P_A(DR)*P(N) + (1-P_A(DR))*P_n(N)$$

where $P_A(DR)$ is the probability of the diesel running at the start of that hour i.e. the probability of the diesel running at the end of the last hour.

The new $P_A(DR)$ value required, that is the probability of the diesel running at the end of the current hour, is given by;

$$P_A(DR) = P_A(DR)*P(DR) + (1-P_A(DR))*P_n(DR)$$

This is then the $P_A(DR)$ value for use in calculating the next values of $P_A(D)$ and $P_A(N)$. The initial value of $P_A(DR)$ is zero i.e. it is assumed the diesel starts from rest at the beginning of the year.

These values of $P_A(D)$ and $P_A(N)$ are then summed for the whole year, together with each hourly incremental fuel use which is given by $P_A(D)*FU(load)$ where $FU(load)$ = hourly diesel fuel use at the assumed load level. An approximation is therefore given of the number of diesel starts and fuel used over a year for the system configuration and load profile specified.

7 YEARLY PERFORMANCE ASSESSMENT USING THE WEIBULL DISTRIBUTION

As discussed earlier, for generality it is desirable to estimate the system's performance in a variety of different wind regimes. Wind speed frequency distributions may be described by the Weibull distribution and this representation was applied here by means of the following method; Weibull shape and scale parameters, k and c, are specified. From these values and the Weibull distribution equation, a binned wind speed probability distribution is deduced, giving the number of hours per year for which the wind speed is within each one metre per second wide band of wind speeds. However all time series information is lost when using such a distribution and some assumptions must be made in order to use it with the method of hourly performance assessment detailed in section 5.

Firstly, it is assumed that the hours per year at any windspeed are randomly distributed throughout the year and throughout the day; that is it blows for 1/24th of its total duration for each hour within the daily load profile. This is equivalent to assuming no diurnal influence on the wind speed distribution, valid in the U.K. and in many other areas. (The method could easily be extended for sites where information on diurnal trends is available.) Secondly, to calculate appropriate values of $P_A(N)$ and

$P_A(D)$, a value of $P_A(DR)$, i.e. the probability of the diesel running at the start of that particular wind speed and load hour, is required. Thus, information is required about the distribution of wind speeds in the hours preceeding the wind speed and load hour in question.

Figure 3 (a), (b) and (c) shows the distribution of preceeding hour wind speeds for an eight year period of hourly mean wind speed data recorded by the Meteorological Office station at Plymouth (9). Wind speed has been lumped into 3 bands and each preceeding hour average wind speed binned relative to the current hour wind speed. The data is recorded in integer knots, so bin width is 0.515 m/s. Superimposed on each of the figures is a normal distribution with the same standard deviation as the data in each wind speed band. It can be seen that for the lower and medium wind speed bands, the distributions are quite similar to the normal distribution apart from a rather high peak value for the measured data, indicating a tendency for a given wind speed to persist. For the higher wind speed band, the distribution is markedly skewed, indicating a tendency for such hours to be preceeded by lower wind speeds, as might be expected. The standard deviation varied slightly across these bands from 0.885 to 1.015 m/s. Examination of data from a number of different Meteorological Office sites gave very similar results, with the standard deviation being generally between 0.9 and 1.05 m/s.

For the majority of wind speed and load combinations, the values of $P(N)$ or $P(D)$, and $P_n(N)$ or $P_n(D)$, are only slightly different. This is particularly true at high wind speeds. Thus, the final values of $P_A(N)$ and $P_A(D)$ are relatively insensitive to the value of $P_A(DR)$ used for interpolation. A normal distribution with σ =1 m/s was therefore assumed for preceeding hourly wind speed values for all wind speed and load combinations and a six part approximation of this distribution used to derive the average $P_A(DR)$ value. This was then used to derive $P_A(N)$ and $P_A(D)$ values which were summed for all wind speeds and loads. Figure 4 shows the process used in flow diagram form.

8 ACCURACY OF THE PROBABILISTIC WEIBULL METHOD

These assumptions relating the time series structure of hourly averaged wind speeds to the overall frequency distributions were checked by comparing the results obtained in this manner against results obtained, using the method of Section 6, with files of measured hourly wind speed values. These year long files have had best fit Weibull parameters calculated (9) and several files were selected with shape parameter values within ± 0.1 of 1.6, 2.0 and 2.4, and as wide a possible range of scale parameters (proportional to mean wind speed). Data on fuel saving and number of diesel starts obtained by the two methods for a typical system configuration are shown in Figures 5 and 6.

In general, it may be seen that for fuel saving the two methods agree very well, both in terms of the trends with respect to shape and scale parameters and in terms of absolute values. The point at which the three lines cross is significant only in that at mean wind speeds below this point, the system saves more fuel in a low shape factor wind regime due to the broader distribution of windspeeds which therefore include some fuel saving high winds. With mean wind speeds higher than 6 m/s, a high shape factor wind regime saves more fuel since the narrower distribution leads to less time spent at low wind speeds that save no fuel. In practice, the inclusion of a furling speed in the model would bring these curves back together and make fuel saving tend to fall for very high mean wind speeds. For the range of wind speeds considered here, however, this effect is negligible.

Good agreement is also observed for the number of diesel starts data, Figure 6. It is clear that the value of the Weibull shape parameter has a greater effect on this value, leading to noticeable scatter in the measured hourly wind speed data due to the $\pm$0.1 shape parameter bin width used. Adding to this is the effect of deviations in the measured data from the ideal Weibull distribution shape. Again though, the trends with respect to shape and scale parameter are similar for both methods, together with their absolute values.

9 CONCLUSIONS

The method described, in which a representation of the time series structure of hourly averaged wind speeds is combined with various wind velocity duration distributions and detailed second-by-second modelling of the system for periods of less than an hour, appears to provide an accurate and quick way of assessing system performance in a wide variety of wind regimes, and for any load profile within the matrix bounds.

The assumptions used in combining performance values for successive hours become less accurate as the time constant of the store becomes large in proportion to the length of data used to assess short term performance; that is, as the "end effects" of initial and final states of charge become more significant. For use with hour long files of fast frequency wind data therefore, only energy stores having a time constant of less than a few minutes may be examined in this way. The assessment of any system subject to more complex control strategies such as minimum diesel run-time, hysteresis and so on, also presents a problem.

However, the method described does allow accurate estimation of the long term performance of a wind/diesel/hydraulic system rapidly and economically for a range of mean wind speeds and load profiles. The results shown in Figures 5 and 6 are illustrative of the method, but by changing the system parameters - and in particular the capacity of the energy store - the number of starts per day and the fuel saving can be significantly altered. The sensitivity of the system performance to changes in energy store capacity, wind turbine size, turbulence intensity etc is reported in (4) and (3).

Also, though the results indicated in figures 5 and 6 are for a 1 kW average load, this nominal figure was chosen for convenience in presenting the data. The results of the assessment are believed to be appropriate and applicable with reasonable accuracy to wind/diesel/hydraulic accumulator systems with ratings in the range from about 1 kW to about 100 kW. (At higher power levels an optimised system would probably use multiple wind turbines and/or multiple diesel engines.)

10 ACKNOWLEDGEMENTS

The authors acknowledge with thanks the financial support of the SERC, via grant GR/B88105, for the work described in this paper.

REFERENCES

(1) LIPMAN, N. H., DUNN, P. D., MUSGROVE, P. J., SEXON, B. and SLACK, G. Wind generated electricity for isolated communities. Reading University report for Dept. of Energy, ETSU WN3544, 1982.

(2) INFIELD, D., LIPMAN, N. H., BLEIJS, H., FRERIS, L. L., JENKINS, N., TSITSOVITS, A.J. and ATTWOOD, R. Current progress in the development of a wind/diesel system for autonomous electricity production. Alternative energy systems; Electrical integration and utilisation conference. Coventry Polytechnic, September 1984.

(3) SLACK, G. and MUSGROVE, P. J. Hydraulic accumulator storage for use in wind/diesel generating systems. Wind Energy Conversion 1985 (Proc.7th BWEA wind energy conference). 185-192. MEP 1985.

(4) SLACK, G. The integration of a wind turbine and diesel engine with hydraulic accumulator storage to supply electricity in isolated locations. Ph.D. Thesis, Reading University. 1985.

(5) SEXON, B. A. and DUNN, P. D. A wind turbine simulator test rig. Proc. 4th BWEA wind energy conference. 129-135. BHRA.1982.

(6) FREAN, P. B. Optimal characteristics of components for small wind energy conversion systems MSc Thesis, University of Strathclyde. 1983.

(7) BOSSANYI, E. A. and HALLIDAY, J. A. Recent developments and results of the Reading/ RAL grid simulation model. Wind Energy Conversion 1983. (Proc. 5th BWEA wind energy conference.) Cambridge University Press. 62-74 1983.

(8) SOMERVILLE, W. M. and STEVENSON, W. G. An appreciation of the 10 metre Windmatic Aerogenerator. Proc. 3rd BWEA wind energy conference. 120-127. BHRA. 1981.

(9) HALLIDAY, J. H. A study of wind speed statistics of 14 dispersed UK Meteorological stations with special regard to Wind Energy. Rutherford Appleton Laboratory Report RL-83-124. 1983.

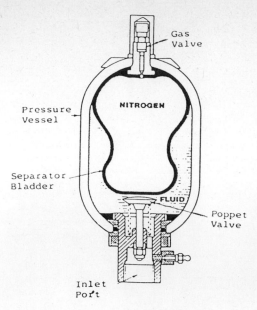

Fig 1   Hydraulic accumulator

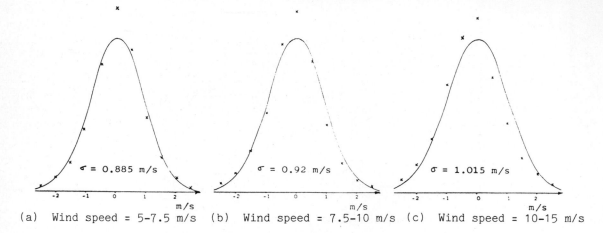

(a) Wind speed = 5-7.5 m/s   (b) Wind speed = 7.5-10 m/s (c) Wind speed = 10-15 m/s

Fig 3   Distribution of preceeding hour wind speeds
(Data used: Plymouth 1970—77)

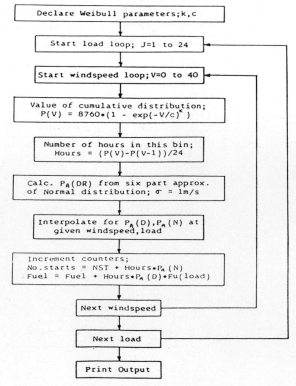

Fig 4   Flow diagram — Weibull method

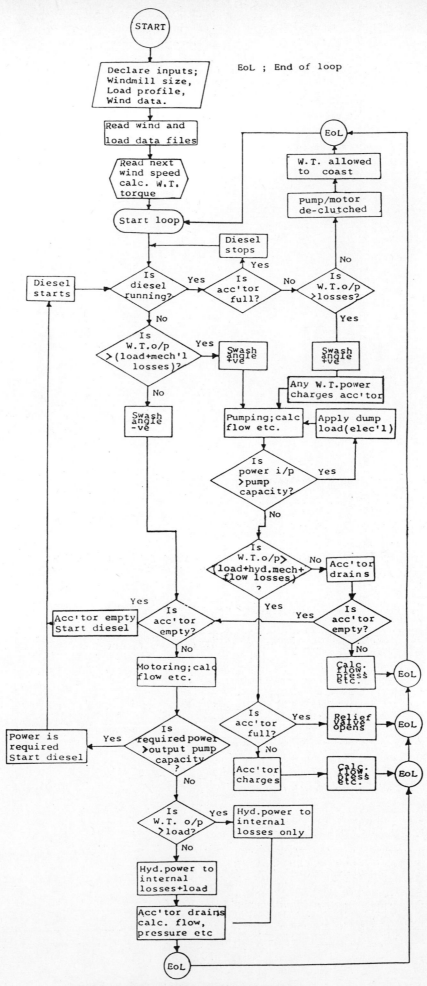

Fig 2   Flow diagram of system operation

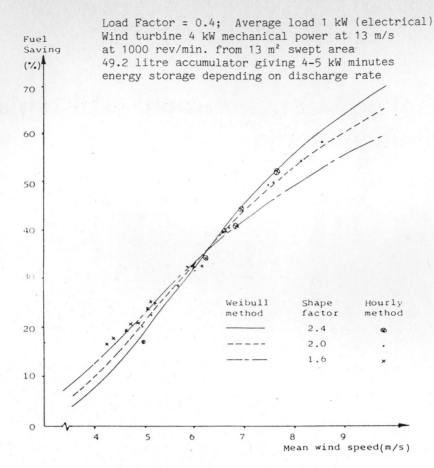

Load Factor = 0.4; Average load 1 kW (electrical)
Wind turbine 4 kW mechanical power at 13 m/s
at 1000 rev/min. from 13 m² swept area
49.2 litre accumulator giving 4-5 kW minutes
energy storage depending on discharge rate

Weibull method	Shape factor	Hourly method
——	2.4	⊗
— — —	2.0	·
— · — ·	1.6	×

Fig 5   Fuel saving versus mean wind speed

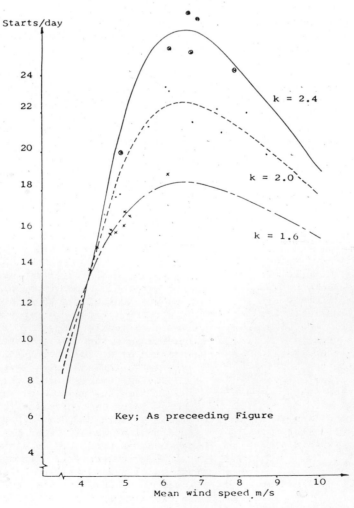

Key; As preceeding Figure

Fig 6   Diesel starts versus mean wind speed

Wind/diesel power generation – strategies for economic systems

J H BASS and J W TWIDELL

ABSTRACT

Experience gained during experimental studies of a diesel based, combined heat and power unit, and the monitoring for a one week period of the Lundy wind/diesel system is described. Techniques to statistically describe load data are reported and conclusions made about their variation with increasing numbers of consumers.

Finally, computer simulation models of stand-alone, energy supply systems, and their use to examine different operating strategies, are described. The modelling results indicate that the use of load management is one of the better strategies and lowest predicted energy costs of about 4 pence/kWh. The most effective strategies are similar to those now operating on Fair Isle and Lundy Island.

1. INTRODUCTION

This paper investigates control strategies for diesel/CHP and wind/diesel energy supply systems. Practical laboratory tests on a small diesel generator and field studies of a working wind/diesel system on the Island of Lundy were performed to support detailed computer simulation work. Statistical data obtained from the long term monitoring of individual household's electricity usage has been used as input data. The results show that energy supply costs can be reduced by up to six times compared with conventional diesel generation using, (i) load management control, (ii) waste heat recovery and (iii) wind turbine generation.

We distinguish between three forms of energy supply for the control strategies, (a) essential electricity (service power), (b) heating power (i.e. electricity used for heating), and (c) heat (i.e. water and air heated directly without electricity). These definitions allow us to discuss the 'priority' of particular energy demands. 'High priority' demands are those that require electricity (i.e. essential electricity) and that must be satisfied at all times, i.e. lights, T.V., stereo etc. 'Low priority' demands can be satisfied with either heating power (i.e. electricity) or heat, and, because of the large thermal capacities usually associated with them, can be rescheduled without reduction of the consumer's perceived service. Examples are space and water heating. These definitions are used in the further sections.

Before describing the computer modelling work in more detail, we discuss in the following three sections some of the supporting experimental work.

2. THE LUNDY ISLAND WIND/DIESEL SYSTEM

Lundy is a small island in the Bristol Channel with a successful autonomous wind/diesel installation serving twelve permanent residents and additional summer tourists on a local grid. The 50kW capacity aerogenerator with its load control system was supplied and installed by the International Research and Development Co Ltd in November 1982 (1, 2, 3).

Electrical power is available at two levels of priority:

(1) Service Power for 'high priority' demands only, and not usually for heating. This is always available during 'guaranteed periods' each day, either from the aerogenerator if it is windy or the diesel generators if not, and may also be available at other times if there is sufficient wind.

(2) Heating Power for 'low priority' demands. This is usually only available when there is sufficient wind power to exceed the total service power demand. However, some households are allowed to received a limited amount of heating power from the diesel sets in calm periods.

To distribute the variable amount of heating power available, each household has its own frequency sensitive load control unit. These units contain three frequency controlled relays which each control separate circuits. As the wind changes, the aerogenerator's rotor speed and hence the grid frequency change also. In response to these frequency variations these circuits are enabled or disabled so that (i) the load follows the power available, and (ii) the aerogenerator is controlled to maintain a grid frequency close to 50Hz. In practice it has been found that at least 70% of the entire network load must be connected in circuit and therefore available for control. The system has proved successful and is praised by the islanders. Standards of comfort have improved greatly and the extra energy has encouraged growth in the island's tourist trade.

Several aspects of the system were studied during a week spent on Lundy in September 1984. We present only selected results here; further information is available elsewhere (4).

2.1 Barton Two Cottage

It is interesting to see how the load control system affects individual houses and in particular how service and heating power are distributed on a 'day-to-day' basis. Fig 1 shows the variation in service power (shaded block) and heating power (line), for one dwelling (Barton Two cottage) over a 48 hour period (24/9/84 - 26/9/84) during which the winds changed from strong (>5m/s at 4m) to calm (<2m/s). We see that:

(i) Heating power is highly variable, often ranging between 0 and 2kW in less than a minute.

(ii) Service power varies slowly in time and changes in a step-wise fashion.

This is as expected; heating power is centrally distributed in response to wind power availability and service power, subject only to guarantee periods, is solely consumer controlled. The abundance of heating power at the beginning of the period corresponds with a period of strong local winds. These winds gradually fade out and the amount of heating power supplied decreases to zero in response. After 6 pm on the 25th, little further heating power is available and the cottage receives only service power subject to the guarantee periods (denoted 'G' on the figure). We note that this particular cottage is untypical in that heating power can be received from the diesels (i.e. from service power) during calm periods and at 10 pm on the 25th this can be seen to occur.

Over the 48 hour period the cottage receives 16.4kWh for heating (low priority) and 7.7kWh for essential services (high priority).

2.2 Aerogenerator Performance

The performance of the aerogenerator was assessed by monitoring its three phase power output, system frequency and hub weight wind speed at the power house. Data was recorded continuously for 4,000 seconds at a rate of 1Hz and the first 100 values of each of these quantities are shown in Fig 2. A fairly high degree of correlation is apparent between each pair of variables, with cross correlation coefficients ranging from 0.68 to 0.97.

Some houses on Lundy are directly connected to the aerogenerator's busbars, so that the power they consume goes unmetered in the power house. Allowance was made for this and each value of measured power had the expected extra contribution from these houses added.

The power performance curve was obtained by the 'method-of-bins' (5, 6). Where possible the recommendations of the IEA were followed, although this was not true for the 'in situ' anemometer mounted at 5m on the aero-generator tower.

Windspeed was corrected to hub height using a wind shear correction factor (5). While a 10 minute averaging time for wind speed and power data is usually recommended (5), a 30 second averaging time was used here. This shorter averaging time enables a greater number of performance data points to be obtained from a given sample and it has been found that if used to generate a performance curve, the curve is typically within 5% of that obtained from 10 minute averaged data (6). Fig 3 shows the performance curve obtained from the 150 pairs of 30 second averaged data. No data was obtained around the 'cut-in' or 'cut-out' speeds due to the shortness of the data set. For comparison the performance curve of the similar machine on Fair Isle is also shown (7, Fig 3).

The final report of the monitoring experiments (4) describes other features of the Lundy system; for example the use of waste heat from the diesel sets and the attempts to incorporate phase change heat storage. The experience gained from the monitoring supported the modelling exercise discussed in section 5.

3. SMALL DIESEL COMBINED HEAT AND POWER UNIT

A small diesel generator based combined heat and power unit has been built and tested at the University of Strathclyde, and considerable experience has been obtained with this (8, 9, 10, 11,). The unit is designed as a 'retrofit' and, being constructed from cheap and easily available parts, cost only £950 to build, at March 1985 prices, (exclusive of the diesel itself). The 4kVA diesel generator is now over twelve years old and has seen heavy use. Its condition is considered typical of those used to generate power in grid remote areas. The CHP Unit is formed by mounting the diesel set in a combined thermal/acoustic enclosure. This both 'traps' heat and provides attenuation of sound. Waste heat is reclaimed in two ways:

(1) An electric fan force ventilates the enclosure and cools the engine. The effluent warm air is suitable for space heating.

(2) An air/water heat exchanger recovers heat from the exhaust gasses and provides hot water suitable for domestic use.

When being operated in a CHP mode the main energy flows are therefore:

(i) The energy input as chemical energy in the fuel.

(ii) The energy output as:

(a) electrical energy, (b) sensible heat in air, (c) sensible heat in water.

These energy flows were measured with the diesel set loaded in steps across its entire operating range, so that the performance of the unit could be determined. Since previous studies have shown that transient effects have no measurable effect on diesel fuel consumption (12, 13), the experiment was conducted in steady state conditions, i.e. with static loads. Diesel performance tests are usually made in accordance with BS 5514/1 (1977) which defines a set of standard reference conditions (14). These conditions were reproduced in all respects, save only that the temperature of the air inside the enclosure was necessarily 15 – 25°C higher than that recommended. This causes a slight derating (4 – 8%) of the diesel set (15).

Fig 4 shows the magnitudes of the various energy outputs as functions of the chemical energy input. The peak electrical efficiency of the CHP unit was found to be (23 ± 1)% at a load of 2.75kW(e), although the expected position of peak efficiency was 3.2kW (15). The discrepancy is due to a combination of effects, these being:

(i)   the temperature derating effect
(ii) worn bearings, piston rings etc
(iii) the power factor <1
(iv) an overlong exhaust pipe with three right angle bends. Although necessary for safety purposes this caused a large 'back pressure'.

The overall efficiency of the unit was found to remain constant at (58 ± 4)%, irrespective of load, at 2.75kW(e) a (35 ± 5)% increase on the peak electrical efficiency alone. Note that the relative contributions to this total figure of 58% from electricity, hot air and hot water vary depending on diesel loading.

It has been demonstrated that waste heat recovery from small diesel sets is both possible and fairly inexpensive. Obviously many improvements could be made for more efficient CHP operation, in particular, the remaining unrecovered heat in the exhaust gasses suggest this as the most profitable avenue for further reclamation.

4. STATISTICAL ANALYSIS OF CONSUMER ELECTRICITY DEMAND

Just as it is necessary to obtain representative wind speed data for use in wind/diesel modelling (16, 17, 18) so it is necessary to obtain load data appropriate to the application for which the system is designed. Unfortunately there is a shortage of such data, so that often inappropriate load data has been used.

We can identify three ways in which the load data chosen for use in modelling may be inappropriate. These relate to:

(a) the wrong type of consumer, eg industrial/agricultural/domestic. Each have their own characteristic patterns of energy use.

(b) artificial smoothness. This might be introduced as a result of averaging over time, and tends to leave only 'trend' information. Short term, stochastic variations tend to get 'filtered out'.

(c) the wrong number of consumers. This has a similar effect to (b) above. Increasing consumer diversity with large numbers of consumers tends to produce smooth, trend-only, profiles.

The assumption that consumer load profiles (data) scale linearly with the number of consumers is almost certain to produce misleading results. To illustrate this Figs 5 and 6 show a 'typical' daily demand profile of a single consumer, and of a group of about 40 consumers respectively. The single consumer's demand is composed of a short term, cyclic variation (probably a 'fridge or freezer) and irregular, fairly large peaks to

around 2.5kW (probably either an electric kettle or an induction motor). There is little apparent 'trend' in the data and the profile is dominated by the short timescale variation. The profile of the group of forty consumers is very different and a typical 'low daytime', 'high evening' trend is obvious. There are small, short timescale variations about this trend but no prominent peaks as in Fig 5. Thus it would be quite misleading to use a scaled-down form of Fig 6 as a representative data set for a single consumer.

Thus before selecting appropriate load data for use in modelling one must first:

(i) decide precisely what sort of data is required, e.g. number and type of consumers

(ii) investigate the variation in the data available and identify a representative sample. Here we confine ourselves to systems designed for individual remote domestic consumers.

4.1 Statistical description of load data.

The load data used in the analysis comes from the long term monitoring of a group of forty houses in Abertridwr, Wales, (19, 20). The data is 'essential electricity' use only, none being used for domestic heating.

A total of 94 days of 5 minute load data was used for each of the 40 houses monitored in the statistical analysis. The 94 days were composed of five periods, of about 20 days each, taken at different times of the year. For each house, and for each period, we consider the load data to be a discrete time series $\{X(t)\}$. The following statistics are computed for each data set:

(a) Mean load, $\bar{X}$
(b) Maximum and minimum load, Xmax, Xmin
(c) Standard deviation from the mean, Sm
(d) Load factor, $\alpha = \bar{X}/Xmax$
(e) Coefficient of Variation, $\beta = Sm/\bar{X}$

The average values of these statistics over all the data sets are shown in Table 1, together with the range of values. This enabled a representative sample for use in modelling to be identified. The most important characteristic of this single consumer 'high priority' demand data is the very low load factor of 0.07, i.e. a peak demand of twelve times the average value. Note that there was no significant 'seasonal' variation in any of these statistics.

Finally, group profiles of a variable number of consumers from one to forty were created and these statistics computed for each, to see how they are related to group size. Several permutations of each grouping were made (typically 20) and the average value of the statistics determined for each group size. Fig 7 shows the variation of $\bar{X}$, Xmax, Sm, α and β as functions of group size. Whilst the linearity of mean load, $\bar{X}$ with group size might be expected, the linearity of peak load, Xmax and standard deviation, Sm, with group size, are not. Both load factor and coefficient of variation show rapid variation initially but finally level out with increasing group size. The sensitivity of these two functions to group size in the range 1 – 15 show how significant the effects of consumer diversity can be.

5. COMPUTER SIMULATION MODELLING

So far this paper has (i) investigated a working wind/diesel installation, (ii) considered practical heat recovery from small diesel sets, and (iii) obtained statistical information on representative sets of load data for a varying number of (i.e. 40) domestic consumers. This final section deals with the results of a series of computer simulation models of stand-alone energy supply systems, ranging from diesel only systems to wind/diesel based CHP systems; all of these are sized at a level to meet the demands of individual remote households. The main objective is to identify those systems that meet the demands most cheaply.

The modelling discussed here is different from other wind/diesel models in its approach to consumer demand. Previous studies (16, 17, 18) have assumed that consumer demand cannot be interfered with and must be met as demanded. Here we consider the consumer's demand to be the sum of two components, a 'high priority' demand and a 'low priority' demand, as defined in section 1. The 'high priority' demand requires 'essential electricity' and must be met at all times, whereas the 'low priority' demand, which requires either heating power or heat, can be interfered with, i.e. rescheduled as appropriate.

The constraints on the simulation modelling here, are:

(a) essential electricity must be available to meet 'high priority' demand at all times.

(b) the total amount of energy supplied for heating purposes must be of at least a specified amount over 24 hours, but may be rescheduled freely within this period. The strategies considered allow this heating load to be satisfied by heating power (electricity) and/or by direct heat (air or water) according to the particular model. Implicit in the modelling is the knowledge that commercial control systems exist for such rescheduling, and that houses and devices may be adequately insulated and have sufficient thermal capacity to maintain temperatures over such periods.

The modelled demand consists of:

(i) a 'high priority' demand profile from a typical, single household on the Abertridwr estate (section 4), with a total demand of 6.8kWh/day.

(ii) a 'low priority' demand of 42 kWh/day, typical of a rural house in parts of Scotland (21), which can be rescheduled as required. The integrated total demand is therefore 48.8kWh/day.

The relative cost of the different systems and strategies are assessed using a net present value type analysis (4). The following assumptions are made:

(a) discount rate of capital, 10%/y

(b) inflation, 12%/y and diesel fuel escalation rate 5%/y above inflation (NB These are means of the figures for the last eleven years.)

(c) diesel fuel price, 0.43 £/litre (March 1985)

(d) equipment lifetime, approximately 20 years

(e) operational and maintenance cost of all equipment, 5% of the initial capital costs each year

(f) capital costs of diesel generators taken from a survey of June 1985 prices (4).

5.1 Diesel only – systems options.

Table 2 contains a summary of the modelled results for some of the options considered in our work.

(a) diesel electricity – no control. The simplest option is to supply the entire 48.8kWh/day total demand with diesel generated electricity. Fig 8 shows an example of a 'typical' total daily demand profile (i.e. the sum of components (i) and (ii) above) of a consumer who uses electricity for all heating purposes but does not control his load. To meet this demand, a diesel set rated at 7.2kW(e) is required. However, the load factor is only 0.28 so that the diesel often runs on low load. This option requires the use of 37 litre/day of diesel fuel, the average conversion efficiency is 12.5% and the unit energy cost is 21.5 pence/kWh. These results are reasonable compared with those from field studies (21).

(b) diesel electricity – load management control of 'low priority' demand. This option is similar to that above except that the 'low priority' demand can now be rescheduled; the aim is to reduce the peak demand whilst improving load factor, giving reduction's in both the capital cost of the plant needed to meet the load and its O & M costs. The resultant, controlled total demand can be met with a 4kW(e) diesel set and the load factor is increased to 0.50. Fuel use is decreased to 25 litre/day, average conversion efficiency increased to 18.5% and the unit energy cost decreased by 32% to 14.7 pence/kWh. This result agrees with experimental results from a laboratory system (8).

(c) diesel electricity for 'high priority' – bottled gas for 'low'. We now treat the heat and electricity supplies separately, as do perhaps a large majority of actual installations. However, it is apparent from Fig 5 that 'high priority' demand profiles have large peaks and poor load factors, so that they present very unfavourable operating regimes for diesel generators. To meet the 6.8kW/day of 'high priority' demand, a 4kW(e) diesel set is required and the average conversion efficiency is only 4%. Even though bottled gas is priced around 4 pence/kWh (NB A combustion efficiency of 60% is assumed) this strategy is dominated by the diesel fuel costs. Nevertheless, the average unit energy cost is 12.9 pence/kWh overall, which is better than option (b) above. We note that this option would almost certainly increase (O & M) costs to above the 5%/y assumed.

(d) diesel electricity - battery store. One possibility is to use an appropriately sized diesel generator running continuously at it's rated power (here 2kW(e)) into a battery storage bank. An inverter supplies mains 240V, 50Hz power to the consumer who draws power as required. Unit energy costs are assessed at 11.5 pence/kWh. However, the costing of this strategy is very uncertain since no practical installations are known. Further, there are likely to be restrictions on maximum power input/output.

(e) CHP diesel - no control. The diesel generator supplies only 'high priority' demand and the waste heat recovered is used to meet the 'low priority' demand. Realistic figures for expected waste heat recovery are obtained from experiment (section 3). To meet the 'high priority' demand the diesel has to be rated at 4kW(e) and uses 16.4 litre/day. Whilst the CHP unit operates at about 60% efficiency overall, producing 6.8kWh of electricity and 93.4kWh of sensible heat, only 42kWh/day of this heat is required and the remainder has to be dumped. Despite this wasted heat, the unit energy cost drops to 10 pence/kWh.

(f) CHP diesel - load control. Our scenario does not allow us to interfere with the demand for essential electricity. Control is only possible if electricity is needed for heating power (i.e. low priority demand). However, with CHP, there is an abundance of heat, see (e) above and therefore no requirement for heating power from electricity. In this situation there is no benefit in electrical load control, and in general it seems that load control is most beneficial when CHP operation is not practical.

Summary of diesel only strategies. Analysis of the models above shows that fuel costs dominate the total costs for every option. This suggests incorporating a renewable energy device having zero fuel cost.

5.2 Wind/diesel systems options. The only way to significantly reduce costs with diesel sets is to turn them off completely so that diesel fuel usage is reduced. The following strategies consider 'either-or' generation from an aerogenerator with a capacity between 4 and 8kW in average wind speeds of 3 to 9m/s (at 10m), and a diesel generator. The contribution from the aerogenerator has been modelled as the power generated assuming a simple performance characteristic and artificially generated wind data, this being random deviates sampled from an appropriately chosen Rayleigh distribution. No attempt was made to reproduce the 'time series structure' of real wind data. Aerogenerator capital costs are assumed at £1000 per kilowatt of maximum rating (4). Firstly we describe the strategies and then we give a summary of the conclusions.

(g) wind (no control)/diesel (no control). As in (a) but wind generated electricity is used whenever it exceeds the total, uncontrolled demand. When this occurs the diesel is switched off.

(h) wind (load control)/diesel (load control). If the aerogenerator can meet the 'high priority' demand the diesel is switched off. Any excess wind power goes to 'low priority' demand. The diesel only comes on if the 'high priority' demand is not met. To keep the diesel operating in it's most efficient operating region, it provides some heating power.

(i) wind (load control)/diesel (no control). As in (h) but the diesel is not loaded to obtain optimum fuel efficiency and meets only 'high priority' demand.

(j) wind (load control)/CHP (load control). As in (h) but waste heat is recovered from the diesel set when it has to operate.

(k) wind (load control)/CHP (no control). As in (i) but with waste heat recovery.

Summary of wind/diesel strategies. Each wind/diesel strategy was modelled for wind speeds averaging 3, 5, 7 and 9m/s at 10m and for aerogenerators of capacity 4, 6 and 8kW (i.e. 60 combinations). The best strategies for giving lowest fuel use and lowest overall energy costs are given in Table 3. Figs 9 and 10 show these results graphically. Note that the choice and performance of the best options are expected to depend on the average wind speed.

The strategies that always give the cheapest energy involve load management control for the aerogenerator. At the lower average wind speeds of 3 and 5m/s, option (k), with heat recovery from the diesel, was cheapest. At 7 and 9m/s, option (i) with no heat recovery, was best. This is because in the latter case the diesel was not operating frequently enough to make heat recovery viable. An energy flow diagram for option (k), with a 4kW aerogenerator in 9m/s average winds, is shown in Fig 11.

The lowest energy costs range from 3.7 pence/kWh at 9m/s average windspeed, to 9.2 pence/kWh at 3m/s. In all cases it appears best not to add load to the diesel generator to optimise fuel efficiency, i.e. it is best to allow the diesel set to meet just the 'high priority' demand, even though this may be small at times.

The most expensive option is (g), when neither the aerogenerator nor the diesel have their load controlled. Such an arrangement is always more expensive than the cheapest diesel-only option, option (e), i.e. CHP.

6. CONCLUSIONS

The strategic modelling of small energy supply systems, sized to supply the demands of individual, remote households, has had practical input from (i) field monitoring on Lundy Island, (ii) laboratory CHP plant, and (iii) monitoring of individual consumer's essential electricity (high priority) demand. Our conclusions are:

(a) The cheapest diesel-only option is CHP, giving energy at 10 pence/kWh overall.

(b) Wind/diesel is likely to provide cheaper power than diesel-only systems at average wind speeds above 3 - 5m/s. In all cases it is best to use load management control for the aerogenerator, but not for the diesel. CHP from the diesel is worthwhile at average wind speeds below about 6m/s. In favourable wind speeds above 7m/s average, energy costs can drop to 4 pence/kWh using load management control for the aerogenerator.

Finally we note that our best options from the modelling closely parallel those installed on Fair Isle and Lundy Island, and our worst options are those without load management control.

7. ACKNOWLEDGEMENTS

We thank those of our group mentioned in the references, who gave much by way of discussion. Thanks also to Adam Pinney, Dave Infield, Jim Halliday and Paul Gardner. We would also like to express our thanks to the SERC for allowing access to the Abertridwr data and to Professor O'Sullivan and his team at UWIST, Cardiff, who collected it as part of an SERC funded 'Energy in Buildings' project.

REFERENCES

1. SOMERVILLE W M & PUDDY J (1983). Wind Power on Lundy Island. In Proceedings of 5th BWEA Annual Conference (ed P J Musgrove) pp 185 - 197, Cambridge University Press.

2. INFIELD D G & PUDDY J (1983). Wind Powered Electricity on Lundy Island. In Energy for Rural & Island Communities III (ed J W Twidell, F G Riddoch, B Grainger) 1st ed, pp 137 - 144, Oxford, Pergammon Press.

3. SOMERVILLE W M (1984a). Operating Experience and Developments on Some British Wind Turbine Installations. Proceedings of BWEA Practical Experience and Economic Aspects of Small Wind Turbines. 27th September 1984.

4. BASS J H (1986). PhD Thesis, University of Strathclyde (to be submitted).

5. FRANSDEN S, TRENKA A K & MARIBO PEDERSEN B (1984). IEA's Recommended Practices for Wind Turbine Testing. 1. Power Performance Testing. WindMatic Sales Manual for WM 12S.

6. AKINS R E (1982). Method-of-Bins Update. Presented at Wind Energy Expo '82, Amarillo, Texas.

7. STEVENSON W G & SOMERVILLE W M (1983). The Fair Isle Wind Power System. In Proceedings of 5th BWEA Annual Wind Energy Conference. (ed P J Musgrove) pp 171 - 184, Cambridge University Press.

8. WYPER H (1982). The Use of Load Control in Small Electricity Systems. M Sc Thesis, University of Strathclyde.

9. BARBOUR D (1983). Energy Supply for the Island of North Ronaldsay, Orkney. M Sc Thesis, University of Strathclyde.

10. BASS J, BARBOUR D, GRAINGER W, TWIDELL J & WYPER H (1983). Combined Heat and Power with Load Management for an Isolated Dwelling Using a Small Diesel Engine. In UK-ISES Rural Power Sources Conference Proceedings, March 1983, pp 69 - 80.

11. BASS J H, & TWIDELL J W (1984). Small Diesel Generator Based Combined Heat & Power. Research Report to the Orkney Island Council, Department of Applied Physics, University of Strathclyde.

12. STILLER P H, SCOTT G W & SHALTENS R K (1983). Measured Effect of Wind Generation on the fuel consumption of an isolated diesel power system. IEEE Trans Power Apparatus & Systems, Vol PAS - 102, No 6, June 1984.

13. BLEIJS H, JENKINS N, TSITSOVITS A & INFIELD D G (1984). Some Aspects of Wind Diesel Integration. In Proceedings of the 6th BWEA Conference. (ed P J Musgrove) pp 382 - 401, Cambridge, Cambridge University Press.

14. BRITISH STANDARD INSTITUTION (1977). British Standard 5514/1: Specifications for Reciprocating Internal Combustion Engines. London, British Standards Institution.

15. PETTERS LTD (1981). PH Range Workshop Manual. 1st ed, Hamble, Petters Limited.

16. TSITSOVITS A J & FRERIS L L (1985). A Statistical Method for Optimising wind Power contribution in a Diesel Supplied Network. IEE Proceedings, Vol 132, Pt C, No 6, November.

17. SLACK G W (1983). Small Diesel Generation Systems: The cost and fuel savings possible by the addition of a wind turbine with and without short term storage. Internal Report, The Energy Group, Department of Engineering, University of Reading.

18. BANDOPADHAYAY P C (1983). Economic Evaluation of Wind Energy Applications for Remote Location Power Supply. Wind Engineering 7, No 2, 1983, 67 - 78.

19. SMITH W A (1979). Report on the Abertridwr 'Better Insulated Houses' Project. Colloquium on Static & Dynamic Processes in Exterior Walls.

20. SLACK G (1983). A Preliminary Analysis of Abertndw Load Data. Internal Report, The Energy Group, Department of Engineering, University of Reading.

21. TWIDELL J W & PINNEY A A (1985). Energy Supply and Use on the Small Scottish Island of Eigg. Energy, Vol 10, No 8, pp 963 - 973.

22. SINCLAIR BA, STEVENSON W G & SOMERVILLE W M (1983). Wind Power Generation on Fair Isle. In ERICIII (ed. TWIDELL, RIDDOCH, GRAINGER) pp 155-162, Oxford, Pergammon Press.

23. TWIDELL J W & WEIR A D (1985). Renewable Energy Resources, London, E & FN Spon. Ltd

Table 1 Single consumer demand statistics for
Abertridwr demand data.

Quantity	X̄/W	Xmin/W	Xmax/W	Sm/W	α	β
Range	62–817	0–102	372–9432	61–1077	0.01–0.22	0.5–2.8
Average	275±5	0.6±0.2	4270±70	350±5	0.072±0.002	1.27±0.04

Table 2 Summary of diesel only options

Strategy	Minimum diesel rating/kW	Load factor	Diesel fuel use/(litre/day)	Primary energy use/(kWh/day)	Extra * cost/£	Average conversion efficiency/%	Unit cost of energy/(pence/kWh)
a	7.2	0.28	37.1	390	–	12.5	21.5
b	3.9	0.52	24.6	260	1,000	18.5	14.7
c	3.9	0.52	16.2ƒ	245	1,000	20.0	12.9
d	2.1	1.00	18.2	190	2,500	25.5	11.5
e	3.9	0.52	16.4	170	950	28.0	10.0

* Cost for extra equipment, other than diesel set and ancillaries.   ƒ Plus 9.1 litre/day of propane gas

Table 3 Summary of optimum wind/diesel options

Average wind speed / (m/s)	Optimum strategy *	Optimum WTG rating/kW	Primary energy use/(kWh/day)	Unit energy cost/(pence/kWh)	Criteria
3	k	4	129.2	9.2	1
	k	"	"	"	2
5	k	8	49.8	6.6	1
	k	"	"	"	2
7	i	8	32.9	5.7	1
	k	6	35.7	5.0	2
9	i	6	23.0	4.4	1
	k	4	26.3	3.7	2

*There are two criteria for determining 'optimum' strategy.
1. Minimisation of primary energy use.   2. Minimisation of unit energy cost.

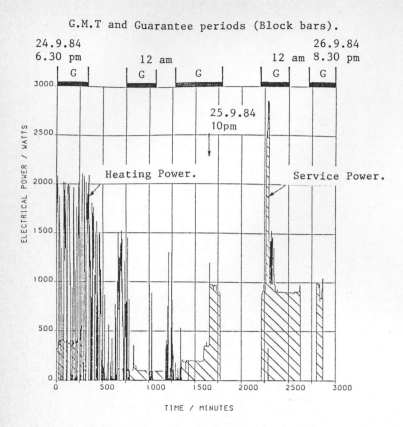

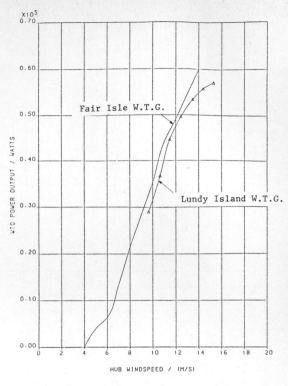

Fig 1   Service (essential) and heating power supply to a single house on Lundy Island, 24–26 September 1984

Fig 3   Comparison of Lundy WTG's power performance with that of an identical machine on Fair Isle (see text for details)

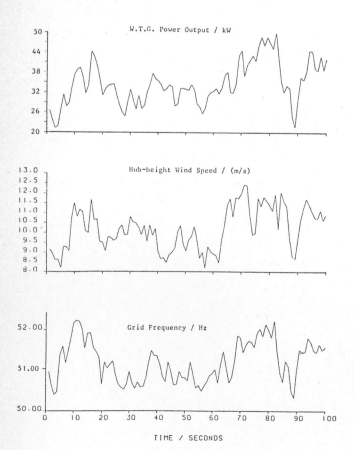

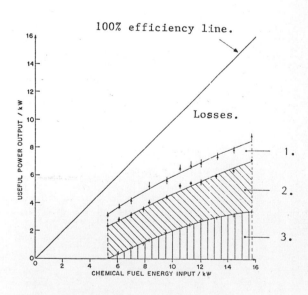

1-Contribution from sensible heat in water
2-Contribution from sensible heat in air
3-Contribution from electrical energy

Fig 4   Sensible heat and electricity output from the CHP unit as functions of the chemical energy input rate

Fig 2   Aerogenerator power, wind speed and electrical grid frequency for a 100 sec period, logged at 1Hz

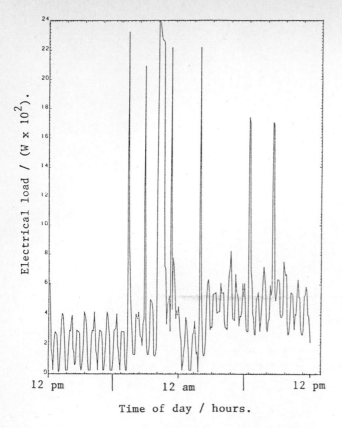

Fig 5   Typical daily demand profile of a single consumer

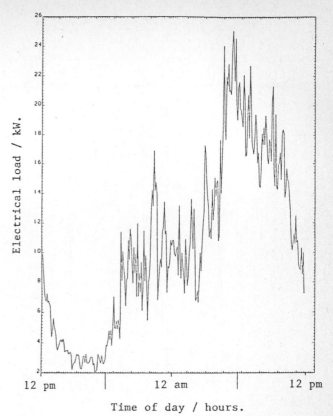

Fig 6   Typical daily demand profile of a group of forty consumers

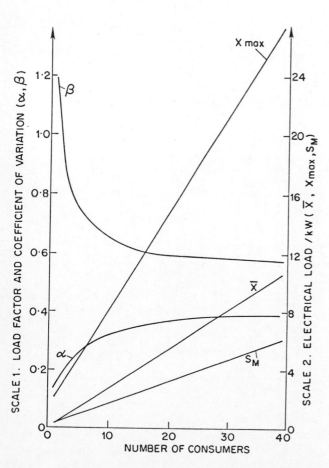

Fig 7   Statistical parameters plotted as functions of the number of consumers

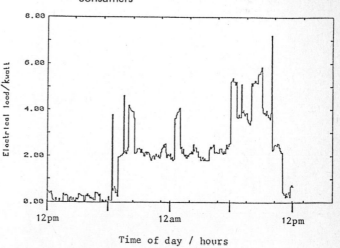

Fig 8   Total uncontrolled daily demand profile for single household (as used for modelling)

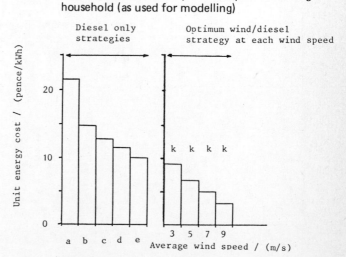

Fig 9   Comparison of unit energy costs of the diesel only options and the best wind/diesel options (see text for details)

59

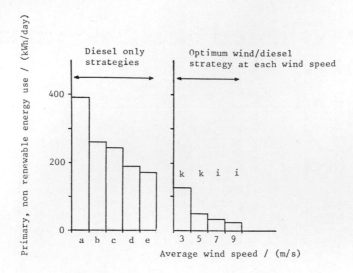

Fig 10 Comparison of primary energy uses of the diesel only options and the best wind/diesel options (see text for details)

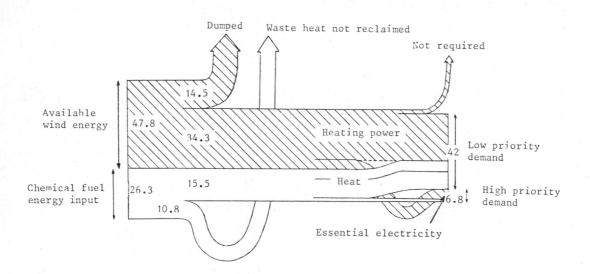

Fig 11 Energy flow diagram for option (k), with a 4 kW aerogenerator in a 9m/s average wind speed (in units of kWh)

Analysis of wind and turbulence measurements on Shetland

E A BOSSANYI and J A HALLIDAY
Rutherford Appleton Laboratory
P GARDNER
University of Strathclyde

SYNOPSIS   Two hill-top sites on Shetland have been instrumented as part of a joint project between the North of Scotland Hydro Electric Board, the University of Strathclyde and Rutherford Appleton Laboratory.   The project aims to evaluate the potential contribution of wind power to electricity supply for the islands of Shetland.   This paper presents some results of the analysis which has been carried out so far on the data recorded at the two sites.   As well as long-term mean wind speeds, wind shear and wind roses, a limited amount of rapidly-sampled data has been analysed to yield some information on turbulence intensities and spectra.   Attention is drawn to the inhomogeneity of the terrain, in contrast to the flat, uniform land to which much of the micrometeorological literature refers.

1. BACKGROUND

The project began in September 1984 as a collaborative study funded jointly by SERC and NSHEB.   The project had three principal aims:

1)   to assess the potential of two hill-top sites previously selected by NSHEB as possible sites for wind turbine generators, by means of measurements at the sites and comparison with data from nearby Meteorological Office recording stations,

2)   to evaluate the potential contribution of wind power to the Shetland electricity supply system, by constructing and running a computer simulation model of the grid system and diesel power station, and

3)   to investigate any possible adverse or beneficial effects of wind power on the quality of electricity supply and on the operation of existing plant.

Since the start of project however, NSHEB have decided to proceed with the installation of a 750kW wind turbine on one of the hill-tops, subject to statutory consent.   Consequently an additional aim has emerged, that of providing site-specific wind information which may be of use to the designers of the proposed wind turbine.

1.1   The existing system on Shetland

Shetland is a group of islands situated some 200 km to the North of the Scottish mainland.   The total population is about 23000, much of it in the main town, Lerwick.   The maximum demand on the electricity grid system is in the region of 30MW, although it appears to be declining slightly since the Sullom Voe oil terminal was completed in 1982/3.   At present the electricity is supplied from a diesel power station in Lerwick.   This has 13 diesel generators ranging from 1.5MW to 8MW, a waste heat boiler driving a steam turbine, and two gas turbines which are infrequently used at present.

Electricity is currently sold to consumers at mainland rates although it costs considerably more than this to produce.   The high generation cost coupled with very high average wind speeds make Shetland a prime candidate for the deployment of wind power systems.

1.2   The wind turbine sites

Two hill-top sites were chosen by NSHEB after consultation with the island authorities and owners and users of the land, taking into account factors such as existing land use, ease of access, topography and proximity of existing power lines.   Scroo Hill (altitude 248m amsl) is a relatively exposed, rounded hill about 11 km South of Lerwick.   The Hill of Susetter (170m) is a flat-topped ridge running North-South and situated some 20km North of Lerwick.

1.3   The instrumentation

Both sites are similarly instrumented.   Each installation consists of a 45m high guyed mast with booms mounted at 5m intervals starting from the 10m level.   Anemometers are mounted at the end of each boom, which point into the prevailing wind to minimise tower shadow effects.   All anemometers are cup anemometers with a response length of 5m, with the exception of the 30m level where two propeller anemometers (response length 3.3m) are mounted at right angles to each other so that their outputs can be resolved to give wind speed as well as direction.   Additional direction measurements are provided by wind vanes at 15m and 45m.   Temperature is recorded at 10m, 25m and 45m, enabling atmospheric stability to be estimated.

All the instruments are normally sampled every minute by a data logger connected to a cartridge recorder, but faster sampling rates (up to 1Hz) can be used for short periods, limited by the need to change the cartridge more frequently.

An earlier paper by Halliday et al (1985) gives further details of the instrumentation and the problems which have to be faced in setting up such installations in such a remote environment.

2 EXTENT OF AVAILABLE DATA

The installation at Scroo Hill has been operational since December 1984, and Susetter since August 1985. However there has been some down time at both sites due to problems with the instrument power supplies, neither site having any mains power available. In addition the 25m anemometer at Scroo was damaged during installation and could not be repaired for some months. Problems are now being encountered with the propeller anemometers at 30m, as will be described later.

Nevertheless a considerable amount of data is now available, covering the periods 1 December 1984 to 2 December 1985 at Scroo and 30 August 1985 to 21 January 1986 at Susetter. In addition some fast data was taken at Scroo when researchers were available on site to change the cartridge every few hours. This consists of nearly 24 hours of 1-second data and 12 hours of 2-second data. This data was used to produce the turbulence information presented below. It is hoped that rather larger amounts of fast data will be collected later this year at Susetter.

3 HOURLY DATA

The instruments are normally sampled once a minute, however to date much of the analysis has been carried out using hourly mean values in order to determine underlying trends. In the course of the analysis a computer routine was developed to find the average of wind direction readings.

3.1 Averaging wind directions

Care is needed in averaging wind direction because of the discontinuity at 0/360°. All directions are measured in the range 0-360°, so directions of 10° and 350° would give an average of 180° rather than 0° or 360°. The correct result depends on whether the change in direction really was -020° rather than +340°. With digital sampling it is impossible to know but assuming that directions do not change rapidly with respect to sampling interval the following scheme can be used. If the difference between two successive readings is greater than 180° in either direction, 360° is added to or subtracted from the later reading until the difference falls in the range -180° to +180°. These corrected readings will lie in the continuous range -∞ to +∞, and can therefore be averaged normally. If the final result lies outside the range 0-360° it can be corrected by multiples of 360° until it is inside.

The method is simple, reliable and easily implemented. Fig 1 shows a section of FORTRAN code which averages the N directions stored in the array D.

3.2 Wind Roses

Wind roses have been calculated for both the hills, the one for Scroo Hill (35m boom) being plotted in figure 2. The form of wind rose presented shows not only the distribution of wind directions but also how the wind speed distribution varies for each direction. The dotted circles represent 100 hours and the solid lines wind speed contours. Thus it can be seen although the wind direction at Scroo Hill is between 165° and 195° for a total of 977 hours, for 111 hours it is less than 5.0 m/s and only between 10 m/s and 15 m/s for 330 hours. The wind rose of the Hill of Susetter has a similar shape even though it is calculated from a much shorter data sample. The long term wind rose at the Lerwick Meteorological Office station is shown in figure 3. It should be noticed that although the Scroo Hill and Hill of Susetter wind roses have a similar shape to the long-term (1970-1980) Lerwick wind rose, marked annual changes are possible as shown by the 1983 Lerwick wind rose (figure 3). When the characteristic of the 750 kW Howden wind turbine (figure 4) is applied to the hub-height (35m) speeds, a wind power rose can be plotted (figure 5).

3.3 Means and extreme values

The mean of the available hourly means for each site has been calculated and is shown in figure 6. It appears that the Hill of Susetter has a higher mean than Scroo Hill. However, this is an illusion caused by the different sample lengths. When a detailed analysis of the period during which data was being recorded at both sites was performed it was found that on average the 35m Scroo winds were about 1 m/s higher than the 35m Susetter winds. The long-term (1941-1970) mean recorded at 10m height at the Lerwick meteorological station is 7.3 m/s. Figure 6 also shows the highest one minute readings recorded at each height for both the sites, which occured on the same day. The highest gust recorded at Lerwick at 10m height for the period 1921-1983 was 49 m/s in January 1961.

3.4 Wind Shear

The average vertical wind shear may be seen in Figure 6, the shear exponents between 10m and 35m being 0.067 at Scroo and 0.098 at Susetter. Figure 7 shows the shear at Susetter for different wind directions and speeds. It can be seen that there is a marked negative shear between the 40m and 45m when the wind is blowing from between 337° and 067°. It seems likely that the wind flow at 45m is being disturbed by a guy wire at the top of the mast. There is also evidence of a slight retardation at 15m, for certain directions, though the cause is less clear.

Additionally it can be seen that for directions between 337° and 022° there is a more marked shear between 10m and 35m than for other directions. This is probably due to the shape of the hill; the mast is on the Southern tip of a ridge. Winds from the north would reach the mast after blowing along the ridge for some distance, by which time a stronger shear will have built up, more like the wind shear found on flat land. However, it can be seen that for the predominant wind directions the wind shear is uniform and not speed dependent.

A similar result was found for the Scroo Hill data (without the ridge effect).

3.5 Diurnal trends

The average diurnal variation of wind speed was examined for both sites: at Scroo Hill the amplitude of the variation decreased from 0.75 m/s at 10m to 0.38 m/s at 45m and at Susetter from 1.07 m/s at 10m to 0.81 m/s at 45m. The hourly wind power data (calculated with the characteristic of figure 4 and the 35m speeds) was analysed for a diurnal pattern - the amplitude was only 19 kW around a mean of 376 kW.

3.6 Weibull Parameters

Weibull parameters have been calculated for the 2 sites using a graphical method (see Halliday (1983) for a description of methods available). As is usual in wind energy, the tails of the wind speed distribution have been disregarded. The values found were: Scroo Hill shape parameter (k) = 1.89 scale parameter, (c) = 9.83 m/s; Hill of Susetter shape parameter(k) = 1.91, scale parameter (c) = 10.04 m/s.

4 GILL ANEMOMETERS - PRACTICAL RESULTS

Overall, the Gill anemometer pair at 30m has been found to be giving significantly lower wind speed readings than adjacent cup anemometers. The discrepancy varies with wind direction in a significant way. Figure 8 shows the difference between hourly means measured by the 35m cup anemometer at Scroo and the 30m Gill pair, binned by direction (as measured by the vane at 45m) over the period January to July 1985. The dotted lines, showing the standard deviation in each bin, give an indication of the scatter. Given the 5m height difference one would expect a slight positive difference due to wind shear, but the large (up to 3 m/s hourly average) direction-dependent differences clearly indicate other effects.

Tower shadow on the cup anemometer probably accounts for the negative difference in a Northerly wind. The Gills would experience tower shadow effects a few degrees either side of North because of the way they are mounted (see diagram in fig 8). This could explain the large positive differences either side of North, although at about ± 20° they would appear to be a little too far away from North. This could indicate that the mounting boom could be causing substantial shadowing. This is not unlikely as the boom is in the same horizontal plane as the Gill rotors, whereas the cup anemometer rotor is 20cm above its mounting boom. Together with the possible shadowing of one Gill by the body and rotor of the other, this might explain the considerable discrepancy which occurs in all directions except in the Southern quadrant, where there would be no such interference. One conclusion is that perhaps the two Gills and the boom should each be fixed at slightly different levels, perhaps one anemometer above and one below the boom.

Another feature is also apparent in fig 8, namely that the difference between the cup and the Gill pair shows definite dips at around 45°, 135°, 225° and 315°. In each of these directions one of the propellers is sideways on to the wind, and there is a narrow range of angles over which that rotor will be stalled, leading to uncertainty in the actual speed and direction. More importantly, since the response of the propeller at angles near to 90° is very small, a small error in the assumed directional response curve could lead to significant errors in these regions.

4.1 Reliability of Gill anemometers

These instruments are intended to be continuously fed with dry air, which is passed through the mounting and out to atmosphere via holes close to the front bearing. The purpose is to prevent the penetration of dust and moisture. The flow rate is not specified by the manufacturer: by experiment a figure of around 5 litres per minute was thought to be the very least that would be satisfactory (ie 10 litres per minute for a pair). As the installations are remote, this quantity of purging air could not be provided. This is probably the reason why three out of the four propeller anemometers have started sticking and are no longer producing meaningful signals. The failures occured after 3,5 and 9 months operation. The instruments will be removed and examined at the next opportunity.

Directional response of Gill anemometers

Gill propeller anemometers have a near-cosine response, so that if two are mounted at 90° the vector windspeed can be calculated by:

$$V = (V1^2 + V2^2) \qquad (1)$$

$$\theta = \tan^{-1}\left(\frac{V1}{V2}\right) \qquad (2)$$

where V1, V2 are the two measured speeds, V is the resultant velocity and θ the direction.

However the response actually differs from a cosine law in practice (see fig 9): the manufacturers supply a response curve which deviates in the y direction significantly from the cosine law in places. If the response is given by a function f(θ) (≅cosθ), then

$$V1 = Vf(\theta) \qquad (3)$$

$$V2 = Vf(\theta-90°) \qquad (4)$$

Hence V1/V2 = f(θ)/f(θ-90°) = g(θ) (≅cotθ) (5)

Given f(θ), g(θ) can be calculated. Hence, given V1 and V2, this can be used to calculate θ, and hence also V from (3) or (4).

This procedure was carried out using linear interpolation: firstly to find g(θ) for a number of θ values, again to determine θ from g(θ), and then again to find f(θ). The first step needs only to be done once: values of g(θ) are stored in an array for use when required.

One problem is that g(θ) is sometimes infinite. Consequently a modified function is actually used, defined by

$$g'(\theta) = g(\theta) \quad \text{for } -1 < g(\theta) < 1$$

$$g'(\theta) = 1/g(\theta) \quad \text{elsewhere}$$

so that $g'(\theta)$ is always between -1 and $+1$, and is calculated as

$$g'(\theta) = V1/V2 \quad (|V2| > |V1|)$$

$$g'(\theta) = V2/V1 \quad (|V1| > |V2|)$$

In order to be able to calculate θ from $g'(\theta)$, the latter must be single-valued. This is only true within each of the four 90° quadrants. However it is easy to determine the correct quadrant by examining the signs and magnitudes of V1 and V2.

Finally, to ensure the best accuracy, V is calculated from either (3) or (4) as follows:

$$V = V1/f(\theta) \quad (|V1| > |V2|)$$

$$V = V2/f(\theta-90°) \quad (|V2| > |V1|)$$

5 TURBULENCE ANALYSIS

As explained above, only a limited amount of fast data for Scroo Hill has so far been collected, so the results presented here on turbulence cover a limited range of wind speeds (around 7-14 m/s) and directions (60°- 150°) during 2 days and the intervening night in July 1985.

5.1 Turbulence intensity

Fig 10 shows turbulence intensities of ten-minute data samples, i.e. standard deviation of 1 or 2 second data about the 10-minute mean (after removal of a linear trend fitted by least squares). There is only a small difference in using 2-second rather than 1-second data: to be strictly comparable, the night-time turbulence intensities should be increased by a few per cent.

The turbulence intensities in fig 10 are mostly in the range 0.05 to 0.08, and apparently not correlated with wind speed at all. These figures are for the 35m anemometer, which is the hub height of the proposed 750kW turbine.

The mean turbulence intensity over the period was 0.064 at 35m, and 0.076 at 10m. Increasing the sample length from 10 minutes to one hour increases these figures to 0.074 at 35m and 0.086 at 10m.

Fig 11 shows the variation of turbulence intensity with height for three direction bins, again using 10-minute samples. Above 30m direction has little effect, but near the ground there is a much greater increase in the 120°-150° direction bin. There is no obvious explanation for this, but clearly more data covering a wider range of directions would be useful.

Note that the Gill readings are kept separate because (apart from the problems mentioned above) they are not strictly comparable to the results from the cup anemometers, since the Gill readings are analogue signals sampled instantaneously while the cup anemometer outputs are averaged over the sampling interval (by counting pulses). The Gills would therefore be expected to give higher apparent turbulence intensities.

5.2 Turbulence spectra

A turbulence spectrum calculated from 5½ hours of 1-second data is shown in fig 12. A clear peak around 30 seconds can be seen. At lower wind speeds (not shown) the peak moves to lower frequencies as expected. The smooth line is a Kaimal spectrum for neutral conditions fitted using a least squares algorithm. The parameters of the fitted spectrum are given in the figure. The expected parameters would depend on the roughness length which is unknown, and the applicability of such "standard" spectra on top of a steep hill is in any case open to doubt. The logarithmic plot clearly shows the -5/3 asymptote which is expected at high frequencies from theoretical considerations. Note, however, the considerable deviation at low frequencies, which might be caused by topographic effects or thermal instability. Note also that there is no drop-off from the -5/3 response at high frequencies, indicating that the response of the cup-anemometer is adequate for a one-second sampling rate. The anemometer is a Vector A100M.

5.3 Instantaneous wind shear

It may be of interest to know the variation of wind speeds in space as well as in time, for example to give an idea of the difference in wind speed experienced by two points on a wind turbine blade at the same instant. Fig 13 shows this information in histogram form between pairs of points separated by 5m and 25m respectively. A difference of 4 m/s between the speeds at the blade tip and the hub of the 750 kW turbine is not unexpected. The graphs also show that the differences can reasonably be approximated by a normal distribution.

6 CONCLUSIONS

The overall results presented cover almost a year at Scroo Hill but only 4 months at Susetter so a full comparison of the sites is not yet possible. However it does appear that Susetter has slightly lower wind speeds than Scroo, by about 1 m/s on average at any given time. Turbulence information is very preliminary as it only covers one 36 hour period at Scroo.

However, the following general conclusions seem to hold:

Wind shear is smaller than that found above flat terrain presumably due to the exposed hill-top locations, although a Northerly wind at Susetter does blow along the ridge for some distance which probably accounts for the more marked shear in this direction,

Turbulence is low (at Scroo Hill at least). Again this is probably due to the hill-top location, as the wind some way above

the summit experiences relatively little
influence from the ground. Nearer the ground
the turbulence does increase,

   Mean wind speeds are high: over 10 m/s at
35m at Scroo, with a Weibull distribution shape
factor of 1.9.

   Wind roses are generally in agreement with
the figures for the Lerwick Meteorological
office station.

   Although the average diurnal variation of
hourly windspeed appears significant, there is
virtually no average diurnal effect when the
mean hourly wind turbine output is computed.

7 ACKNOWLEDGEMENTS

The authors gratefully acknowledge the help and
advice given by other project members (J W
Twidell of the University of Strathclyde, N H
Lipman, M Collier of RAL, W G Stevenson, G A
Anderson, N L Holding of NSHEB). In addition,
the staff of NSHEB Lerwick Power Station are
thanked for their great efforts in the operation
and maintenance of the two sites, in particular
J Cousins and J Watt. The project also
acknowledge the funding given by the Science and
Engineering Research Council under grant
GR/C/82022 and the North of Scotland Hydro
Electric Board.

8 REFERENCES

Halliday, J A, Gardner, P, Bossanyi, E A (1985).
Wind Monitoring for Large Scale Power Generation
on Shetland. In: Twidell, J W, Lewis, C,
Hounam, I (Eds), Proceedings 4th Conference on
Energy for Rural and Island Communities,
Pergamon Press (1986).

Halliday, J A (1983). A Study of Wind Speed
Statistics of 14 Dispersed UK Meteorological
Stations with Special Regard to Wind Energy.
Energy Research Support Unit, Rutherford
Appleton Laboratory; Report RL-83-124

Shetland Island Council (1985). Shetland in
Statistics. Number 13 (1984). ISBN
0-904562-20-4.

```fortran
      REAL D(N)

      AVE=D(1)
      XVAL=D(1)
      DO 10 I=2,N
      VAL=D(I)
5     IF(VAL-XVAL.GT.180.0)THEN
         VAL=VAL-360.0
         GOTO 5
      ELSE IF(VAL-XVAL.LT.-180.0)THEN
         VAL=VAL+360.0
         GOTO 5
      END IF
      AVE=AVE+VAL
10    XVAL=VAL

      AVE=AVE/FLOAT(N)
15    IF(AVE.GE.360.0)THEN
         AVE=AVE-360.0
         GOTO 15
      END IF
20    IF(AVE.LT.0.0)THEN
         AVE=AVE+360.0
         GOTO 20
      END IF
```

Fig 1   Section of FORTRAN code for averaging wind
directions

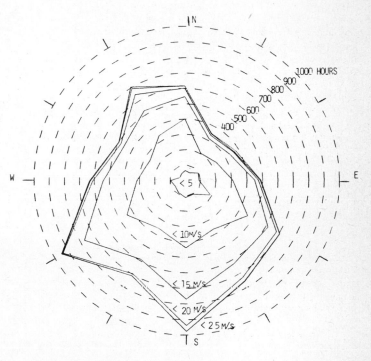

Fig 2   Wind Rose for Scroo Hill (1 Dec. 1984 to 2 Dec.
1985) (7778 hours). Wind speed at 35m, wind
direction at 15m

Wind Direction and Speed: 1983

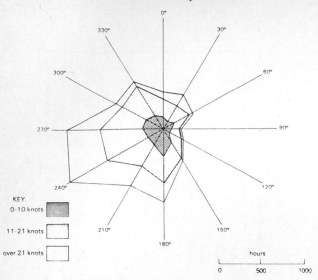

KEY:
0-10 knots
11-21 knots
over 21 knots

hours
0   500   1000

Wind Direction and Speed: 1970-1980

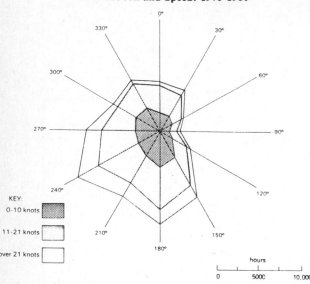

KEY:
0-10 knots
11-21 knots
over 21 knots

hours
0   5000   10,000

Source: Shetland Islands Council (1985)

Fig 3   Wind Rose Lerwick Meteorological station:
(a) 1983, (b) 1970—1980

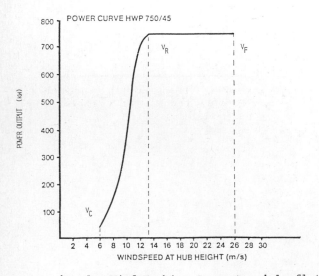

Source: 'Howden Wind Turbine Generators' leaflet
James Howden & Company, Glasgow, UK.

Fig 4   Characteristic of the Howden HWP-750 wind turbine

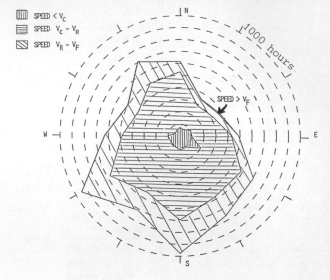

SPEED < V_C
SPEED V_C - V_R
SPEED V_R - V_F

SPEED > V_F

1000 hours

Fig 5   Wind power rose for Scroo Hill (1 Dec. 1984 to
2 Dec. 1985) (7778 hours)

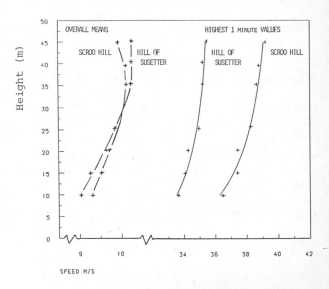

Fig 6   Mean and extreme wind speeds at Scroo Hill (1 Dec.
1984 to 2 Dec. 1985) and the Hill of Susetter
(30 August 1985 to 21 Jan. 1986)

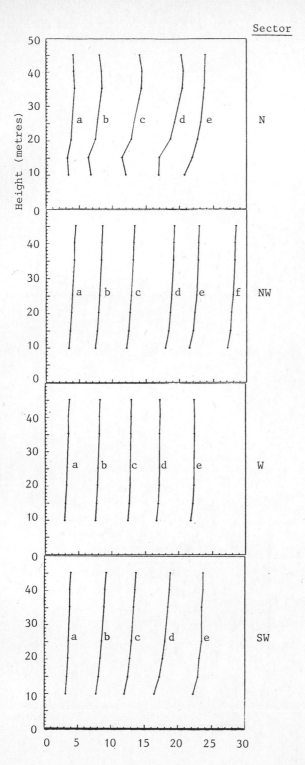

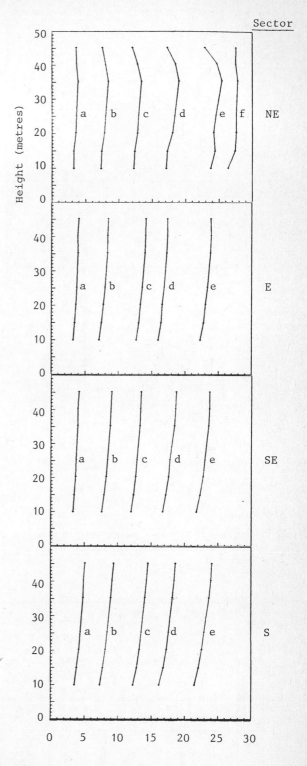

Speed Bins

a   0- 5 m/s at 10m
b   5-10 m/s at 10m
c   10-15 m/s at 10m
d   15-20 m/s at 10m
e   20-25 m/s at 10m
f   25-30 m/s at 10m

Fig 7   Wind shears at the Hill of Susetter — 30 Aug 1985
to 21 Jan. 1986

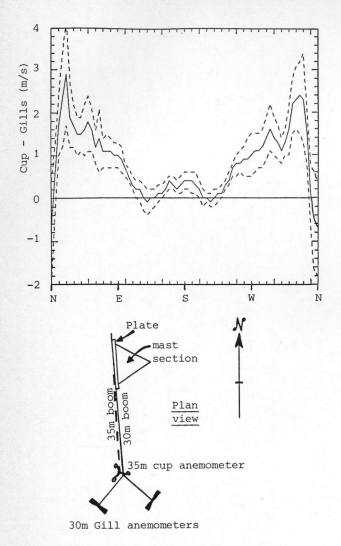

Fig 8   Discrepancy between results from propeller anemometer pair and nearby cup anemometer at Scroo Hill

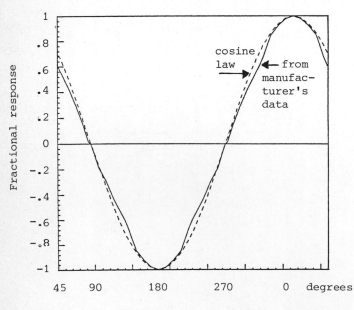

Fig 9   Directional response of a Gill propeller anemometer

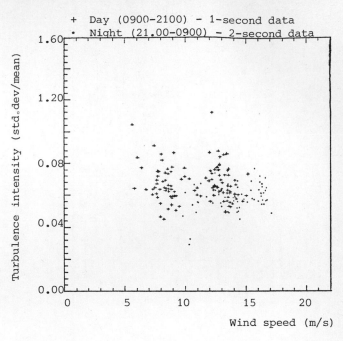

Fig 10   Turbulence intensities of 10 minute samples at Scroo Hill, 35m height

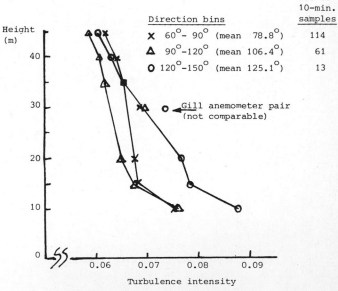

Fig 11   Variation of turbulence intensity with height and direction at Scroo Hill

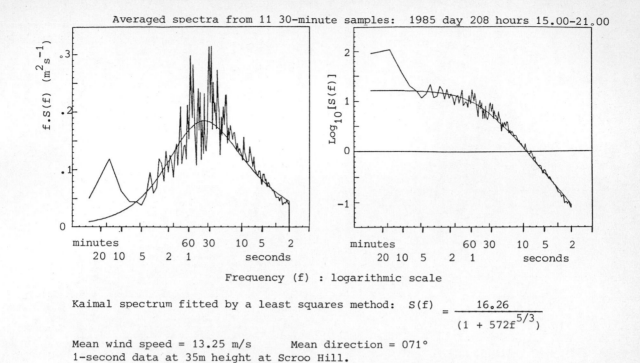

Frequency (f) : logarithmic scale

Kaimal spectrum fitted by a least squares method: $S(f) = \dfrac{16.26}{(1 + 572f^{5/3})}$

Mean wind speed = 13.25 m/s   Mean direction = 071°
1-second data at 35m height at Scroo Hill.

Fig 12   Typical turbulence spectrum at Scroo Hill, 35m
          height

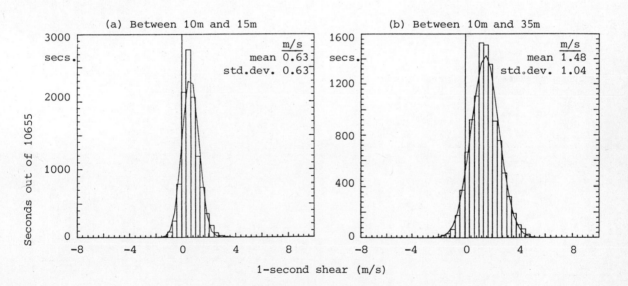

Fig 13   Instantaneous wind speed difference between two
          points on a notional rotor blade

Meteorological and statistical characteristics of high and low windspeed months over the British Isles: a preliminary analysis

J P PALUTIKOF, T D DAVIES, P M KELLY
Climatic Research Unit, School of Environmental Sciences, University of East Anglia
J A HALLIDAY
Energy Research Group, Rutherford Appleton Laboratory, Chilton, Didcot, Oxon.

SYNOPSIS Earlier work has shown that temporal variations in annual and monthly mean windspeeds over the British Isles are on a sufficient scale to merit consideration in the estimation of wind power potential at candidate WEC sites. In order to allow for this temporal variability in the wind power planning process, it is necessary to investigate the underlying meteorological and statistical characteristics. The results of a preliminary analysis of these characteristics are presented in this paper.

A set of high and low windspeed months were isolated as the two samples for analysis. This was done by performing a principal components analysis on a matrix of 42 station records by 240 monthly observations (ie 42 twenty-year time series). The first component of this analysis summarizes the general wind field. Thus months with high/low scores on this component were those when relatively high/low windspeed regimes for that month prevailed over the U.K. in general. Two samples, of six high windspeed (scores >+10) and eight low windspeed (scores <-5) months were drawn. Using a gridded (5° lat. x 10° long.) data set of Northern Hemisphere monthly mean sea level pressure, composite pressure anomaly maps were drawn up for each sample. These maps show that high windspeed months are associated with strong westerly flow and intensification of the two major pressure systems which dominate the area: the Iceland Low and the Azores High. Low windspeeds are related to a weakening in these two systems.

To investigate the statistical characteristics of high and low windspeed months, hourly windspeeds for a high windspeed month (December 1974) and a low windspeed month (December 1976) were analysed for four anemometer records. First, the frequency distributions were examined. Second, the mean windspeed at each hour of the day was calculated and the diurnal variations plotted. Third, a study was made of run lengths of hourly windspeeds above a range of threshold values. All three characteristics display within-sample similarities and between-sample differences. The results of this preliminary analysis will assist in decisions relating to future research directions.

1   INTRODUCTION

In an earlier project funded by the Central Electricity Generating Board (October 1983 – December 1984 inclusive), we investigated temporal and spatial variations in monthly mean windspeeds over Britain. This work clearly demonstrated that fluctuations in time series at this scale are sufficient to merit consideration in the estimation of wind power potential at candidate WEC sites (1,2).

As an example of the possible range of variation, we may examine the record from Southport Marshside. The anemometer and site characteristics for this record are described in Palutikof et al (3). Over the period 1898 to 1954, windspeeds at this site ranged between a high of $7.3ms^{-1}$ in 1923 and lows of $5.2ms^{-1}$ in 1899 and 1945. If these extremes were random occurrences in a stationary series (in which the mean and variance were stable with time) then they would be of only minor importance. However, this is not the case. Rather, they form part of a strongly auto-correlated time series with well-marked phases of high and low windspeeds which persist for up to two decades. In the case of the Southport annual record, windspeeds were low at the beginning of the series but gradually increased until 1910. From 1910 to 1919 there was only one year at this site when the annual mean fell below $6.5ms^{-1}$.

From then until the early 1930s windspeeds declined, remaining low from 1930 to the end of the record. To illustrate this, from 1930 to 1954 (the last year of observation) there was only one year with an annual mean windspeed greater than $6.5ms^{-1}$, and only six years exceeding $6.0ms^{-1}$. Palutikof et al. (3) in an inspection of seven long-term (at least fifty-five years) windspeed records noted that such marked temporal instability was a characteristic of northern sites.

The power in the wind is given by the expression:

$$E = \rho AV^3/2 \qquad \ldots 1$$

where E is the available kinetic energy per unit time, ρ is the air density, A the projected area swept by the turbine blades, and V the wind speed. Thus, ignoring turbine characteristics and any changes in air density, the available power is determined by the cube of the windspeed. Using an air density of $1.1kgm^{-3}$ and an aerofoil diameter of 60m in Equation 1 it is possible to arrive at some estimate of the implications of instability in windspeed time series. The extreme values of $7.3ms^{-1}$ and $5.2ms^{-1}$ in the Southport record represent a change in the available ´annual mean power´ of from about 600kW to about 220kW. However, these are the extremes, and it is perhaps more realistic to examine the background decadal values in which the extremes are embedded. These vary from around 5.5ms^{-1} at the beginning and end of the record, to 6.8ms^{-1} for the decade 1910-19, corresponding to a change from about 280kW annual mean value to about 480kW. The noisier monthly series produce an even greater variation, particularly in the winter half-year. The range in each winter month was found to be from a maximum of around 9ms^{-1} at some point in the decade 1910-19 to a minimum in the region of 4ms^{-1} in the decade of the 1930s. Such a change represents a variation by a factor of 11, from over 1,100kW to around 100kW.

Such temporal instability has important implications for the wind power planning process. If estimates for the power available from a proposed wind turbine at Southport had been based on data for the decade 1910-19, then the output over the decade 1930-39 would have been very disappointing.

The above analysis was performed using monthly, seasonal and annual data. It pinpoints the potential importance of climatic variability, but the results cannot be directly utilized by the wind energy industry. A new programme of work, funded by the U.K. Department of Energy, will examine the impact of climatic change on windspeeds at time intervals of more direct relevance for wind power production. This involves the analysis of hourly and daily data. The results of some preliminary analyses of hourly windspeed data are presented in this paper.

One problem in analysing data at this level is the large quantity of observations. For the earlier work on monthly windspeeds a network of 52 stations over 27 years was employed, giving a possible maximum (ignoring missing values) of some 16,800 observations. For the same network over the same period, an analysis at the hourly level would involve some eleven million values. Particularly for a preliminary analysis, such a large number of data items is unmanageable. Thus the first step was to create subsamples of data for exploratory analysis. This was done on the basis of the results of a principal components analysis (PCA) performed on monthly mean windspeed data, as described in the next section.

2   THE SAMPLING TECHNIQUE

It was decided that the exploratory analysis should be performed upon a set of high and a set of low windspeed months. We have demonstrated in the analysis of monthly data that coherent high and low windspeed phases, with a persistence of up to two decades, exist (Palutikof, 1986). What is required now is further insight into the morphology of the two phases. First, how are they constructed from the hourly data: do they, for example, reflect a reduction in the number of calms, replaced by an increase in the number of occurrences at the low windspeed end of the frequency distribution, or do they arise from a small number of extremely high windspeed observations? Second, are there associated, but separate, changes in the statistical characteristics (such as persistence, or the shape of the diurnal cycle) of hourly windspeeds during high and low windspeed phases?

In order to answer these questions, as a first step, two samples were drawn: one of high windspeed months and one of low windspeed months. The population from which they were drawn was a subset of the 1956-82 network of 52 stations used in the earlier, monthly, analyses, and as described in (4). It was not possible to select months by simple inspection of this data set. The set of, for example, five windiest months at one station bore little if any resemblance to the set of five windiest months at another. Instead we identified the two samples using the results of a PCA performed on monthly mean windspeed data. PCA is a statistical technique which identifies statistically independent linear combinations of variables which account for decreasing amounts of the total variance in the data set. The first combination, or component as it is more generaly known, accounts for the maximum amount of the total variance, the second for the maximum amount of the residual variance, and so on. For general information on the application of PCA in climatology, reference may be made to (5) and (6).

The original data matrix consisted of 52 stations by 27 years (1956-82). However, two problems arose from this. First, the network was unduly weighted towards the south of England, and therefore ten records were arbitrarily removed from that region. Second, there were a large number of missing values at the beginning and end of records and thus we curtailed the record length to twenty years (1961-80). The data were converted to deviations from the monthly mean for each station; this effectively removes the seasonal cycle factor. The final PCA was performed on the correlation matrix for 42 station records by 240 monthly observations. (In fact, since the products of the standard deviations were almost always close to unity, there was little difference in the results from the correlation and covariance matrix analysis).

The first combination of the station records (or PC1) accounted for 51% of the total variance in the data set. The contribution of each wind record to each component is given by the _loading_ of each station on that component. The loadings for PC1 were, apart from the extreme northern stations of Lerwick, Kirkwall and Wick, of the same sign and of approximately the same magnitude at each station. Thus the first component, which summarizes the broad-scale characteristics of the windfield over Britain, is related to all stations (except those in northern Scotland) in a similar way.

The strength of each component from month to month can be expressed as a series of amplitudes or _scores_. Since the first component summarises the general wind field, a month with a large positive score will be a high-windspeed event for Britain as a whole, and a month with a large negative score will be a low-windspeed event. We used this relationship to select the two samples. The sample of high-windspeed months was taken to be all occurrences with scores greater than or equal to +10, and the sample of low-windspeed months all those with scores less than or equal to -5. Table 1 lists the months in each sample.

Table 1

High-windspeed months

    February 1962
    March 1967
    October 1967
    January 1974
    December 1974
    January 1976

Low-windspeed months

    March 1962
    November 1962
    February 1968
    September 1971
    March 1973
    March 1974
    April 1974
    December 1976

In order to test the internal homogeneity of the two samples, and also to determine their underlying meteorological characteristics, composite pressure anomaly maps were drawn up for each set of months. Measurements of sea level pressure at a spacing of 5° latitude by 10° longitude were taken for the area 10^ON to 75^ON by 70^OW to 80^OE from a data set supplied by the U.K. Meteorological Office. At each grid square the pressure anomaly for each month was calculated by subtracting the long-term mean (1866-1980) from the observations. Then, for the high-windspeed sample and the low-windspeed sample, the anomalies were summed algebraically for all months and then divided by the number of months in the sample to produce two composite pressure anomaly maps.

These maps show that the strength of the wind field over Britain, as reflected in the first component of the PCA, is closely related to features of the large-scale atmospheric circulation. The major features of the pressure field over the study area are the Azores anticyclone and the Iceland Low. Pressure patterns in high windspeed months are dominated by an intensification of both systems, with a +5mb anomaly west of Spain and a -9mb anomaly west of Norway. The associated strong pressure gradient over Britain must obviously lead to the generalized high (westerly) windspeeds of this sample. The low windspeed sample is characterised by a slackening of the general circulation, as shown by a -3mb anomaly in the Azores anticyclone and a +7mb anomaly in the Iceland Low. These results justify the sampling technique used here, and suggest that the two samples are internally consistent with respect to the underlying atmospheric circulation patterns.

3   THE ANALYSIS OF HOURLY WINDSPEED DATA

It is intended to base much of the early work on hourly windspeeds on samples drawn using the technique described above. Clearly the size of the sample can be adjusted by changing the threshold value of the amplitude of the PCA, to admit fewer or more months as desired.

In this paper we are only able to examine and compare data from one low-windspeed month (December 1976) and one high-windspeed month (December 1974). For these two months we have used hourly windspeed data for four stations: Benbecula, Tiree, Aberporth and Cranwell. The basic siting information for these stations is given in Table 2. Three of the stations are located at western coastal sites: Benbecula in the Outer Hebrides, Tiree in the Inner Hebrides and Aberporth in Cardigan Bay. Cranwell is chosen to be representative of the eastern part of the country. Table 3 shows the mean monthly windspeeds (MMW) for these stations for the two sample months.

Hourly windspeed regimes for these four stations and for the months December 1974 and December 1976 were analysed with respect to three characteristics. First, the frequency distributions were examined with respect to the mean and shape of the distribution curve. Second, the mean windspeed at each hour of the day was calculated and the diurnal cycle plotted. Third, a study was made of run lengths of hourly windspeeds above specified critical values. The results of each analysis are described below.

3.1   The Frequency Distributions

The Weibull distribution (WD) is commonly used in the literature to describe windspeed frequency distributions. It is a special case of the generalized Pearson Type III (gamma) family (8). In this project it is intended to use the parameters of the WD as the main summary statistics of windspeed. However, here we simply present histograms of the frequency distributions of the eight station-months to allow qualitative comparison (Figure 1).

The following points of interest may be noted in Figure 1:
1. As expected, the peak (or mode) of the frequency distribution shifts towards higher windspeeds in December 1976 at all four sites.
2. With respect to the kurtosis of the curves, it appears that the two northern sites, Benbecula and Tiree, display flatter shapes than the two southern sites in both months. Furthermore, there is a change in the kurtosis between months, although this is not in the same direction at all sites. Thus at Aberporth the curve for December 1974 has the higher kurtosis of the two sample months, whereas at Benbecula and Cranwell the reverse is true.
3. In December 1974 the shape of the frequency curve at all four stations tends towards a Gaussian distribution. A WD with a shape parameter of 3.6 approximates very closely to a Gaussian distribution (9). In December 1976 the four frequency distributions display a positive skew. Thus in this month the distributions conform more towards the expected shape (see, for example, (10)) with a lower value for the WD shape parameter.

There is clearly much scope for further work in this area, particularly to determine whether the differences noted in points 2 and 3 above apply generally or are simply a characteristic of the two sample months at these four sites. Furthermore, it is intended to determine what contrasting features emerge when windspeeds are binned according to direction at different sites.

Table 2: Site details for the four windspeed records

Station	Location	Height of site above MSL(m)	Estimated Effective height (m)	Latitude	Longitude
Benbecula	Outer Hebrides	16	10	57°30N	7°24W
Tiree	Inner Hebrides	21	11*	56°30N	6°54W
Aberporth	Cardigan Bay	147	13*	52°66N	4°36W
Cranwell	Lincolnshire	68	13*	53°0N	0°30W

*Adjusted to an effective height of 10m for this analysis, using an exponent value in a column of 0.16 in the power law equation following (7).

Table 3: Comparison of monthly mean windspeeds (MMW) for December 1974 and December 1976, (ms^{-1})

	1956-82 December MMW		MMW December 1974	MMW December 1976
	Mean	S.D		
Benbecula	8.15	1.45	10.4	5.4
Tiree	9.31	1.22	11.3	7.2
Aberporth	7.85	1.26	9.4	5.6
Cranwell	5.55	1.22	8.5	5.1

3.2 Diurnal Cycles

The diurnal cycle of windspeed in the boundary layer is generally described as varying between a maximum in the mid-afternoon and a minimum at night (11). Soon after sunrise solar heating generates a surface radiative surplus. Convective warming produces an unstable layer next to the ground which gradually deepens through the morning. This allows vertical exchanges to penetrate well up into the atmosphere, which in turn permits greater momentum transport from faster-moving air aloft and increases surface winds. The depth of the mixed layer thus created may reach 0.5 to 2km by mid-afternoon. At night surface cooling produces a low-level temperature inversion which in turn dampens vertical exchange and momentum transport. Therefore nocturnal winds are generally light.

The diurnal patterns for the eight station-months under analysis are shown in Figure 2. Clearly they do not show any simple trends between higher daytime and lower nocturnal values. This is to be expected in the high windspeed month (December 1974) when strong regional flow should break down any boundary layer constrasts produced by the effects of insolation. What is, perhaps, unexpected is the amount of variation that does occur in this month. One might expect that a strong regional flow would give steady windspeed values throughout the day, whereas in fact there is substantial variation. For example, the range of hourly means at Cranwell is from $15.5ms^{-1}$ at 1900h to $19.1ms^{-1}$ at 1200h.

In the low windspeed month (December 1976) it might be expected that with a weak mesoscale flow the boundary layer would be sufficiently well-developed to envelop an anemometer located at a standard height of 10m. However, for December 1976 this is not the case. It should be recalled that the sampling technique selects months which have low (or high) windspeeds relative to the same months in other years, and not to the total population. Thus December 1976 has low windspeeds relative to all Decembers, not to all months. Had we selected a low windspeed summer month, it is possible that it would demonstrate a true diurnal cycle.

The suggestion from this preliminary analysis is that the calculation of 31-day means of hourly windspeeds to demonstrate daily variations obscures important features of the data. High-windspeed months are thought to arise from spells of high values caused by storms passing through the area. It is likely that diurnal patterns during these spells will be dictated by the characteristics of the storm and will differ fundamentally from the diurnal patterns at other periods of the month. It is not possible at present to suggest the underlying characteristics of low-windspeed months, and more work remains to be done, particularly bearing in mind the low-windspeed seasons of the year.

3.3 Persistence

The third characteristic to be analysed was the duration of runs of hourly windspeeds greater than specified values. For each month and each station a search was made for runs of windspeeds equal to or greater than a range of threshold values, from 4 to $35ms^{-1}$. Thus thirty-two passes were made through each monthly file, one for each threshold. The pattern that emerges is as follows. The number of runs in each month is small for the low threshold values, since these are exceeded by almost all the observations. Towards the middle range of threshold windspeeds the number of runs increases, as the few long runs are gradually broken up by intervening spells of lower windspeeds. And finally, towards the upper end of the threshold range, the number of runs drops back again because there are only a few (short) spells of high winds. A typical example is for Cranwell, December 1974. For a threshold of $4ms^{-1}$ there are only two runs, of 45 and 698 hours. The number of runs increases until for a threshold of 19ms-1 there are 57 runs. At the final threshold value of 35ms^{-1} there is only one run, of duration just one hour.

Although this broad pattern prevails in most of the cases analysed here, there exist between-station and between-month differences superimposed upon it. In order to illustrate these, frequency surfaces were constructed, as illustrated in Figure 3. The run length counts were binned as shown in the axis notations of this figure, using a linear division for the windspeed threshold, and a logarithmic division for the run length. The results were used to construct an isometric projection of a single-valued three-dimensional surface Z(X,Y) where X is the windspeed threshold and Y is the run length. Thus, for example, at the grid point (X=16-19ms^{-1}, Y=1-3 hours), the number of occurrences for Cranwell, December 1974, is Z=108. The projections in Figure 3 are viewed from (M,N) looking towards the origin. No scale for Z is provided on these diagrams, since the shape is the most important consideration, but it should be noted that the vertical scale is the same in all eight diagrams.

The following points may be made about Figure 3:

1. The threshold values with the greatest number of runs are lower in December 1976 than in December 1974. This is particularly noticeable at Benbecula and Tiree. At Benbecula the threshold range with the maximum number of runs shifts from 24-27ms^{-1} in the high windspeed month to 8-11ms^{-1} in the low windspeed month. At Tiree the shift is from 20-23ms^{-1} to 4-7ms^{-1}.

2. The commonest run duration is 1-3 hours. However, a submaximum exists in the range 10-31 hours, and indeed at Tiree in December 1974 this is the largest category for the threshold range 20-23ms^{-1}, with 64 observations. The existence of this submaximum probably reflects a diurnal cycle in the data. This diurnal cycle is least pronounced at Benbecula in both months and at Aberporth in the high-windspeed month.

3. Long runs, of over 100 hours, are only found at low windspeed thresholds. Although it might be expected that they would be restricted mainly to the high windspeed month, in fact at Aberporth and Cranwell a substantial number of long runs are found in December 1976 (eleven and nine respectively).

4 CONCLUSIONS

We have shown in the above discussion that time series of monthly mean windspeeds are strongly autocorrelated, with well-marked phases of high and low windspeeds which persist for up to two decades. This clearly has important implications for the wind power planning process, since if estimates of the available power at a candidate WEC site are derived from data for a high windspeed phase, the subsequent actual output could well be disappointing.

A sample of high windspeed months and a sample of low windspeed months were drawn from the 1956-82 time series of the wind field over the British Isles, using the technique of principal components analysis. We were able to show that these two samples were internally consistent with respect to their underlying atmospheric circulation patterns, and displayed distinct between-sample differences.

A preliminary analysis was performed of hourly data from one month of the high windspeed sample and one month of the low windspeed sample. Four stations were used, giving a total of eight station-months. The characteristics considered were the frequency distribution, the diurnal cycle and the persistence (or run length). There were quite distinct differences both between months and between stations. However, the samples were too small for any generalized inferences to be made.

There is clearly much scope for more detailed analysis of windspeed data at the hourly level. Further information is required on the temporal and spatial variations in the statistical properties of hourly windspeeds which are of direct relevance to the wind power industry. Accurate estimation of the likely wind power potential from a candidate WEC site is impossible without this knowledge.

ACKNOWLEDGEMENTS

This work was supported by the Central Electricity Generating Board, the U.K. Atomic Energy Authority and the Science and Engineering Research Council.

REFERENCES

(1) PALUTIKOF, J.P., KELLY, P.M. and DAVIES, T.D. An analysis of the spatial and temporal variations of the wind field over the British Isles. Proc. European Wind Energy Conference, 1984, W. Palz (ed.), 1985a, H.S. Stephens and Associates, Bedford, pp.65-70.

(2) PALUTIKOF, J.P., DAVIES, T.D. and KELLY, P.M. Windspeed variations and climatic change. Wind Engineering, 1986, (forthcoming).

(3) PALUTIKOF, J.P., DAVIES, T.D. and KELLY, P.M. An analysis of seven long-term windspeed records for the British Isles with particular reference to the implications for wind power production. Wind Energy Conversion 1985, A. Garrad (ed.), 1985b, Mechanical Engineering Publications Ltd., London, pp.235-240.

(4) PALUTIKOF, J.P., DAVIES, T.D. and KELLY, P.M. A databank of wind speed records for the British Isles and offshore waters. Wind Energy Conversion 1984, P. Musgrove (ed.), 1984, Cambridge University Press, Cambridge, pp.414-425.

(5) MITCHELL, J.M., DZERDZEEVSKII, B., FLOHN, H., HOFMEYR, W.L., LAMB, H.H., RAO, K.N. and WALLEN, C.C. Climatic change. WMO Technical Note 79, WMO No. 195, 1966, World Meteorological Organization, Geneva.

(6) STIDD, C.K. The use of eigenvectors for
     climatic estimates. J. Appl. Meteor 6,
     1967, pp.255-264.

(7) HALLIDAY, J.A. and LIPMAN, N.H. Wind speed
     statistics of 14 widely dispersed U.K.
     meteorological stations.
     Proc. 4th BWEA Conference, 1982, BHRA Fluid
     Engineering, London, pp.180-188.

(8) HENNESSEY, J.P.Jr. Some aspects of wind
     power statistics. J. Appl. Meteor. 16,
     1977, pp.119-128.

(9) CONRADSEN, K., NIELSEN, L.B. and PRAHM,
     L.P. Review of Weibull statistics for
     estimation of wind speed distributions.
     J. Appl. Meteor. 23, 1984, pp.1173-1183.

(10) PETERSEN, E.L., TROEN, I., FRANDSEN, S. and
     HEDECAARD, K. Windatlas for Denmark, 1981,
     Risø National Laboratory, Roskilde.

(11) OKE, T.R. Boundary Layer Climates 1978,
     Methuen and Co. Ltd., London.

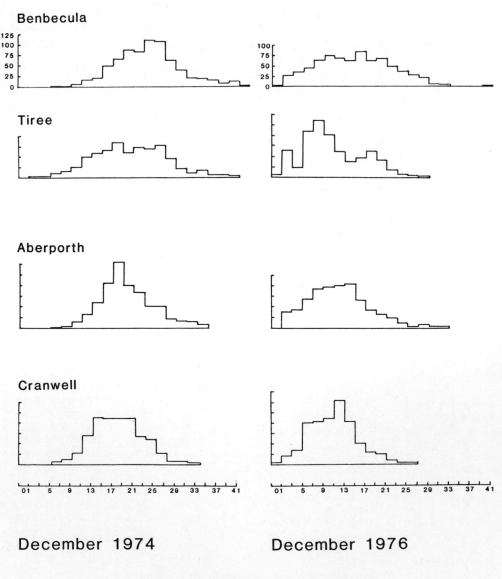

December 1974   December 1976

Fig 1   Frequency distributions of hourly windspeeds.
         Vertical axis shows the number of observations;
         horizontal axis shows the wind speed in ms^{-1}

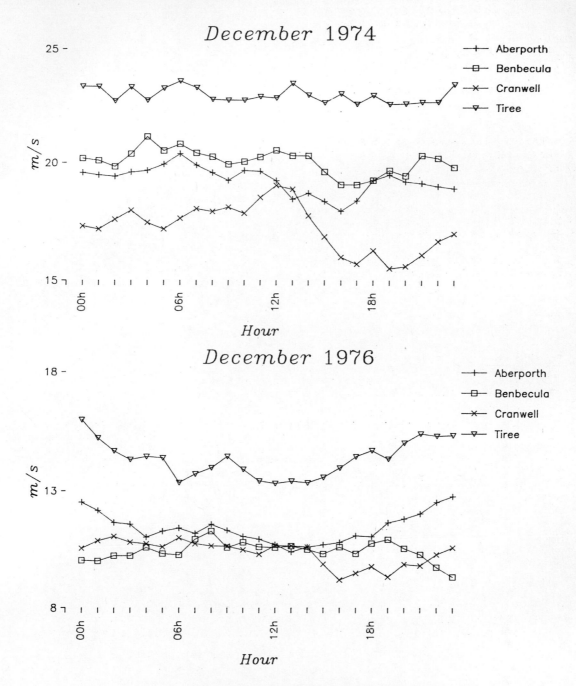

December 1974

m/s

Aberporth
Benbecula
Cranwell
Tiree

Hour

December 1976

m/s

Aberporth
Benbecula
Cranwell
Tiree

Hour

Fig 2   Diurnal variation of hourly windspeeds based on
the 31-day mean at each hour

Fig 3   Frequency surfaces of run lengths (hours) above
specified threshold windspeeds (ms⁻¹)

Applied wind generation in small isolated electricity systems

W M SOMERVILLE
International Research & Development Co. Ltd

SYNOPSIS   Some problems in the operation of a small electricity supply system for an isolated community are discussed in relation to the application of wind energy.   The choice of aerogenerator, the use of load management techniques to balance supply and demand and possible benefits of energy storage systems are examined.   The importance of community involvement in load control is stressed and some results given.

1   INTRODUCTION

The provision of an electricity supply to a remote or isolated community by a connection from a national network may, by reason of distance, mountainous terrain or some other natural barrier, be totally uneconomic or technically impractical.   To provide a supply by a diesel generator will require the lowest capital investment, but the running of such an installation will involve the provision of non renewable fuel and the transport of fuel to the generator.

Small remote communities persist because the people love their homeland.   Only few are prosperous, but many can only provide subsistence sufficient to maintain the population with little, if any, surplus for export to provide the purchasing power to maintain a diesel generated electricity service unless subsidised by a much larger community or as a matter of national policy.

It has been shown that a wind turbine generator, in conjunction with diesel generation, can, when suitably proportioned to the needs of a community, show considerable benefits both reducing the unit cost of electricity and improving the standard of life by providing dry heat to dwellings when the turbine is able to produce more power than required by essential services.   Dry heat is particularly important in providing comfort for the very young and elderly as it combats the all pervading dampness typical in island and mountain communities.

2   DIESEL GENERATION

The operating cost of small diesel generating plant can be relatively high, figures of 15-30p/kWhr are typical.   The main element of cost is the fuel, but a significant portion of cost is needed to provide lubricating oil, replacement parts, labour to service the plant and insurance against disaster.   These latter elements are fixed costs directly linked to running hours.

When generating, fuel is used to overcome engine friction, windage and auxiliary loads such as fuel pump, oil pump, dynamo, water pump and cooling fan and provide for heat losses.   The alternator too has friction and windage losses and, in addition, power is required to provide excitation and to overcome iron losses.

If the community to be served by the system is scattered, it may be necessary to transmit power at high voltage to minimise line volt drop and reduce cable costs.   Here additional power will be absorbed in transformer iron losses, with a small copper loss, before power is delivered to the consumers. The balance of the fuel consumed is converted to usable electrical power.

The operating efficiency of a diesel engine is acceptably high at rated load, even allowing for the engine losses, but the additional losses from the generator and system energisation cause an even more rapid fall in efficiency as load is reduced.

3   LOAD FACTOR AND PRIORITIES

In the operation of a small system with a limited number of consumers, it is extremely difficult, if not impossible, to maintain a high load factor.   The demands on an electricity system in a small, isolated community will vary considerably from one situation to another, but from our experience, it would seem that the priorities are as follows:

(a)   electric lighting,
(b)   power for freezers,
(c)   entertainment,
(d)   water heating,
(e)   power tools, appliances etc,
(f)   space heating.

These various appliances, preferred time of use, thermostatic controls etc., make for a daily cycle of demand that is only broadly predictable, with a considerable random element.

The widespread availability of standard appliances at reasonable cost encourages a steady growth in demand, and a major problem to be considered is the rating of plant for a given community to suit the accepted pattern of living.

Large elements of load, used in a random manner, can make this calculation very difficult. Typical 'problem' loads in a small system are electric kettles, immersion heaters, washing machines (with heaters), tumble driers, instant electric showers and cookers. Most of these appliances are rated for speed and convenience, and, on a large network of several megawatts capacity, switching any single unit makes an insignificant load change. This is not the case on a small network, and, for example, the author has experienced island blackouts caused by the common kettle when, after a social event, too many islanders decided to have a cup of tea when returning home. 16 kettles will quite effectively shut down a 40 kW generator - an ever present danger when every house has one. This is the problem of maximum demand, and if the generator is rated to handle such loads, the normal operating efficiency will be very low. Rules of diversity are applied based on experience but it is important to remember that a situation which produces a temporary local overload on a large network will blackout a small system.

4 COMMUNITY INVOLVEMENT

From the foregoing comments, it will be understood that, in small system operation, there is greater need for individual responsibility in the use of appliances if the operating costs and charges to the consumers are to be kept acceptably low. This may also require that the system be shut down for agreed periods each day when the demand is low. In a small system, the use of several generating units operating in parallel for high demand and shutting down units in sequence as demand falls, is not necessarily a viable option as the smaller units are usually less efficient, and the automatic start up and synchronisation of sets in sequence can add appreciably to the plant cost and the maintenance burden.

The problems of controlling maximum demand and, at the same time, operating at a high load factor requires a degree of involvement by all members of the community, and one method of organisation which proved effective in a small community was to have all consumers represented on the operating committee. The main advantage being an open forum to discuss problems, promote clear communication and develop a community awareness of the consumers role in the effective running of the system.

5 WIND GENERATION

Using an aerogenerator as a prime source of electrical power for an isolated community introduces new problems. Firstly, it is a new and unfamiliar apparatus and, at present, experienced operators are few and far between. Thus, the first requirement is that the machine must have a low maintenance requirement and, further, the maintenance that is required must be compatible with skills available within the community if down time is to be minimised.

To be compatible with commonly available appliances, the wind turbine must produce power at conventional voltage and frequency. For simplicity, this will involve a fixed ratio power train, between turbine and generator, and constant speed operation to generate constant frequency. A self excited alternator with automatic voltage regulator is required for constant voltage output.

There are many ways of controlling turbine speed, such as varying the pitch of the whole or part of the blades of the turbine, and by using spoilers or similar lift control devices, but reflection on the variable nature of both the wind speed and the applied load will show that these control devices will be in continuous motion which, in addition to the added complexity and cost of providing control systems of sufficiently fast response, will increase the need for maintenance and competent adjustment.

The simplest option would seem to be the use of rotor blades designed to be 'stall regulated' to limit the maximum power, and to use electrodynamic braking to control the turbine speed. This option allows the use of fixed pitch rotor blades, and the controls may be completely static semi-conductor elements free from fatigue, wear and tear. Additionally, since electronic devices can respond nearly instantaneously, speed of response is unlikely to be a problem.

Modern electronic control systems, when built on a modular basis, although very complex in circuit detail and function, can be maintained by unskilled staff replacing plug in units in response to a written fault finding routine. Modules suspect and faulty can be discarded, or returned to the manufacturer for repair as appropriate.

Operationally, the variable power output from an aerogenerator, as wind speed varies, adds to the complexity of matching supply and demand. By reference to a basic power performance curve, Fig. 1, there are three distinct operating conditions:

(a) below cut in wind speed there is no power available,

(b) between cut in and rated wind speed the power is highly variable and,

(c) above rated wind speed the power available is substantially constant.

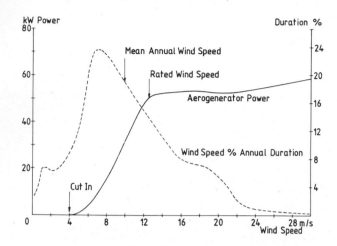

Fig 1   A stall regulated aerogenerator performance curve and the windspeed duration curve for Fair Isle

In high winds, the aerogenerator will deliver power on demand, subject to maximum demand not exceeding rated power. In low winds, an alternative source of supply is required, but the most difficult range of operation is clearly between cut in and rated wind speed. The importance of this operating region is clearly seen when relating the power performance curve to the wind duration curve. The effectiveness of an aerogenerator in meeting load demand in the most frequently occurring wind speeds is critical to the economic success of a wind and diesel co-generation scheme if fuel costs are to be minimised.

The stall regulated wind turbine is designed to be most efficient in the lower operating wind speeds, in order to produce most benefit from the most frequently

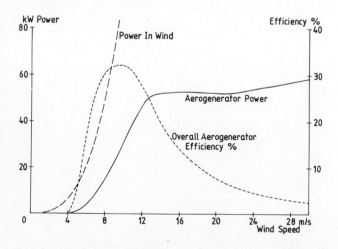

Fig 2   The relationship between wind energy, aerogenerator output, and aerogenerator overall efficiency

occurring wind speeds. At higher wind speeds the overall efficiency falls rapidly, as the 'stall regulation' of the blades sheds progressively more and more of the energy passing through the rotor circle in order to limit the captured energy to a safe level. This is shown in Fig. 2. This design feature makes for the best possible performance in this critical region of operation, whilst retaining simple rotor construction with low maintenance requirements.

Overlying this problem of reduced power output are the rapid changes in wind speed which occur continuously, second by second, due to local airflow disturbances created by local topography, obstacles such as trees and buildings, and thermals generated by sunlight and cloud cover on varying surface conditions.

To illustrate the problem Fig. 3 has been drawn to show the effect on the power output of ±15% variation in wind strength at three mean wind speeds of 5, 8 and 11 m/s with a typical power performance curve for a 50 kW wind turbine. The wind speed variation has been assumed sinusoidal to show more clearly the changing nature of the power fluctuations. This shows that in the critical range even small wind speed variations can cause a considerable change in the output power. In the case illustrated, where the wind turbine is governed by a local dump load, the maximum power available to serve the network, if network load is manually controlled, in a mean wind speed of 8 m/s ±15% is only 18 kW although a maximum power of 38 kW is available periodically and the mean power is 28 kW.

6   LOAD MANAGEMENT

Since most of the electricity used is for heating, the author's company was prompted, in considering this problem, to look at the feasibility of automatic control of heating appliances, and develop a system which could respond to the available wind energy continuously adjusting a large proportion of the available consumer load as required. This system can fully utilise the available energy, largely avoiding waste to dump. Thus, in the example given, the consumers can benefit from the mean output of 28 kW. This makes the system operation more cost effective as more energy is delivered for the same operating costs.

A second benefit arises from this system in that, given automatic load control, the utilisation of electricity may be re-examined and use categorised into essential and non-essential groups.

These categories can be further divided where some essential services may be considered essential within time limited bands, i.e. freezers – require power for, say 2 hours twice a day as a minimum, and entertainment limited to evening hours.

To keep the essential service demand to a realistic minimum, the following loads are assumed for each household:

Lighting — subject to need and available power. 200 W

Radio/Television — normally evenings only or when power available. 250 W

Fridge/Freezer — supply 2 hours in every 12 hours 300 W

Total variable 100 W – 750 W.

This total, Spartan by mainland standards, is acceptable in conditions when the power available is strictly limited.

The time dependent essential loads can be effectively served by adopting a policy of guaranteed service hours for example by ensuring a service for say two hours each morning and from dusk until 11.15 pm. When the wind turbine cannot meet the demand during these periods, the diesel generator may be run either alone or assisting the wind turbine as required. By such means, the essential services load, subject to demand, may be reduced to a minimum, thereby extending the number of hours annually that the aerogenerator can sustain the system.

On its own, this system is highly restrictive on the use of appliances – for example a washing machine with built in heater could not be used on the automatically controlled circuit, as a rapid sequence of starts would overheat the motor. It could be run by splitting the drive and heating circuits, allowing heating to be controlled and proceeding with the wash programme when the necessary temperature was reached. However, such modifications repeated many times would be costly and, for this reason and considerations of safety after modification, a simpler solution, involving voluntary regulation, is preferred.

By providing a simple form of available power indication at the automatic load control units, each consumer can judge the output being fed to controlled heaters and decide whether there is sufficient to run a particular appliance without overloading the demand for services. This precaution, possibly linked to an agreed rota for use, allows high load appliances to be used when the energy is available.

To encourage and reward consumers for co-operative operation of the system, a suitable tariff structure is desirable. This must encourage the use of controlled appliances in favour of directly switched units, and further apply a higher tariff when the diesel generator is in use, provided this fact is also signalled to the consumers. This may, or may not, be combined with a quarterly or annual standing charge. The tariff structure used on Fair Isle is as follows:

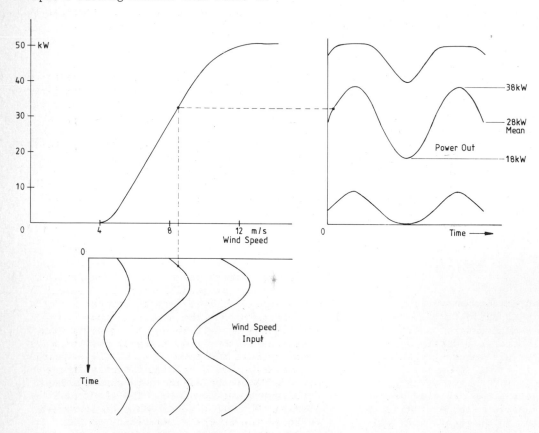

Fig 3   The effect of varying windspeed ±15 per cent on output power at mean windspeeds of 5, 8, and 11 m/s

Quarterly connection charge £20
Power to controlled appliances 2p/kWhr
Power on demand from aerogenerator 3p/kWhr
Power on demand from diesel generator 7p/kWhr

This tariff generates sufficient funds to meet all running costs of the system, including a sinking fund for the eventual replacement of the major components of the system.

Having made provision for load management to match the power fluctuations at mid range on the power performance curve, the same techniques will handle the dips in power when operating in the transition region close to rated power.

However, at the lower transition point close to 'cut in' wind speed, a different situation exists which also differs from the case of the network connected machine. The network connected machine produces power by trying to run faster than the speed dictated by the network frequency, so, as the wind speed dips, the turbine is automatically unloaded and it can remain energised, waiting for the wind to freshen at normal speed or until such time as the wind fails and the machine trips out on drawing motoring power.

The stand alone machine, on the other hand, must at all times provide power for excitation and auxiliary loads as well as the minimum demand on the network. Thus when the wind speed drops the controlled loads are shed automatically but, when all controlled loads have been switched off, the machine will continue to decelerate making up the difference between supply and demand from the kinetic energy in the rotating machinery.

To protect the system and appliances connected to it, there is a lower limit to the supply frequency which can be tolerated even transiently and should the speed drop by 10%, the system is automatically disconnected. The power loss due to reduced airspeed over the rotor blades as they slow down is partly compensated by increased angle of attack. Experience has shown that the system can frequently recover from a momentary reduction in wind strength below that needed to sustain load. Normally these lulls in the wind are of short duration typically 4-6 seconds before the wind strength recovers.

7 ENERGY STORAGE

A critical dip in wind strength may occur long before the mean wind speed falls below the sustaining level and the use of short term energy storage systems to sustain output during such brief intervals is being studied by several groups.

Long term storage systems may also play a useful role in conjunction with a wind turbine, both to assist in marginal winds and to provide for essential service demand for limited periods to minimise the use of diesel generation.

For an isolated community to survive, a fresh water resource must exist and be sufficient to provide potable water for the population, animals, hygiene and cultivation. In some cases, the water supply may be abundant, and, where sufficient head exists, a small hydrogenerator may be sufficient to meet the basic needs of the community for at least part of the year. There are many more cases where there may be an abundance of water but insufficient head for a small scale hydrogenerator.

Where the local terrain is suitable and an adequate supply of water exists, although insufficient to provide for hydroelectric plant, there is a case for re-examining the hydroelectric resource as a means for storing energy to provide an essential service to cover limited periods of low wind energy. Where the natural flow is insufficient then, subject to the cost of providing storage capacity, the surplus wind energy could be used to pump water to high head storage. Such a system is to a degree self compensating for seasonal variations since in the cold wet season less pumping will be required and power will be available for heat, whereas in the warm dry season less heat is needed for comfort and surplus wind energy can be used for pumping. It is conceivable that a high head reservoir may offer additional benefits, in some cases, by adding irrigation potential not previously available.

To be effective a combined system operating in a small community must be run automatically but should be programmable to allow for changing needs and conditions in a simple and readily understood way.

Given suitable conditions, it should be possible in some cases to dispense with diesel generation entirely and the difference in generation costs may be sufficient incentive to ensure that the system is managed in this way or at least to minimise diesel running time. It is, however, essential to retain some diesel generating capacity to ensure that a minimum service can be guaranteed during plant outages for essential maintenance and to cater for rare but possible climatic conditions such as unusually long periods of calm dry weather.

8 DIESEL SUPPORT AND CHP

The use of diesel generation to support the aerogenerator is possible but, for reasons given earlier, it is largely counter productive due to the low efficiency experienced on part load. If, however, the diesel generator is an integral part of a combined heat and power system, the economics of operation would be significantly improved provided location adjacent to sizeable heat requirement was possible. The electrical loading could then justifiably be increased to raise the energy availability to the community at large. Such operation would be relatively efficient in cold weather but there would be a high risk of inefficient operation in warmer conditions in Summer so annual operating costs would be greater but less so than using a plain diesel unit in this manner.

9 FUNDING

The capital cost per consumer of establishing a small electricity scheme will be much higher than for a large urban scheme. Even where there is an existing system, the costs of developing the system to utilise increased output from an aerogenerator will be costly as dwellings tend to be more widely spaced than in a typical urban situation, increasing the investment in distribution works. Thus, the cost of the aerogenerator may well be less than half and possibly as low as 10% of the total cost of a new scheme.

It is improbable that a small community can fund or support full interest payment on the capital required and the establishment of operational systems of this type must depend on grant aid to a large degree.

Small systems provide invaluable experience as pilots for larger schemes in the future, both in establishing hardware requirements and operating techniques, and there would seem to be a good case for national support for such schemes in the light of the social and technological benefits gained.

10 BENEFITS

The social benefits to be derived from wind and diesel co-generation schemes are primarily in reduced operating costs, producing electrical power at an affordable cost to the customer. This is enhanced by taking a systems engineering approach to the operation and management of the whole system when wind generation is introduced and further augmented by enlisting consumer involvement in overall management of energy use.

On Fair Isle, this type of system provides more than three times as much energy, and more than two and a half times the hours of supply but consumes only 40% of the diesel fuel that was used before wind generation was introduced at approximately the same total annual cost to the consumers. The aerogenerator provides about 90% of the energy consumed.

The growth in the use of electrical energy on Fair Isle in the three years the system has been in operation is shown graphically in Fig. 4. The reduced output in the last year of operation was the result of abnormally low winds at approximately 85% of the normal mean annual wind speed.

The community have been adept at using the additional energy but even so some energy

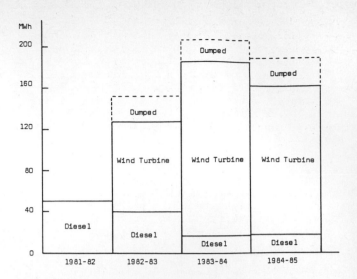

Fig 4 Growth of electrical energy consumption on Fair Isle

is lost to dump, particularly during the warm Summer months. Had the topography and water resource on Fair Isle been adequate this lost energy could have been stored for use to extend the hours of generation.

The operating record of the system is summarised in Table 1. The islanders have appreciated the greatly extended hours of generation, particularly in Winter, and the improvement in comfort afforded by the 'dry' heating from wind energy surplus to service demand. A hidden benefit in the operation of the system has been a reduction of 40% in coal imports and a reduction of 34% in heating oil used. Previous papers and articles describing this scheme in more detail are listed in the Bibliography below.

11 ACKNOWLEDGEMENTS

The author wishes to thank International Research & Development Co Ltd. for, permission to present this paper, and aid in its preparation and also the Fair Isle Electricity Committee for information used.

12 REFERENCES

(1) SOMERVILLE, W.M. 55 kW aerogenerator formally commissioned on Fair Isle. NEI Review 4, February 1983, No. 4, pp. 15-19.

(2) STEVENSON, W.G. and SOMERVILLE, W.M. The Fair Isle wind power system. Musgrove, Wind Energy Conversion, March 1983, pp. 171-184. ISBN 0521 26250X.

(3) SINCLAIR, B.A., STEVENSON, W.G. and SOMERVILLE, W.M. Wind power generation on Fair Isle. Twidell, Energy for Rural and Island Communities III, pp. 155-162. September 1983. Pergamon Press. ISBN 0-08-030580-6.

(4) STEVENSON, W.G. and SOMERVILLE, W.M. Optimal use of wind and diesel generation on a remote Scottish Island. Proceedings European Wind Energy Association Conference, Hamburg, October 1984, pp. 681-684. ISBN 0-9510271-0-7.

(5) SOMERVILLE, W.M. and STEVENSON, W.G. A wind turbine greatly improves the power supply on Fair Isle. Modern Power Systems incorporating Energy International, June 1985, pp. 19-23.

Table 4 Overall summary of performance of Fair Isle wind diesel system

Operating year	1981-82		1982-83		1983-84		1984-85	
	kWh	%	kWh	%	kWh	%	kWh	%
WTG output for year		N/A	119,465	100	189,829	100	172,280	100
Discarded to dump		N/A	26,838	22.5	20,934	11.0	26,460	15.4
Used on network		N/A	92,627	77.5	168,895	89.0	145,820	84.6
Island consumption	49,872	100	129,922	100	185,024	100	162,409	100
Provided by diesel	49,872	100	37,295	28.7	16,147	8.7	16,589	10.2
Provided by WTG	N/A		92,627	71.3	168,895	91.3	145,820	89.8
WTG availability		N/A		64.4		94.5		97.3
WTG load factor		N/A		27.3		43.3		39.3
Diesel fuel used	4,138 gal	100	3,517 gal	85.0	1,513 gal	36.6	1,595 gal	38.5
kWh sold/gallon diesel fuel	12.05	100	36.94	307	122.29	1,014	101.82	845

A 15 kW stand-alone windpower system

R TODD and **T KIRBY**
Centre for Alternative Technology, Machynlleth

SUMMARY   A 15 kW TCR/Polenko (Technische Combinatie, Rheinen) wind generator was installed at the Centre in 1984.   The paper describes the controls and the electrical system which provides for variable or 'fixed' rotor speed, direct or AC-DC-AC feed of the site electrical load and the sharing of surplus power between battery charging and a heating load. An account of the installation of the wind generator and our operating experiences is included together with some results of performance monitoring.

INTRODUCTION

The Centre is based on a 30 acre site with commercial buildings, workshops, educational facilities and accommodation for 30 permanent residents and around 40 students or temporary workers. A main aim is the demonstration of renewable energy technologies and to this end the site electricity supply is not grid-linked. All the electricity used is produced on site mainly from a combination of micro-hydro and windpower.

The load has been kept very low by the use of high efficiency/low power lighting and appliances and high 'energy awareness'. Electricity for heating is confined to controlled loads using the surplus energy after priority loads have been met. The 'on-demand' site load usually ranges from 0.2 to 4 kW, and daily consumption varies from 20-50 kWh, summer to winter. The site loads can be flexibly grouped into several different priority loads. This allows for selective load shedding when necessary.

The TCR/Polenko wind generator was added to the system to a) reduce the need to run the standby generator in long dry periods when insufficient hydro-power is available, and b) to provide a heating input to the main building complex both for hot water supply and space heating.

A site for such a machine had been identified some years earlier and windspeed measurements indicated an annual mean of about 5 m/s at 10m. The site is on a hill over-looking the Centre, some 800m from the main building. Capital availability limited the size of the machine to the smallest which would be capable of meeting a large fraction of the site electricity demand during long dry spells. Given the expected wind speeds, a machine of 10m diameter was indicated.

The TCR/Polenko was chosen after a survey of other similar size machines, because it was one of the few offered as a stand-alone machine, construction was simple and robust, and the price was very competitive. This machine was purchased on the basis that we - through our associated Company, Dulas Engineering Ltd. - would develop and manufacture the control system necessary for our application. The machine was supplied with a basic 'heating only' controller.

WIND GENERATOR DATA

Type   : TCR/Polenko WPS 10
Rotor   : 3 blade, steel, 9.6m diameter
Hub Height   : 20m
Total Weight   : 7.5 tonnes
Gearbox   : Shaft mounted, 30.4:1
Rated Power   : 15 kW
Cut in   : 4.1 m/s at 50 Hz
Best C_p : 0.28 at 7 m/s and 50 Hz   = 3.7
Speed Control   : Load control and tip flaps electrically operated by hub mounted actuator.
Overspeed   : Two spring operated brakes, one each side of gearbox, held off electrically.   3 independent overspeed sensors. Stop time <2 secs.
Shutdown Speed: Approx. 27 m/s

SYSTEM DESIGN

Controls
In periods of no hydro-power, consistency of supply to the site load from the wind system is important. Because of the limited storage capacity available, low windspeed performance is particularly important. The existance of an invertor/battery system would have allowed the option of supplying all site power (during no hydro-power periods) via an AC-DC-AC route, using the wind generator for battery charging. However, the conversion losses are significant

and wasted energy can be minimised by feeding the load directly from the wind generator when possible. Direct powering also has the advantage of making the wind generator's high peak power capability available to the site, ensuring for example, much better motor-starting than is possible with the present invertor. The wind generator is fitted with a 3 phase synchronous alternator, so output frequency is directly proportional to R.P.M. One phase can easily supply the site load given adequate wind.

Rotor speed and hence frequency can be held within a narrow range by electronic load control, however allowing a wide range of speed (approximating to constant tip-speed-ratio operation) improves the power output particularly in low wind speeds. To investigate the importance of this, a simple computer model of the machine was used to produce the family of curves shown in Fig. 1. It is clear that below about 6 m/s there is considerable loss of output if the frequency is held up to 50 Hz. This is borne out in practice. For best energy output the variable wind speed curve shown in the figure should be followed.

A further computer programme simulating Rayleigh wind distributions was used to compare the daily average power output of the machine using fixed 50 Hz control with that achieved using variable speed. Fig. 2 shows the results of this and indicates a large advantage to variable speed on low wind speed days. On such days with non-Rayleigh speed distributions the advantage of variable speed can be even greater.

The site load requires a fairly constant frequency, although we find in practice that a 45 to 55 Hz range is quite acceptable. Voltage is limited to 220 volts rms at low frequencies to avoid possible saturation effects in transformers. Most equipment happily tolerates 220V when nominally rated for 240. Frequencies higher than 55 Hz can cause difficulties with fluorescent tube starting and equipment incorporating thyristor/triac controls. One disadvantage of the 45-55 Hz range is the inability to use conventional time switches. However, quartz controlled units are readily available. Very few modern domestic appliances are particularly frequency sensitive.

When supplying the site, the controller therefore must hold the frequency within the 45-55 Hz range with some sacrifice of low wind-speed output, or switch the site load to the invertor with the same inherent AC-DC-AC conversion losses, which are likely to be in the range of 20-50%.

The chosen strategy operates the turbine in variable speed (30-55 Hz) mode when supplying battery charging and heating loads. The alternator AVR is modified to progressively reduce the voltage with frequency down to 150V at 30 Hz; this avoids transformer saturation problems and protects the alternator excitation windings.

If the frequency exceeds 45 Hz and the power available exceeds the site load consistently for 15 seconds, the site load is switched directly to the wind generator. This minimises conversion losses and maximises battery charging capability. Surplus power is initially absorbed by the controlled battery charger. If the windspeed increases further and the charger reaches full power or the batteries become fully charged, surplus power is fed to the heating load.

If the frequency falls towards 45 Hz and the site load is above a preset threshhold, all the charging and heating load is removed to help maintain adequate frequency for direct supply with some sacrifice in turbine output (see unloader - Fig. 3). This minimises the frequency of switching between the direct and invertor supply. From an efficiency point of view, computer modelling suggests that, in low windspeeds, there is little to chose between under-loading the turbine (to maintain frequency and direct supply) and switching to invertor supply with the inherent conversion losses.

If the frequency falls below 45 Hz, the low priority site loads are disconnected. A further fall switches the high priority loads to the invertor. Speed control remains in the 45-55 Hz mode for a short period, but if the available power stays below the site load, control reverts to the 30-55 mode.

As the site is fed from one alternator phase only, an additional control loop (single ∅ limiter - Fig. 3) is used to protect the alternator by limiting the heating load applied to that phase as the phase current approaches its maximum safe value.

The controller block diagram is shown in Fig. 3 and the loading curves used shown in Fig. 4.

The remaining controls are associated with the tip flaps, braking, remote starting and stopping of the machine. These are split between a modified Polenko control box in a shed adjacent to the turbine and a remote control and monitoring unit in the main building. Communication is via 800m of 20 pair armoured flaps.

Electrical System

The complete electrical system is shown in Fig. 5. The 800m power cable was sized for approx. 2% annual energy loss. The auxiliary supply cable feeds power from the invertor (or other source) to the turbine shed for controls, lighting, power tools, test gear, etc.

Lightning strike precautions take the form of :

1) Foundation slab reinforcements bonded to tower and earth rods beneath.
2) Earthing counterpoise of 4 x 30m buried conductors radiating from tower base.
3) Bare 25 mm^2 earth conductor buried with armoured cables.
4) Surge arrestors between all other conductors and earth at the turbine, half-

way down the hill, and in the main building.

5) Earth plates at both the halfway point and adjacent to the main building.

Switching the site load between direct and invertor supply is at present carried out asynchronously by de-rated electromechanical relays. Despite the potential problems this seems to work well in practice; changeover causes barely noticeable lighting flicker.

The heating load consists of 5 x 3 kW 3-phase immersion heaters mounted in a 250l tank. Hot water from this supplements space heating and the building hot water supply. A 5 x 3 kW convector heater load is mounted adjacent to the main building to act as a dump load when the tank thermostat cuts off. These loads are controlled by a step and phase device. One 3 kW 3-phase load has full range triac phase angle control. The others are switched in as 3 kW steps. The fully controlled load is used to fill in the steps so that the total heating power can be varied smoothly from zero to 15kW. This approach is preferable to complete phase control of all 15 kW, as it causes much less waveform distortion. It is also superior to the zero voltage switching system supplied by Polenko; this causes voltage flicker which is not tolerable when the machine is simultaneously feeding a lighting load.

Battery Chargers

The requirements here are somewhat unusual. The power level must be controllable in response to an external control signal and to battery voltage. They must also accept voltages in the range 150 to 240 and a frequency range of 30-55 Hz. Efficiency should be as high as possible over the whole power range and the input current waveform should be as close as to sinusoidal as practicable with a high power factor. These are important to minimise losses in the cable and alternator and to minimise waveform distortion.

An off-the-shelf unit could not be found so a charger has been designed and built from scratch. Toroidal transformers feed to a 3-phase bridge followed by an inductive input filter. The resulting DC is fed to the batteries via a 3-phase high frequency transistor chopper which allows full control of the power level whilst maintaining a reasonable AC input current waveform. The charging requirements can be met by two 100 Amp 24V nominal output units. Efficiencies between 80 and 90% can be achieved.

INSTALLATION

The first stage of the installation was the excavation and laying of the foundations, completed in Feb. 1984. The concrete pad is approximately 6m square by 650 mm thick, reinforced top and bottom with 200 x 10 mm ∅ BRC mat. 48 off M30 hookbolts were set into the concrete using a ring template supplied by Polenko to match the base flange of the steel tower. Soil cover is low at the hill top site, so the pad sits on bedrock (slate).

The next stage was the laying of the electrical cables from the C.A.T. control room to the concrete pad; 3 AEI armoured cables (4 x 16 mm^2 power cable, 20 pair control cable and 2 x 2.5 mm^2 auxiliary power), each 800m long and climbing approx 100m. The cables were buried by means of a mole-plough, drawn by a 4 wheel drive tractor.

The windmill and tower arrived at C.A.T. in Machynlleth in August 1984 in one container. The container was unloaded onto 4 heavy duty agricultural trailers, which were then pulled up almost 2 km of steep agricultural track to the machine site, again by 4 wheel drive tractors. The heaviest trailer, loaded with the two tower sections packed one inside the other, weighed almost 6 tonnes and had to be pulled by 2 tractors in tandem.

Erection was fixed for 1.9.84, arrangements were made for the hire of an Iron Fairy 'Cairngorm' 12 tonne 4 wheel drive crane. The crane was first used to offload the tower sections from their trailer, then to draw out the smaller diameter top section from inside the lower section. Wheels had been fitted to one end of the upper section at the factory to facilitate this. The internal ladder sections were then fitted into the two tower halves, then the first section was lifted and positioned on the foundation bolts. A theodolite was used to establish vertical, adjustment being made on 4 orthogonal studs before final tightening of the rest. An extension fly jib was then fitted to the crane to allow the 20m plus lift required for installing the upper tower section. The two sections mate by means of internal bolted flanges.

The next lift, the nacelle, was by far the most difficult to position, despite the use of tapered dowel bolts. The main difficulty arises from the limited orifice (750 mm) through the yaw gear, making it difficult and potentially dangerous for an engineer to assist with engagement. Once alignment on the taper dowels was achieved, it was simply a matter of fitting and tightening 30 x M10 bolts.

The rotor was assembled on the ground, using the crane and a set of purpose built frames to support the blades during fitting. As the light faded at 9.00 p.m. the rotor was lifted, engaged and bolted into place. The crane had only arrived on site at 1.00 p.m.

It was nearly 4 weeks later when the machine first operated - most of that time being taken by completing the electrical connections and the switchgear in the C.A.T. control room.

MONITORING

The only continuous monitoring at present is the regular recording of energy production and wind-run. A Vector Instruments anemometer mounted at 10m height adjacent to the turbine is linked to the main building, where it drives a km counter and a m/s speed speed indicator. The speed output signal has also been used for some short term measurements of power vs. windspeed. The resulting curve is shown in Fig. 6, with the manufacturers curve superimposed, allowing for the difference in heights.

Fig. 7 shows the daily average power outputs plotted against daily mean wind speeds. Superimposed is the expected performance from the computer model. Agreement is good on low windspeed days, but problems with tip flaps control and unnecessary shut-downs causes considerable departures on windier days.

Figure 8 shows the monthly energy production since May '85. Availability (brake-off/total hours) has been 75% over this period. Output over the last year has been 17,500 kWh. We expect this figure to reach 28,000 kWh in a typical year (5 m/s at 10m). The present shortfall is due in part to the inadequate flap control strategy (as indicated by Fig. 7). Also, the benefits of the variable speed 30-55 Hz) control system has only been seen in the last two months, since it was completed in December 1985.

OPERATING EXPERIENCE

Most of the loss of availability so far has been caused by two problems :

1. Although the machine is supplied as a stand-alone device, it needs a 240V 50 Hz supply for its control gear. The brake reset motor alone needed 10 amps reactive to start up - particularly difficult with an 800m x 2.5 mm^2 cable! This motor was replaced with a low voltage DC unit, powered by a small battery at the base of the tower. In the very early days the machine was very prone to mysteriously stopping - seemingly in response to transients or 'glitches' on the auxiliary power (the fail-safe brakes automatically activate in the event of power failure). Several latching relays were converted to capacitor-backed DC operation, and the micro-processor was given a back-up NiCad battery supply. These measures have reduced stoppages considerably, though there are still occasional unexpected halts.

2. There have been problems with wear in the rubber bushes of the peg and block drive coupling linking the high-speed gearbox shaft to the alternator. Although the service factors applied to the coupling appear generous, 5 sets of rubbers have now been consumed. Correspondence with the coupling manufacturer's agents suggests that the device may be intrinsically unsuitable for the application, though changing to a different type is not easy within the constraints of the nacelle layout. It is felt that more investigation is needed before replacing the coupling, although we do intend to change when the best solution is identified.

Both of the above problems are associated with Polenko's low experience with stand-alone machines - ours being only the 3rd out of 200 plus production W.T.G.s not to be grid-linked. In the main we are very happy with the performance and reliability. The machine is almost unique in C.A.T.s collection in providing a power/windspeed performance slightly beyond manufacturers claims.

Local acceptance is good, despite visual obtrusion being high. Noise is not normally perceptable from any dwellings except when the coupling bushes become too worn.

In the nuisance category, the tip flap control is crude, the flaps being fully extended at a preset r.p.m., returning after a set time delay, subject to r.p.m. falling below the preset. This causes persistant cycling in moderate winds, leading to loss of energy and wild swinging of frequency. An improved proportional controller is planned. There is also an annoying voltage flicker at certain frequencies, though this probably arises from resonances affected by the coupling problems.

COSTS

The total project cost is expected to be approximately £20,000. This includes the machine, foundations, cabling, anemometry, control and protection switchgear, the 15 kW electrical boiler, emergency dump-loads and the battery charger. The costs of equipment hire and sub-contract labour (e.g. foundation digging) has been included, though internal C.A.T. labour has not been recorded.

In recognition of the demonstration aspects of the installation, Polenko and AEI gave discounts on the machine and cabling respectively. Eurodrive Ltd. kindly donated the gearbox free of charge. Local farmers and earth-moving contractors were very interested in and supportive to the project, an enthusiasm reflected in the bills for their services.

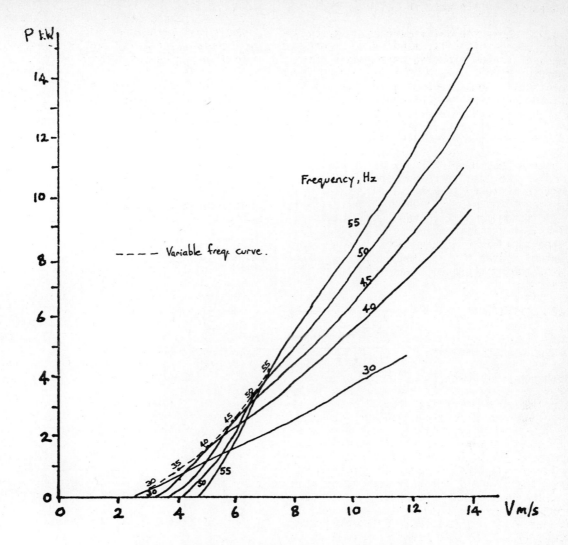

Fig 1   Turbine characteristics from computer model

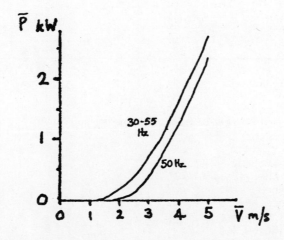

Fig 2   Average output on low windspeed days from model

91

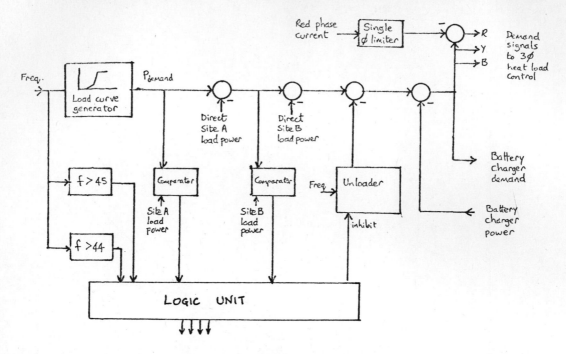

Fig 3　Controller block diagram

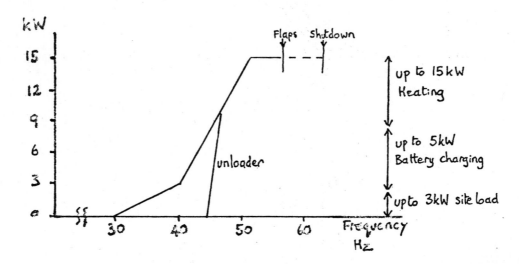

Fig 4　Controller loading curves

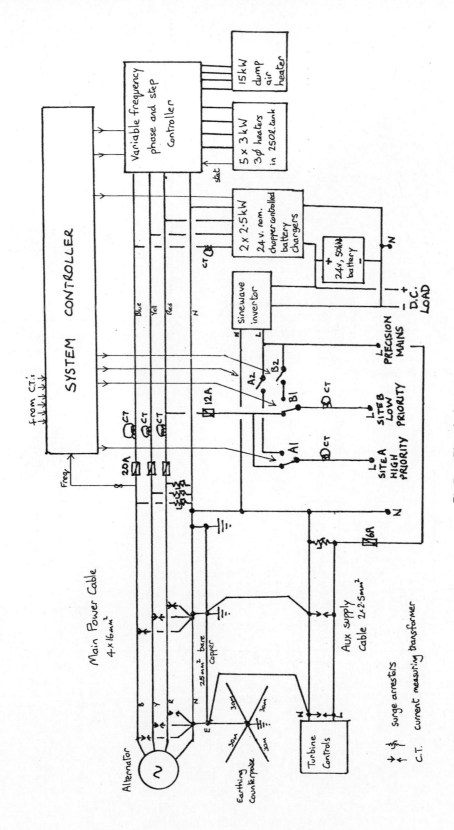

Fig 5   Electrical system

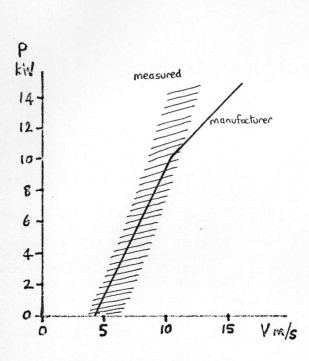

Fig 6　　Power vs windspeed

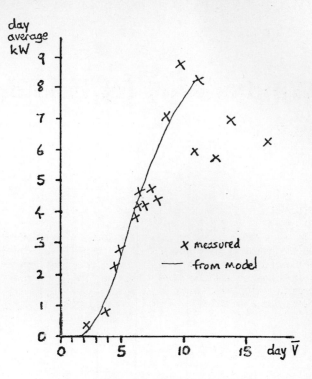

Fig 7　　Daily performance

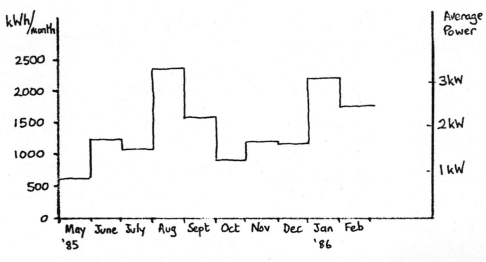

Fig 8　　Monthly energy output

Wind energy for land drainage in Ireland

BRIAN HURLEY
Dublin Institute of Technology, College of Technology, Bolton Street, Dublin 1, Ireland

SYNOPSIS

The available wind energy resource is examined and the impact of height variation is assessed. The water pumping performance of a typical windpump with a 7m tower, varies by a factor of 2.6 due to variations in the wind regime from one region to another. Rainfall and evaporation data are investigated as the basis for defining the maximum pumping requirement. The technology of water pumping wind machines, and recent developments are reviewed. Three models are developed using wind resource and rainfall data to estimate the area of land a wind machine would drain.

1. AVAILABLE WIND ENERGY.

On the basis of international comparison Ireland is well endowed with energy available from the wind. On the island itself, the resource is greatest in the coastal regions with greater availability on the west coast. (1)

The resource is four times larger at 20m hub height in the extreme west than it is in the southern region of the midlands. This variation is a little less marked for hub heights closer to the ground where at 4m the variation is three and one-half times.

The variation in expected pumping performance from one region to another is less than that implied by the availability of the resource but is still of great significance. The performance of a typical windpump at a hub height of 7m varies by a factor of 2.6 due to variation in the wind regime (See figure 1).

2. CLIMATE AND DRAINAGE.

There are no pronounced wet and dry periods in the Irish climate. Rainfall is higher in the August through January period. The ratio for this period to the Spring/Summer period is about 7/5. If surface run-off is neglected, a moisture deficit does not usually occur. In practice, surface run-off is appreciable in all months on average, except July, when a lowering of the water table does take place.

A drainage criterion of 7mm/day is widely used in Western Europe where yearly rainfalls of 750mm/year are common. For Ireland values up to 20mm/day are used. (3)

3. RAINFALL AND EVAPORATION.

The other key variable besides the available wind resource is annual rainfall. This gives a measure of the expected maximum drainage requirements. If mountainous regions are excluded, most parts in the eastern half of the country have between 750 and 1000mm of rainfall in the year with less than 750mm at a number of places in the Dublin area. (1) Rainfall in the west generally averages 1000 to 1250mm and up to 1500mm in some coastal areas. In most parts, April is the month with least rainfall and December the month of greatest rainfall. Evaporation reduces these figures when the drainage requirement is being calculated. Its main impact is between april and September. If it is neglected it will have the effect of making the calculation of drainage required more conservative.

4. CORRELATION OF WIND AND RAINFALL.

It is important to establish some estimates of the correlation between periods of rainfall and the availability of wind, and also to establish the relationship between the intensity of rainfall and wind velocity. This has a major bearing on the amount of storage required in the overall system, either in the land itself or in ditches or sumps associated with the intake of the pump. On a countrywide basis the wettest month on average is December, and April the driest. However, for wind, strong winds are most frequent in the period November to March and the lowest mean windspeed are generally June to September. Preliminary work has been undertaken using hourly windspeed and rainfall data for 1982 for Dublin Airport. On aggregating this data on a daily basis, a correlation is found to exist for January, in the wetter season, but less so for July, when windspeeds are lower. However if the data is aggregated on a monthly basis no obvious correlation is apparent (See fig. 4,5, and 6) The evidence for a daily correlation between wind availability and rainfall should reduce the amount of water storage necessary. However, the only firm basis for sizing an actual system would be to use the hourly average windspeed and water output against windspeed for an actual machine, and the hourly rainfall. This although simple in essence, is tedious in practice unless the weather data is available on computer accessible tape or disc. A simpler procedure is used later.

5.   WATER PUMPING BY WINDPOWER.

In many parts of the world windpower is successfully used for waterpumping. The very long successful history of its suitability for such purposes is brought out by the fact of its continuous use. In the last several decades its use was overshadowed by diesel and electrically powered pumps. But it was still being used in Australia, North America and South America. Windpumps are still manufactured in these areas. In very recent years there has been a great revival in interest. New manufacturers have entered the market. The most commonly used and widely distributed type of windpump is the slow-running multivane metal fan, driving a piston pump. Originally developed in France, the multivane first attained widespread use in the U.S.A.

6.   LARGE VOLUME LOW HEAD PUMPING.

The most widely available windpumps are adapted to high head low volume pumping. These machines are not suited to large volume low head pumping applications of the type required for drainage either of large areas of agricultural land or peat bogs. In recent decades several countries, The Netherlands, Denmark and West Germany, have used specially developed windpumps for land drainage. The common features are higher speed rotors with less blades than traditional multibladed water pumping machines. These rotors do not exceed 4m in diameter. There is no common approach to pump drive or pump type. Reciprocating rods are used with diaphragm pumps, and rotary shafts with centrifugal pumps. More recently in Ireland the classic multibladed machine has been adapted to work successfully with diaphragm pumps.(7) Such windpumps will be referred to as Direct Drive Mechanical Windpump sets. (See Figure 3.)

More recent technical developments offer further options and much increased pumping capacity.

7.   WIND ELECTRIC PUMP SET

In the past decade many wind electric wind turbines have come available. A particular feature is the extension of rotor sizes now commercially available. Unlike direct drive mechanical windpump set rotors, diameters up to 20m or more are now available, although few of these have been applied to water pumping or irrigation. Work has been carried out in the U.S.A. and Holland (4), (5) (6). Problems of matching rotors and pumps

satisfactorily to the particular end use have yet to be resolved.

TABLE 1:   Performance characteristics of a direct drive mechanical windpump set for a lift of 1 metre.

Instantaneous Windspeed at hub height.	Pumping Rate
m/s	m3/min.
1	0
2	0
3	0
4	0
5	1.05
6	1.33
7	1.50
8	1.78
9	1.96
10	2.03
to	
18	2.03

Rotor diameter :   3.14m
Hub height: 7m

(Source: Manufacturer)

8.   WIND ENERGY RESOURCE
       & ANNUAL OUTPUT.

The matching of a windpump with the drainage requirements of a region with a particular rainfall and annual average windspeed is a complex task. One reason why this is so is that the traditional approach of drainage engineering has been to approach the problem using rainfall allowances based on 24 hour or 1 day time periods. On a proportionate basis this leads to pumping capacities 3 to 4 times greater than that based on the expectation using annual rainfall data, which is more suited to the usual way of estimating wind machine performance. An approach based on the pumping capacities usual to drainage engineering practice leads to over sizing of the wind machine, whereas one based on annual rainfall data may underestimate machine capacity. Ultimately a design figure related to both, and incorporating storage capacity of land or drainage system, is needed.

The wind energy resource and also the annual output of typical machines has been calculated using the thirty year mean annual average windspeeds and a Rayleigh distribution. The mean annual windspeeds have been taken from published Meteorological Office records and have been modified for different hub heights using the 1/7 power law.

9.   METHOD USED TO CALCULATE
       LAND AREA DRAINED.

(1)   "mm/Day" Method.

The area of land drained is calculated by using the annual output of the wind machine and matching it with the usual drainage engineering method of specifying the maximum daily pumping requirement in mm per day.

This daily requirement in pumping capacity is multiplied by the number of days in the year, giving an annual pumping requirement. (See Figure 2).

(2) "Annual Rainfall" Method.

The area of land drained is calculated by matching the annual pumping capacity of the wind machine with the annual rainfall expected in the region. Evapotranspiration losses are ignored, since these are very low in the wetter months. In the drier months, where they become more significant particularly in mid-summer, ignoring these losses will lead to a conservative estimate.

(3) "Creagh Equivalent" Method.

The land area drained is calculated on the results of the pilot scheme operated by the Agricultural Research Institute, at Creagh, near Ballinrobe, Co. Mayo (7) This test has been running for the last four years. A Bosman machine drains approximately 12 hectares of peat land. It is estimated that the machine could drain up to 50% more land, i.e. 18 hectares. Creagh is close to the 5m/s annual average windspeed isovent, which reduces to 4.7m/s at 7m from the ground using the power law. The expected annual output at other windspeeds is calculated by reference to the expected annual output in the different wind regimes using the 'annual rainfall method' and estimating the pumping output on a proportional basis.

The three different approaches give very different results. The "Creagh Equivalent" approach is the best founded, since it is based on actual machine performance.

The "mm/day" approach gives a lower limit to the land area drained. It has been used since it is the basis for the traditional means of sizing pumps in drainage schemes used by engineers. A figure of 10mm/day was used, although for the wetter regions 20mm/day would be more appropriate.

The "Annual Rainfall" approach is more suited to the usual methods employed in the evaluation of the output of wind machines. This method is likely to give an upper limit to the expected area of land drained.

TABLE 2: Land Area Drained.

Windspeed	10mm/day	Creagh Equivalent	Annual Rainfall 1200mm.
		Hectares.	
3.8m/s	6.2	11.0	18.9
4.7m/s	10.2	18.0	30.9
5.7m/s	13.6	24.1	41.4

REFERENCES.

(1) P.K. Rohan
     The Climate of Ireland
     Meteorological Service,
     Dublin 1975.

(2) Meteorological Aspects of the Utilization of Wind Energy as an Energy Source, Technical Note No. 175 W.M.O.

(3) Mulqueen, J. (1975)
     Drainage of Deep Peat in Ireland.
     Proc. Inter. Symposium, Peat in Agriculture and Horticulture.

(4) Vosper, F.C. and Clark, R.N.
     Water Pumping with autonomous wind-generated electricity.
     Transactions of the A.S.A.E. (Vol.28, No.4 pp1305-1308,1985)

(5) Goezinne, F. and Eilering, F.
     Water pumping windmills with electrical transmission. Wind Engineering
     Vol. 8. No. 3 1984.

(6) Clark, R.N.
     Irrigation pumping with wind energy, electrical vs. mechanical.
     Transactions of the A.S.A.E. Vol 27, No. 2 pp415-418, 1984

(7) Robinson T. Personal communication.
     Appropriate Energy System,
     Liffey Trust, North Wall, Dublin.

(8) Shouldice C. Dairying and land drainage applications of wind power in Dept. of Energy, National Windpower Programme. S.E.S.I. Windpower Seminar, U.C. Galway, April 1983.

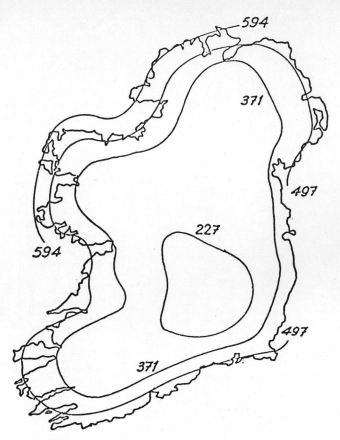

Typical mechanical windpump hub height 7m

Fig 1   Estimated annual output in m³ x 10³

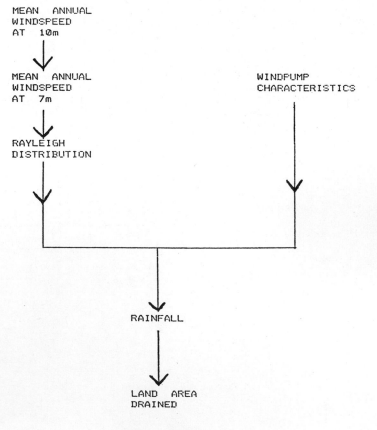

Fig 2   Model

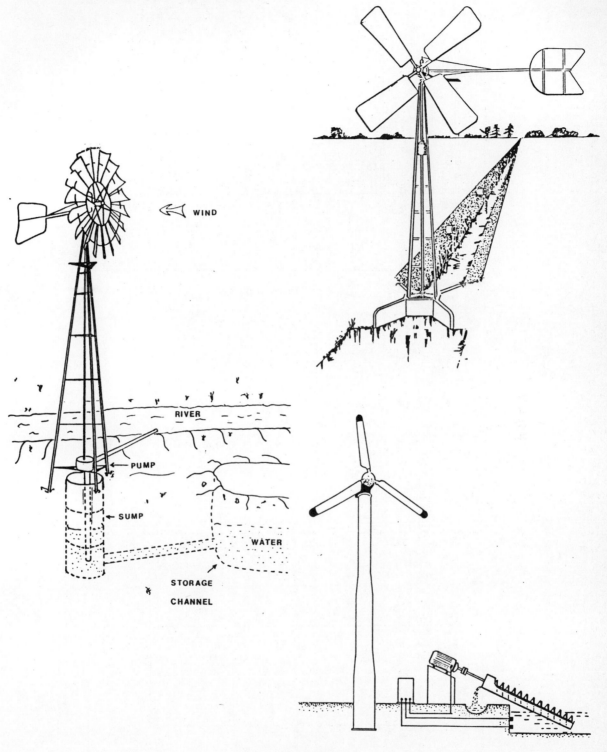

WIND

RIVER

PUMP

SUMP

WATER

STORAGE
CHANNEL

Fig 3   Windpumps

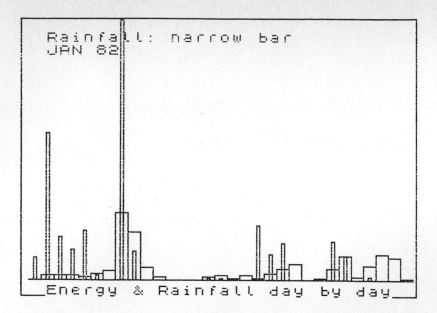

Fig 4

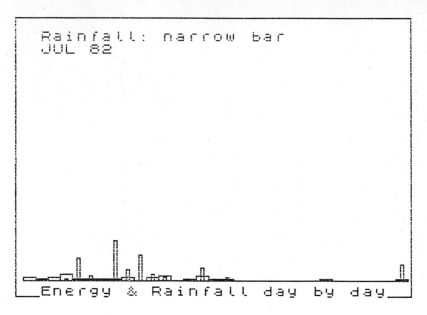

Fig 5

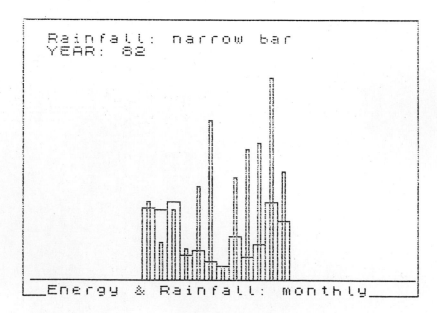

Fig 6

Test at very high wind speed of a windmill controlled by a waterbrake

Ö HELGASON and **A S SIGURDSSON**
Science Institute, University of Iceland

SUMMARY   For the last four years a project on exploiting wind power for househeating, especially at high wind speed, has been carried out on a small island off the Northern coast of Iceland. The mill has a two bladed upwind rotor with a diameter of 5.7 meters. The blades have a fixed pitch and are designed for a tip speed ratio of 5.5. The turbine is coupled to a waterbrake, downstairs, through a heavy duty gear with a fixed gear ratio. By changing the effective blade height in the waterbrake, the loading on the wind rotor can be changed. Thus, the waterbrake serves the dual purpose of transferring rotation power to heat and controlling the rotation speed of the wind rotor. A microcomputer controls the load changes of the brake, according to running conditions, and collects data. Besides giving a short description of the turbine and site conditions, the paper will describe different dynamic aspects of the waterbrake, like how it responds to changes in wind speed and variation of the pressure inside the brake. We will also analyse, to some extent, test runs during three months with wind speeds frequently above 20 m/sec and effective output of more than 40 KW.

1. INTRODUCTION

Many coastal areas in Iceland have average wind speed which exceeds 9 m/s (20 mph) during the wintertime (October-April). Although sufficient hydro and geothermal power is available for the country as a whole, it may be of an economical interest to utilize wind power under certain circumstances. For example, it is useful when combined with other power sources in order to meet high heating demands during the windy winter time. It is of special interest for househeating, which usually exceeds 80% of the total power consumption for a family. In addition to being very important quantitatively, it is not usually vulnerable to minor irregularities, which is the case for electrical power. Also, it is easier to have backup for this type of energy.

During the last four years, a research project on wind power has been carried out at Grimsey Island on the Northern coast of Iceland.   The main objectives of this project are:
a) an investigation of a direct transfer of wind power to heat by using a waterbrake,
b) exploring the ideas of complete control of the whole WEC-system by a variable power load in the waterbrake (this includes an absence of aerodynamic brakes on the wind rotor),
c) designing a low maintenance computerized control system and
d) extracting power from the wind at maximum Cp-value up to wind speed at about 20 m/sec and partly utilize the power up to 25 m/sec by running the turbine overloaded by the waterbrake.

The whole WEC-system has been described in detail elsewhere (Ref. 1 and 2). In this paper only a brief description of the most important features will be given. The main subject this paper discusses will be an analysis of different measurements related to the dynamics of the waterbrake, especially in relation to

using it for overall control of the WEC-system through load changes. It will also evaluate the total energy production for three months (November 85 - January 86), which demonstrates the mill's capability to cope with harsh wind conditions. During this time period, we had an excellent opportunity for exploring different aspects of operating at high wind speeds, since the average wind speed at the site was 10.6 m/sec, based on a 30 minutes average.

2. THE WIND TURBINE AND THE SITE CONDITION

The selected site for the test mill lies on the south tip of Grimsey Island, measuring 5.4 sq.km and situated on the Artic Circle about 50 km north of Iceland. (Fig.1) The annual average wind speed is 8.5 m/sec, and the average

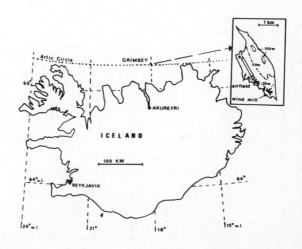

Fig 1   Wind turbine site

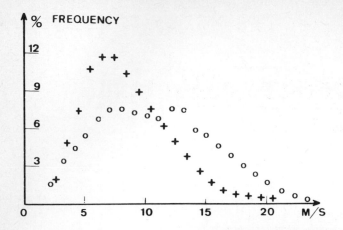

Fig 2   Frequency distribution of the wind speed on Grimsey
        + Indicates the annual distribution
        o Indicates the distribution during Nov. 85 — Jan. 86

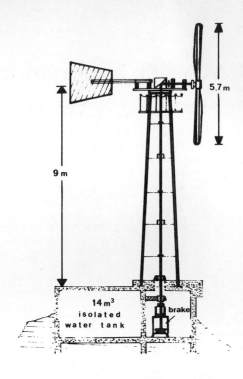

Fig 3   The wind mill

power exceeds 600 w/sq.m or 5300 KWh/sq.m a
year.   During the wintertime, the wind speed
distribution is significantly shifted towards
higher wind speeds as can be seen on Fig. 2,
where speed distribution from a period of three
months (Nov.85-Jan.86) is superimposed on the
annual distribution.   That, increased potential
in wind power, together with high demands on
heating power at the same time of the year, explains our interest in running a WEC-system
capable of following maximum Cp-value up to some
20 m/sec.   At a place with wind power like we
have in Grimsey, a mill with a rotor sweaping
through an area of 25 sq.m. (diameter of 5.7 m)
should produce annually 50 000 KWh, if Cp-value
of 0.4 can be kept up to wind speed of 19 m/sec.
The test mill has a 9 m tower of steel construction which is placed on 6x2x2 m³ concrete foundation.   This foundation houses a 14 m³ hot water
storage tank and a small compartment where the
waterbrake and the computer control system are
located.   The rotor is two-bladed and upwind
with a diameter of 5.7 m (19 ft.).   The profile
is NACA-4412 to 4418 with a tip speed ratio of
5.5.   The rotor blades are made as one unit, not
interrupted at the hub.   Inside is a core of
polyurethane foam, but the shell is made of
sheets (0.4 mm thick) of glass fiber texture em-
bodied in epoxy.   The number of layers increases
from 4 at the end to 30 at the hub.   The rotor
has no aerodynamic brakes and is designed to
cope with more than 500 r.p.m. at full load.
This will be discussed in more detail in section
5.   The rotor is coupled through a heavy-duty
gearbox (with a gear ratio 1:3) to the vertical
shaft which transmits the power down to the
waterbrake.   Fig. 3 shows the outline of the
mill.

3. WATERBRAKE AND CONTROLLING PROCEDURE

Figure 4 depicts schematically the main feature
of the waterbrake.   The vaned rotor swirles the
water around but the stator brakes the angular
motion causing frictonal forces to heat up the
water.   The heating energy is delivered to the
house heating system through a heat exchanger.
In case of using water as a brake fluid this is
not necessary and can even have some disadvantages.
The vanes in the stator go through slots in a
movable plate.   By moving this plate up and down

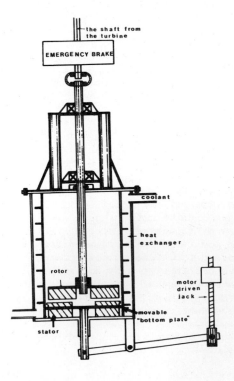

Fig 4   The waterbrake with the 'variable load system'

102

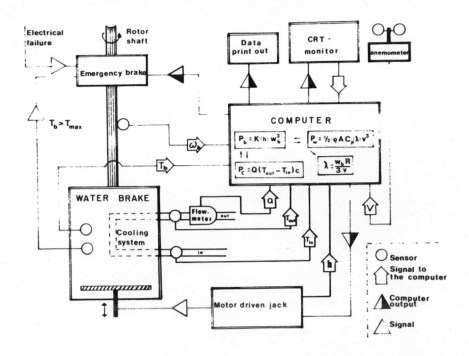

Fig 5   A diagram of the computer control system

the effective height of the stator vanes is altered and hereby the braking power of the waterbrake. The movement of the bottom plate is accomplished by a motordriven jack connected through a 1:3 lever system to an axis at the center of the plate. The movement of the jack is controlled by a computer system (see fig. 5) which simultaneously collects data from different sensors and takes care of the overall operation.

The computer measures:
1. The wind speed, v
2. The angular speed, w_b, of the brakes rotor.
3. The temperature of the brake, T_b
4. The temperature, T_{in}, of the coolant water entering the brake.
5. The temperature, T_{out}, of the coolant water leaving the brake.
6. Flow of coolant water, Q
7. The position of the motor-driven jack which moves the bottom plate and hereby the effective height of the stator vanes.

The computer lists these data and also calculates and lists the power of the brake, P_b, the power delivered to the cooling water, P_c, and the tip speed ratio, λ.

The heating power is calculated by simple caloriemetrics, that is

$$P_c = Q (T_{out} - T_{in}) C \qquad (1)$$

Where C is the specific heat of water. The brake power is calibrated according to the heating power as the brake deliveres almost all it's energy to the cooling water.

According to R. Matzen (Ref. 3), the equation for the brake power is, with some modifications, given by

$$P_b = K(T_b) \ F(h) \ d^5 \ w_b^3 \qquad (2)$$

where d is the diameter and w_b is the angular speed of the brakes rotor. $K(T_b)$ is a coefficient related to the geometrical structure of the brake and the temperature in it as the viscosity of the water changes with temperature. At normal running conditions the temperature changes are not significant but $K(T_b)$ increases considerably when the brake gets cold due to still weather.

F(h) is a function of the height, h, of the vanes in the brake. The brake power can therefore be altered by changing the effective height of the vanes on the stator which is done by moving the slotted plate up and down. The height ranges from 17 mm to 70 mm. In that region F(h) is nearly linear, doubling the height from 17 mm to 34 mm will increase F(h) by approximately 20%, which is less than stated elsewhere (Ref. 2 and 3). Further study of this relation seems necessary.

At wind speed v the power of the wind turbine is given by

$$P_w = 1/2 \ \rho \ v^3 \ A \ C_p(\lambda) \qquad (3)$$

where ρ is the specific mass of air, A is the area which the rotor blades sweep through and $C_p(\lambda)$ is the power coefficient for the rotor, depending on the tip speed ratio, λ. In a normal condition we have:

$$P_b = \beta \ P_w$$

or

$$K(T_b) \cdot F(h) \cdot d^5 \cdot w_b^3 = \beta 1/2 \ \rho v^3 \cdot A \cdot C_p(\lambda) \qquad (4)$$

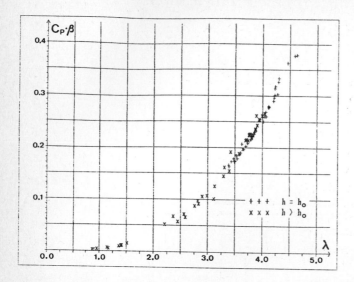

Fig 6   Measurements of the effective power coefficient, $C_p(\lambda) \cdot \beta$, for the Grimsey wind mill

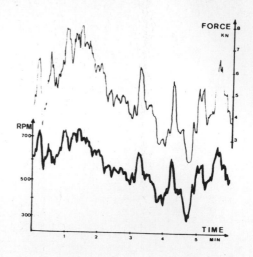

Fig 7   Relation between rpm of the rotor and the force on the bottom plate of the water brake

where ($\beta < 1$) is related to the efficiency of the system.

Since the tip speed ratio is defined by

$$\lambda = \frac{W_w R}{v} \qquad (5)$$

and

$$W_b = 3\ W_w$$

we get the following relation

$$\lambda^3 \cdot F(h) = C_p(\lambda) \cdot G \qquad (6)$$

where

$$G = \beta \ \rho \ R^3 \cdot A/54$$

describes mostly non-varying properties of the system.

Equation (6) shows that the matching between the wind turbine and the waterbrake is independent of the wind speed. This was also to be expected since the power of both systems is related to the cube of the speed.

The $C_p(\lambda)$ - λ relation for the Grimsey windmill is shown on fig. (6), based on data collected in December. (Notice however that the curve on the figure includes β)

From the figure and equation (6) we can see that increasing the blade height of the stator, h, and thereby F(h), results in lowering the tip speed ratio λ, and along with it the efficiency of the turbine. At normal running conditions, up to wind speed of about 18 m/s, the height of the vanes is kept constant, $h = h_o$. Then the mill is running mostly at an overall efficiency of 20 to 28%. For the time being we have a safety stop at $h_o = 17$ mm which is the reason for this low efficiency. Further reduction of the blade height is planned in order to reach tip speed ratio near optimum at 5.5, as is expected of the design of the rotor.

When the rotor reaches 250 rpm (at wind speed of about 18 m/s) the computer starts increasing the blade height of the stator. As F(h) increases the brake overloads the turbine and λ and $C_p(\lambda)$ decreases. The overloading is increased gradually in the region of 18 to 28 m/s. At still higher wind speed a heavy overload

causes the wind rotor to collapse down to tip speed ratio of one. If the wind speed gets over a given maximum the computer will bring the mill to a stop by an emergency brake. An example of this procedure will be described in section 4. In addition to the rpm of the rotor and the wind speed, the computer will overload the mill if the temperature of the brake gets to high, due to insufficient cooling.

The swirling motion of the water in the brake causes substantial force on the bottom plate. This force was measured by mounting a load cell on the motordriven jack. The load cell and a rpm sensor were connected to a plotter and the force and rpm plotted continuously and simultaneously. Part of the results is seen on fig.7. The force seems to be nearly linearly related to the rpm of the rotor of the brake in the range measured. Up to 150 rpm no force was measured but at 800 rpm it reached 7000 N. This equals a mean pressure difference of 0.56 bar over the plate. The pressure under the plate is about 0.2 bar (2 m water head) so the pressure on top of the plate gets down to 0.6 bar$_{abs}$ or less. At that pressure the boiling point of water is only 86°C, which would be maximum temperature of the brake at these specific conditions. It would be of interest to look more carefully at this subject, but for the moment we have no evidence of troubles relating boiling in the brake.

Figure 7 also brings one's attention to reaction time of the system. It shows that the rotating parts react to changes in wind speed in an order of 5 - 10 seconds, similar to the expected reaction time of the wind turbine itself. The reaction of the control-system, that is the movement of the bottom-plate in the brake, is much slower. Doubling the effective height of the stator vanes from optimum at $h_o = 17$ mm to 34 mm and thereby increasing F(h) by approximately 20%, takes about 1 1/2 minute. An estimate of the influence of the brake's slow reaction on the energy production has not been analysed.

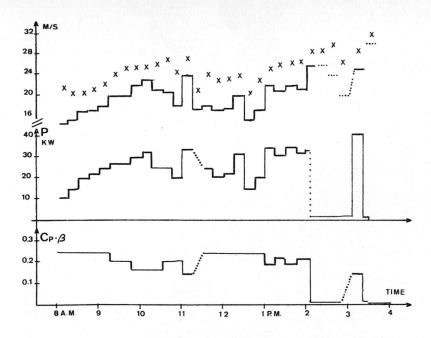

Fig 8   Operation of the WEC-system in a stormy weather, the 17th of
November 1985. The upper part shows the average wind speed
and the highest wind speed for every 15 minutes. The lower part
shows the power production and overall efficiency of the system
for the same time interval

4. OPERATION AT HIGH WIND SPEED AND LONG TERM ENERGY PRODUCTION

In this section we will discuss the operation of the WEC-system in November 85 to January 86. During these three months the average wind speed at the mill site was 10.6 m/sec. The frequency distribution of the wind speed for this period is shown on fig. 2, together with the annual distribution based on data for some ten years. The mill was available for operation 80% of the time during these three months. There was one major stop, for some 14 days due to a break down of an electrical relay in the jack-system of the waterbrake. Of the time available for operation (approximately 1730 hours), there was sufficient wind speed for some 1200 hours for the mill to produce energy. The total amount of energy produced (measured by calorimetric methods as the effective output, see equation (1)) reached 12000 KWh or 500 KWh/sq.m. swept area. Considerable part of the energy was fed directly to house heating system but only minor part stored in the storage tank. Several times the mill produced more than 40 KW or 1.6 KW/sq.m. at wind speeds above 21 m/sec. As might be expected for a three months period with average wind speed of 10.6 m/sec, there have been many days of stormy weather. Thus, we have had excellent opportunities for long term tests of power production and the operation procedure at wind speed above 20 m/sec. Altogether we had more than 40 hours of operation above 20 m/sec. We have detailed data on 4 runs exceeding 6 hours each where the average wind speed is most of the time above 20 m/sec and gusts occurs at more than 30 m/sec. Fig. 8 shows some of the relevant data from one of these runs, the 17th of November when a storm passed Northern Iceland. At normal run, data are printed out every five minutes. However, during

a rough operation the computer prints out data more frequently, even every minute, due to the decision making procedure in the program. In the figure 8, data from intervals of 15 minutes are grouped together for simplifying the graphs. The threefold graph shows the average wind speed, the average effective power and the overall efficiency of the WEC-system as a function of time from 8 a.m. to 4 p.m. At 8 a.m. the wind turbine runs at optimum but at 9:15 due to increasing wind speed, the computer starts overloading and hereby reducing the efficiency. At 10 a.m. the average wind speed reaches 23 m/sec and the load of the waterbrake causes the system to tun at 2/3 of optimum power. At this time the average rpm of the wind rotor was 230, reaching the maximum of 283 rpm for a while. At 2 p.m. the wind increases sharply, frequently passes 28 m/sec. The waterbrake is now driven to full overload which suppresses the rotor down to a tip speed ratio of one (80 rpm) and the mill produces only few KW. Then, for some 15 minutes at 3 p.m., a temporary decrease in wind speed releases the turbine for a while. At 25.2 m/sec it produces 42 KW for quarter of an hour. But the storm goes on and the average of 28 m/secquenches the turbine down again to 80 rpm. A few minutes later, when the computer has measured five times wind speed more than 31 m/sec during an interval of 5 minutes, the emergency brake is turned on and it brings the rotor to a complete stop.

5. DISCUSSION

The last few months are the first period of continueous operation, where we have been able to concentrate on testing the main principle of design in the WEC-system. Problems of different types we came across during the first three years have been described and discussed in earlier papers (Ref. 1 and 2). Most of them were related to structural mistakes in the foundation and a mismatching in the brake-rotor construction. Due to this, we still have a safety margin on the blade height in the waterbrake. The rotor is not allowed, for the moment, to run at designed tip speed ratio, as the Cp graph (Fig.6) indicates. On the other hand, it is the steep part of the Cp-λ curve, where the most important control-operations take place. Therefore, it is natural to focus on this part in our case. As mentioned in the introduction, there were four main objectives of this research project. As the results in section 3 and 4 shows, transferring of wind power to heat by using waterbrake and extracting power effectively at high wind speed have been achieved by this WEC-system satisfactorily. However, it is too early to state anything about the cost of maintenance for system of this type or its economy. Further study on reaction time of the waterbrake, how it's loading depends on the geometrical form of the brake and the forces inside the brake will surely help in designing a more efficient brake. But even with the relatively slow movement of bottom plate in the brake, an operation at harsh weather conditions, like those shown on fig.8, seems to be quite satisfactory. At the present time we are looking at a brake with direct heating, that is without a heat exchanger. Thereby the temperature of outflowing water can be raised some 10-15 degrees. Another interesting matter for discussion is the design of the wind rotor.

By design the mill has no aerodynamic brakes on the wind rotor and therefore relies solely on the variable load of the waterbrake and on an emergency brake on the rotor axis, which transfers the power through a heavy duty gearbox down to the waterbrake. By analysing constructions of woven glasfiber sheets embedded in epoxy we felt confident that a two-bladed rotor constructed as one unit could cope with the most extreme weather conditions, at least to the same extent as the tower itself. In case we are correct, a design of this type has many advantages and this material is not likely to break abruptly (this material is extremely plastic) so danger of

wing part flying away is minimal. In the design it was decided that the rotor should cope with a full load running up to 40 m/sec or 600 rpm. In free running mode, that means without any load of the waterbrake, the rotor should withstand up to 800 rpm. According to this, the rotor was constructed as described in more detail in ref. 2. The rotor has been tested partly loaded for a shorter period up to 500 rpm (the power output then exceeded 90 KW) and unloaded higher for longer time (up to appr. 600 rpm). Then we have had an accident, where in a storm, a safety lock broke while the axis down to the waterbrake was not connected. Detailed information about this case is not available, since after the rpm passed 600 the rpm-detector was damaged and no further data were collected. We do know that the fastening of bearings on the horizontal axis broke and shortly thereafter the rotor had an impact with the tower platform which resulted in damage on top part of the mill, although the wings did not break off. They suffered some damage but could be repaired on the island. For the moment we estimate the upper limit for a construction of this type of blade to be at 8-9 meters in diameter. We see many interesting perspectives in a compact and robust mill, with the main power generation and control down in the foundation and a minimal control mechanism at the top of the tower where major stress exists. Such a wind mill, with a rotor diameter of 8 meters, is capable of producing 100-130 MWh annually, with the wind condition found at Grimsey.

REFERENCES

(1) Helgason, Ö. Exploiting Wind Power of High Wind Speed for Househeating by Waterbrake in Iceland. *Proceedings of AWEA National Conf. Pasadena US 1984,* page 86-112.

(2) Helgason, Ö. Exploiting Wind Power of Heigh Wind Speed by Waterbrake with Variable Load. *Proc. of Delphi Workshop* on Wind Energy Application, *Delphi Greece 1985, p 282-300*

(3) Matzen, R. Wind-Heat Generator, *Second Intern. Symp. on Wind Energy System, H2-17, Amsterdam (1978)*

Aerodynamic interaction between two wind turbines

G T ATKINSON and **D M A WILSON**
Cavendish Laboratory, Cambridge

1. ABSTRACT

This paper reports results from an experimental investigation of the performance of a 5m diameter HAWT operating in the wake of an aerodynamically similar machine ten rotor diameters upstream. The aspects of performance studied included both mean values and excursions in: blade strain; blade coning angle and power output.

Previous studies have shown that the wake of the Cambridge Mk.V Turbine, that upstream in these experiments, suffers a considerable rotor induced sideways deflection. In these circumstances the structure of the wake varies markedly from the axi-symmetric case. In particular a sharp dip in the streamwise velocity appears some distance from the main velocity deficit in the direction of the deflection. Some extra attention is given to the pronounced effect of this novel feature of the wake on a downwind machine.

2. INTRODUCTION

It is important to determine how the power output of a wind turbine and the spectrum of blade strain excursions varies when a number of other machines are placed nearby ; several approaches to this problem have been tried.

1. Model arrays in wind tunnels (For example Ref.1): A relatively complete picture can be obtained and any non-linear effects involved in the addition of wakes are included but there are always uncertainties in scaling up the results and allowing for the unsteadiness and non-uniformity of the natural wind.

2. Two larger machines. (Refs. 2,3) Aerodynamic interactive effects decline rapidly a the wake centreline moves away from the downwind machine so the most natural approach is to measure how power loss,strain excursions etc. depend on the angle of misalignment between the wind direction

and the line between the turbines. In this kind of experiment it is clearly important to control or allow for any yaw error in the turbine upstream and consequent deflection of its wake. The reasons for this are two-fold:

(a) Firstly if no account is taken of wake deflection and the results are presented as averages of, for example, the amplitude of strain excursions across a wake section, the underlying structure that may show strong variation i.e. relatively narrow peaks of large amplitude, is replaced by a weaker more broadly distributed increase in the level of strains. In view of the highly non-linear relation between the amplitude of strain excursions and fatigue damage this may underestimate the problem.

(b) Secondly,the distribution of streamwise velocities in a deflected wake differs significantly from the plane or axi-symmetric case. The effect of these differences on a second machine is the main subject of this paper.

The Cambridge machines are particularly suited to this kind of measurement in that they are free yawing and respond directly to changes in wind direction at the turbine of time scale greater than a few seconds. With a servo-controlled device the measured wind direction and that at the turbine are imperfectly correlated as the point of measurement must be significantly removed from the turbine.

3. DESCRIPTION OF TURBINES

3.1 Common features

The machines have 5m diameter rotors that run downwind of the tower with axes in a horizontal plane 10m above the ground. The machines are separated by 50m (10 rotor diameters) and the fetch is open and uniform when the wind blow along the line of separation. Both rotors comprise two independently hinged blades with a constant twist and linear taper (Ref.4). The nacelles yaw freely and are outwardly identical.

3.2 Dissimilarities

The machines differ in two important respects:

(1) The downwind machine is fitted with a fairing to reduce excitation of strain and cone angle excursions by tower shadow (Ref.5).

(2) For the experiments reported here rubber stops set a lower limit of seven degrees on the coning angle of the blades of the upwind machine. The corresponding value for that downstream was three degrees. This stop curtails blade flapping in the upwind machine particularly at high T.S.R. (where the mean cone angle is low).

The effect of this and the increased tower shadow upon the yaw error is discussed in a separate paper presented at this conference. To summarize the results: the upwind machine runs with a yaw error of similar sense for all T.S.R. (taken to define a positive yaw error); for the downwind machine, where flapping is unconstrained and tower shadow reduced, the yaw is large and negative at high T.S.R., increases with decreasing

T.S.R. to become zero at T.S.R. 10.5, reaches a maximum positive value at T.S.R. 8 and thereafter decreases again rapidly.

In short the action of both stops and tower shadow is to make it easier to follow changes with T.S.R. in the structure of the deflected wake.

The machines also have different alternators .

4. METHODS

Measurements of cone angle and strain were made at a rate of 22Hz and most records were of 1000s duration. Measurements of power were made at 4Hz for 5000s. All data was stored on magnetic discs. The mean cone angle and strain amplitudes are based on 5s intervals.

The form of the profiles obtained is not very sensitive to the period over which the wind direction is averaged so long as this is of the same order of magnitude as the time of passage from the turbine to the second machine (Ref.6).

5. RESULTS AND DISCUSSION

5.1 Introduction

A complete theory of the response of a given turbine to the kinds of incident flow found in a wake depends greatly on the construction of the downwind machine. Therefore, whilst seeking to explain qualitatively the results obtained, we have tried to emphasise simple,generally applicable lessons, intelligible to those not familiar with the dynamics of this kind of rotor (Ref.7).

5.2 Time scales of strain variations.

Fig.1 shows how the standard deviation of strain at the midspan varies with wind direction. All the data is included and the increase in strain is broadly distributed.

By varying the period over which the variance is calculated, some estimate can be made of the frequency of these strain excursions (See Fig.1). It appears that the high frequency variations

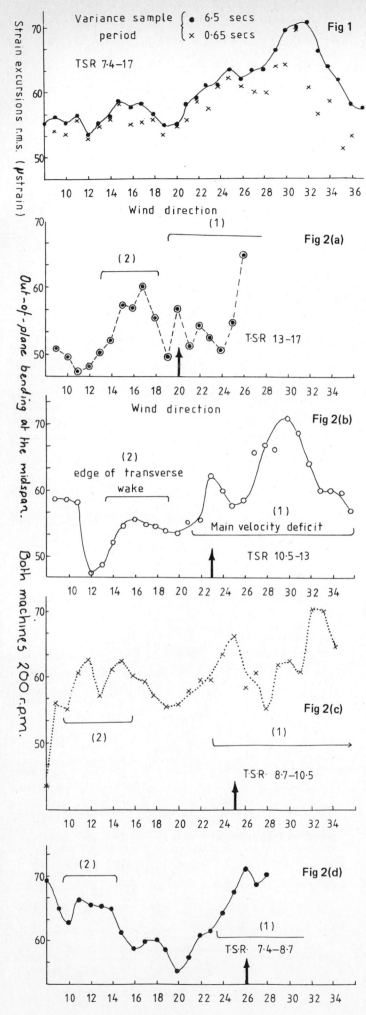

Strain excursions r.m.s. (μstrain)

Out-of-plane bending at the midspan. Both machines 200 r.p.m.

Fig 1

Variance sample period { ● 6·5 secs × 0·65 secs

TSR 7·4–17

Wind direction

Fig 2(a)

(1)

(2)

T·S·R 13–17

Wind direction

Fig 2(b)

(2)
edge of transverse wake

(1)
Main velocity deficit

TSR 10·5–13

Fig 2(c)

(2)

(1)

T·S·R· 8·7–10·5

Fig 2(d)

(2)

(1)

T·S·R· 7·4–8·7

f>2Hz are of much greater amplitude than those at low frequency, although there are such slow variations particularly in the wake.

The average amplitude of high frequency strain excursions is only a first indication of the rate of fatigue damage. Details of the measured probability distribution of the amplitude of strain excursions are omitted for the sake of brevity.

5.3 Increased velocity resolution

The same data is shown in Fig.2(a-d) partitioned according to the mean windspeed.

The double peak in strain excursions associated with the central velocity deficit appears at lower angles with increasing T.S.R. (increasing deflection). This trend is best seen through the position of the peak marked with an arrow in Fig.2 ; this is because the number of data points in each bin is greatest there and because the scales of free stream velocity and direction measured at the anemometer are not distorted by the wake.

5.4 The edge of the transverse wake: Mean values

The effect of the dip in streamwise velocities at the edge of the distribution of sideways velocities (region[i] in Fig.7) on the mean values of coning angle and blade strain during the same experiment are shown in Fig.3. The width of the region in which the measured thrust is reduced is about 5m; a distance which may be determined by the width of the downwind machine and the resolution of this simple method (single point averages used to characterise free stream velocity and direction) rather than the true width of the dip in streamwise velocities.

These results indicate that the edge of the distribution of sideways velocities moves to lower angles with decreasing T.S.R. The increased deflection of wakes at high T.S.R. is thus the result of more effective confinement of sideways velocities as well as increased sideways thrust coefficient.

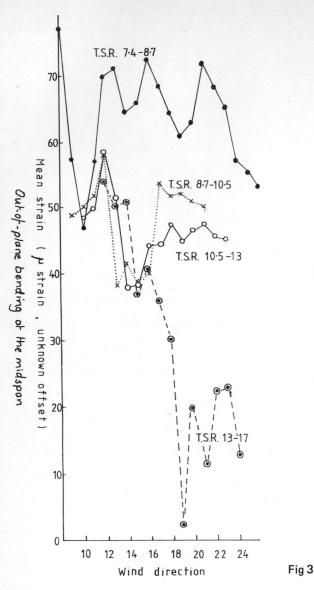

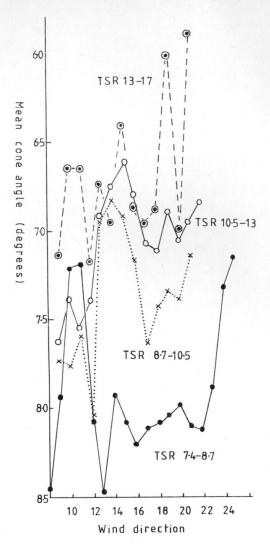

Fig 3

The magnitude of the decrease in mean strain at the edge of the transverse wake is greatest at high T.S.R. ,which is to be expected.

5.5 The edge of the transverse wake: Strain and cone angle excursions

The narrow dip in streamwise velocites will be less effective in exciting cone angle variations when centred on the rotor than when a couple of metres to either side; hence one expects a double peaked structure in the variation of cone angle flapping in this part of the wake (Ref.7). Examples of observations of this effect are shown in Figs.4,5. Note the relationship between the single peaked variation of mean thrust and twin peaks in cone angle variability. The experiment from which the results of Fig.5 were drawn was immediately followed by a control i.e. the upwind machine was stopped and the measurements repeated; no strong variation in mean thrust or

cone angle excitation was observed e.g. Fig.6.

The effect of the edge of the transverse wake on the amplitude of strain excursions in Fig.2 is not so clear cut, although a similar trend in the position and magnitude of the associated structure is discernible. A calculation of the effect of such localised horizontal shears on these highly decoupled rotors has not yet been attempted.

5.6 The central velocity deficit: Strain excursions

A double peaked structure is certainly shown by the data of Fig.2 in the variation of strain excursions across the broad central velocity deficit. The clear asymmetry apparent at high T.S.R. may in part reflect the symmetry breaking effect of a mean wind shear,that yaws the turbine. However the effect can be explained qualitatively

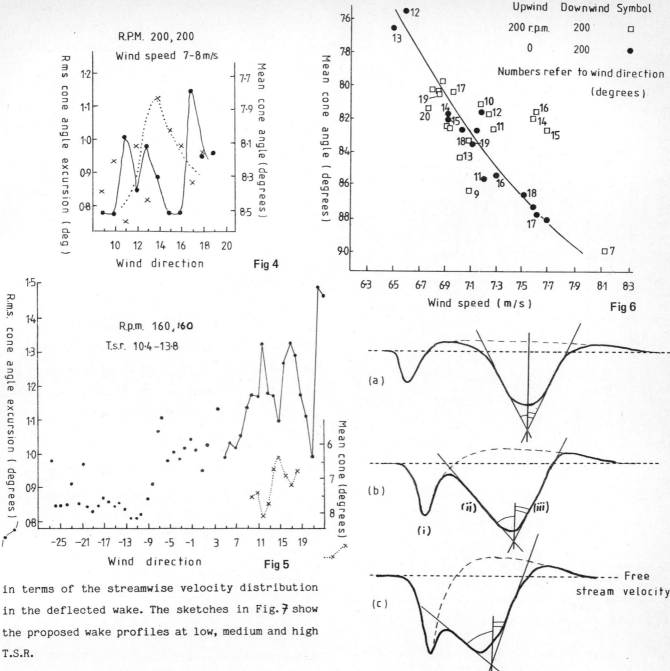

R.P.M. 200, 200
Wind speed 7-8 m/s

Fig 4

R.p.m. 160, 160
T.s.r. 10·4 – 13·8

Fig 5

Upwind Downwind Symbol
200 r.p.m.   200   □
0   200   ●

Numbers refer to wind direction
(degrees)

Fig 6

(a)

(b)

(c)

Free stream velocity

Sketches showing streamwise velocity
T.S.R. Low (a), Medium (b), High (c).

Fig 7

in terms of the streamwise velocity distribution in the deflected wake. The sketches in Fig. 7 show the proposed wake profiles at low, medium and high T.S.R.

The extent to which cone angle variations are excited depends in the first approximation on the gradient of windspeed across the rotor. Reference to Fig.9 shows that this gradient depends on the sum of the gradient associated with the central velocity deficit and that with the edge of the transverse wake. At high T.S.R. these two areas of velocity deficit lie closer together and the latter is relatively deeper, so that the gradient of streamwise velocities is diminished in region (ii) See Fig. 7 and augmented in region (iii). The second effect being small as the gradient of streamwise velocities associated with the deflection declines sharply with distance from the edge of the transverse wake.

5.7 Power Measurements

Fig.8 shows how the power of an upstream machine varies in the wake! Infact, the anemometer used as a measure of the free stream to separate the data is shadowed first by the dip in windspeed at the edge of the transverse wake 26-32 and later by the central wake 41+. Unfortunately the maximum deficit in the power of the <u>downwind</u> machine also lies in the range 27-32 so the position of the anemometer is inappropriate for this kind of measurement.

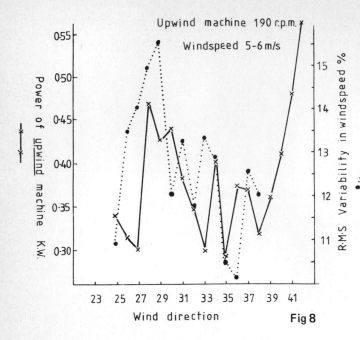

Upwind machine 190 r.p.m.
Windspeed 5-6 m/s

Fig 8

The results of Fig.9 showing the variation of the ratio of the power of up and downwind machines with average windspeed and direction is further complicated by a difference in the performance curves for the two machines - it appears that the upwind machine is less efficient at low wind speeds perhaps because it is in large yaw.

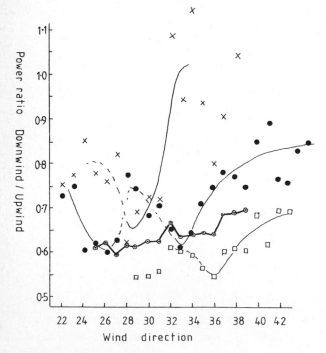

Both machines 175 r.p.m.

T.S.R. 7·6 - 9·1 ☐, T.S.R. 9·1-11·5 ●, T.S.R. 11·5 -15·3 ✗

All data ⊙

Fig 9

The results do however show the upper edge of the wake where the power recovers. The effect of the deflection is to shift this edge to lower

angles with increasing T.S.R. Peaks in the amplitude of strain excursions marked in Fig.2 show a similar shift in the position of the lower edge.

The measurement of wakes for different mean wind directions is particularly important where the terrain is more complex. Changes in the curvature of the undisturbed flow may lead to variations in deflection/asymmetry with windspeed that are not induced by the rotor (Ref.3).

6. CONCLUSIONS

(1) Measurements of the performance of a downwind machine have confirmed that the wake of the MkV turbine is deflected. At a station 50m downstream the displacement of the wake from the line of centres is about 7m greater at high T.S.R than at low T.S.R.

This effect may be important in understanding and perhaps controlling the wind direction for which interactive effects between a number of fixed turbines are operative.

(2) The structure of the deflected flow is more complex than the axisymmetric case. A narrow dip in streamwise velocity associated with the rapid decline in sideways velocities across the steam, appears some distance from the main velocity deficit in the direction of the deflection. The exact position depends on T.S.R. Measurements of the effect of this dip on cone angle variations in a downwind machine and direct measures of the velocity deficit suggest the width of the dip is some 3m and its depth is of order 20% of the free stream (10 rotor diameters downstream ,moderately high T.S.R.).

Wind tunnel experiments are being prepared by R. P. Flaxman from which it is hoped more quantitative information will be gathered about:

(1) The development of the structure of the deflected wake with increasing distance downstream.

(2) The influence of yaw angle and of thrust coefficient on this structure e.g. the position,width and depth of the dip in velocity at the edge of the transverse wake.

(3) The character of the flow outside the hub plane.

7. REFERENCES

1. P.J.H.Builtjes and D.J.Milborrow : Modelling of Wind Turbine Arrays. Proc. 3rd Int. Symp. Wind energy Systems, Copenhagen, BHRA, Cranfield, 1980

2. G.J.Taylor, D.J.Milborrow, D.N.McIntosh and D.T.Swift-Hook : Wake Measurements on the Nibe Windmills. Proc. 7th BWEA Wind Energy Conf. Oxford 1985, MEP.

3. J.W.Buck and D.S.Renne : Observations of Wake Characteristics at the Goodnoe Hils MOD-2 Array. US Dept. Energy Contract DE-AC06-76RLO 1830

4. S.J.R.Powles, M.B.Anderson, D.M.A.Wilson and M.J.Platts : Articulated Wooden Blades for a Horizontal-Axis Wind Turbine. Proc. 4th BWEA Wind Energy Conf. Cranfield, BHRA.

5. D.M.A.Wilson, E.J.Fordham and S.J.R.Powles : A 5m Diameter Horizontal Axis Wind Turbine Using a Commercially Available Alternator. 6th BWEA Conf. Reading, 1984, CUP.

6. G.T.Atkinson and D.M.A.Wilson : Wake Measurements on a Free-Yawing 5m HAWT. 7th BWEA Conf. Oxford, 1985. MEP.

7. G.T.Atkinson, S.M.B.Wilmshurst and D.M.A.Wilson : Blade Dynamics and Yawing Activity of a HAWT in a Sheared Wind. 8th BWEA Conf. Cambridge, 1986, MEP.

APPENDIX: Plan of site.

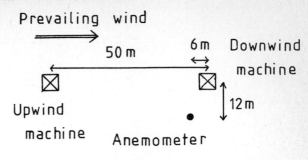

Wake modelling and the prediction of turbulence properties

J F AINSLIE
Central Electricity Research Laboratories, Leatherhead, Surrey

SYNOPSIS This paper describes the development of a numerical wake model first described at BWEA7. In particular, the way in which ambient turbulence should be included in the model is further discussed, and the effects of the boundary layer considered. Comparison with wake data from large wind turbines is very encouraging. Preliminary conclusions on the scale effects for wind turbine wakes are included. The paper also considers ways in which information from the model can be used to estimate turbulence properties in the wake of a wind turbine. Turbulence intensity, lengthscale and shear stresses are considered. It is concluded that the numerical wake model can now be used with considerable confidence to calculate the wake velocity field for wind turbines over a large size range and in various meteorological conditions, and may also give an indication of some of the turbulence properties.

NOTATION

U, V	axial and radial velocities
x, r	axial distance coordinate (downwind from wind turbine) and radial distance coordinate (from wake centreline)
$\overline{uv}$	Reynolds stress cross-correlation
ε	eddy viscosity
ε_a	contribution to eddy viscosity due to ambient turbulence
U_o	free stream windspeed
U_H	free stream windspeed at hub height
U_c	wake centreline velocity
k	Von Karman constant (here taken to be 0.4)
k_1, k_2	dimensionless constants
c	u-v correlation coefficient
L	Monin-Obukhov length
b	full wake width
$F(x)$	filter function to take account of non-equilibrium in near wake region (see Ainslie (1))
K_M	eddy diffusivity of momentum
z	height above ground
z_H	wind turbine hub height
D	wind turbine diameter
u_*	friction velocity
z_o	roughness length
L_T	turbulence lengthscale
k_T	turbulent kinetic energy

1 INTRODUCTION

The generation of appreciable amounts of electricity using wind turbines currently available, or likely to be available in the near future, would require a large number of turbines which would probably be grouped together in some way, either in clusters of various sizes on land, or in shallow offshore waters. It has long been recognised that this grouping together of wind turbines into clusters has associated with it a power loss due to wake effects, and more recently attention has begun to focus on the additional loadings which may be experienced by a wind turbine operating in a cluster. In both cases, predictive techniques of any degree of reliability are still in need of development.

A simplified approach to wind turbine wake modelling was described by Ainslie (1) at the BWEA7 conference, and results of cluster efficiency calculations using this wake model have also been presented by Ainslie (2). The model aimed at predicting the wake velocity field; turbulent mixing was represented by an eddy viscosity technique, and the eddy viscosity had two contributions - the first described the shear generated wake mixing, the second described the additional mixing due to ambient turbulence. The case of an axisymmetric wake in a low ambient turbulence environment was described very successfully by the model. However, there was uncertainty concerning the precise form and scaling factors involved in the ambient turbulence contribution to the eddy viscosity, and two different formulations were presented.

The ambient turbulence contribution to the eddy viscosity is an important term for two reasons. Firstly, it is a dominant term in most situations of practical interest and has considerable influence on the wake velocity field; secondly, it is this term which carries the size-dependent scaling and so the precise form of this term has implications for the scaling up of wind tunnel data to full-scale wind turbines.

In addition to predicting both the wake velocity field in particular cases, and the more general scaling laws to be expected, there is also interest in assessing to what extent simple models can give estimates of turbulence quantities in the wake. Obviously the turbulence information to be gained from a simple model aimed at velocity predictions will be very limited indeed, but is worthy of consideration in the absence of more highly-developed tools for turbulence prediction.

2 MODEL EQUATIONS

The equations used in the model have been described by Ainslie (1), along with a consideration of the simplifying assumptions. The thin-shear-layer equation used is:

$$U\partial U/\partial x + V\partial U/\partial r = -(1/r)\partial(r\overline{uv})/\partial r \quad \ldots (1)$$

The shear stress is described by an eddy viscosity:

$$-\overline{uv} = \varepsilon\ \partial U/\partial r \quad \ldots (2)$$

and the eddy viscosity is found to be given by:

$$\varepsilon = Fk_2 b(U_H - U_c) + \varepsilon_a \quad \ldots (3)$$

where ε_a is the ambient turbulence contribution and k_2 is a universal constant whose value was found to be 0.015 (Ainslie, (1)).

Since most practical situations of interest involve a wind turbine wake immersed in the atmospheric boundary layer, a plausible approach to the ε_a term is to consider the parameter used by boundary layer modellers to describe momentum transfer in the atmosphere. This is the eddy diffusivity of momentum K_M (Plate, (3)), which has the same dimensions as ε and can be defined as:

$$\text{Shear stress } \tau = \rho\ K_M\ \partial U/\partial z \quad \ldots (4)$$

The analogy with equation (2) is obvious and the use of K_M for ε_a follows from this.

3 CALCULATION OF WAKE VELOCITY FIELD IN A BOUNDARY LAYER

3.1 Methods of estimating K_M

K_M can be described in terms of the normal boundary layer parameters. In the surface layer up to 100 m or so in height, in neutral conditions

$$K_M = k\ u_*\ z \quad \ldots (5)$$

and in stable conditions,

$$K_M = k\ u_*\ z/\phi_M(z/L) \quad \ldots (6)$$

where u_* is the friction velocity and $\phi_M(z/L)$ reflects the influence of stability. Furthermore, if a log profile

$$U = (u_*/k)\ \ln(z/z_o) \quad \ldots (7)$$

is taken as being representative of neutral conditions, the non-dimensionalised form of ε_a becomes:

$$\varepsilon_a/U_H D = k_M/U_H D = k^2/\ln(z_H/z_o) \quad \ldots (8)$$

and this equation links the value of ε_a in a simple way to the turbine hub height and the surface roughness.

3.2 Test of the model in neutral conditions

The model equation in non-dimensional form is now:

$$\varepsilon/U_H D = Fk_2(b/D)(1-U_c/U_H) + F\ K_M/U_H D \quad \ldots (9)$$

where $F(x)$ has been retained to take account of non-equilibrium conditions in the near wake (Ainslie, (1)). In neutral conditions this equation can be reduced to:

$$\varepsilon/U_H D = Fk_2(b/D)(1-U_c/U_H) + F\ k^2/\ln(z_H/z_o)\ldots(10)$$

and the ratio z_H/z_o is the only additional input parameter required, apart from turbine thrust coefficient.

Wind tunnel experiments provide a good source of wake data in neutral conditions, although they are themselves subject to modelling uncertainties at small scales. Fig. 1 presents a comparison between the wake model and a CERL wind tunnel experiment using an axisymmetric simulator suspended at one diameter hub height in a turbulent boundary layer flow. The value of $K_M/U_H D$ was calculated as follows. The boundary layer modelled was typical of smooth terrain or water, giving a value of about 0.02 for z_o, and the modelled hub height was about 70 m. Then substitution into equation (8) gives $K_M/U_H D = 0.02$. It can be seen that the agreement with experiment is encouraging.

The Dutch 5 m VAWT wake data (Vermeulen et al., (4)) provides another possible comparison, although the data is of rather uncertain accuracy and the site conditions were not ideal; the thrust coefficient is also uncertain. Assuming conditions to be close to neutral, the roughness length can be estimated to be about 0.1. The hub height of 5 m then gives a value of about 0.045 for $K_M/U_H D$. A comparison between model predictions and the experimental data is shown in fig. 1, and is seen to be reasonable in view of the uncertainties in the comparison.

3.3 A comparison with full scale data

It is tempting to undertake a comparison of the numerical model with available full scale data for neutral conditions, but the difficulties of site measurement must be borne in mind. In particular, measurements on the U.S. MOD-2 turbines (Baker and Walker, (5)) have shown that except in stable conditions wake meandering effects exert a considerable influence on the measured wake deficits. Whilst analytical procedures for estimating these effects must be developed, the wake model – which, at its present stage, calculates instantaneous wake profiles – would be better compared with full scale data in stable conditions where the experimental data are much less influenced by wake meandering effects. Fortunately data has been reported by Baker, Walker and Katen (6) which includes comprehensive measurements of eddy diffusivities in conjunction with wake velocity measurements in stable conditions. The experiments were undertaken on a 170 KW, 17 m diameter Darrieus wind turbine. The eddy diffusivity measurements indicate a value of 0.021 for $K_M/U_H D$. A comparison between model predictions using this value and the experimental data is shown in fig. 2, and the agreement is seen to be encouraging.

Baker et al. (6) also report similar measurements from MOD-2 in stable conditions, although a direct comparison between the MOD-2 data and the 17 m data is not possible on the basis of the results presented because of the difference in thrust coefficient (0.53 for MOD-2 compared with 0.7 for the 17 m VAWT). However, if we assume similar meteorological conditions prevaled for the MOD-2 measurements as for the VAWT measurements (both datasets being recorded in stable conditions), a comparison can be made between the MOD-2 data and the numerical model. Noting that $K_M \propto z$, a value of 0.009 is obtained for $k_M/U_H D$ in the MOD-2 case. Using this value gives very good agreement with the experimental data, as shown in fig. 3.

3.4 Scaling of wake decay

In neutral conditions the scaling law can be calculated as follows:

$$K_M \propto z_H \propto D$$

Therefore,

$$K_M/U_H D \propto 1/U_H$$

i.e.

$$K_M/U_H D \propto 1/\ln (D/z_o) \qquad \ldots (11)$$

This scaling law might be expected to apply in neutral and near-neutral conditions. The influence of $\phi_M(z/L)$ would need to be taken into account in stable conditions. The scale effect given by equation (11) is fairly small and within the range of variation due to meteorological factors.

4 ESTIMATION OF TURBULENCE PROPERTIES

4.1 Turbulent stresses: low ambient turbulence

As with mean velocity information, the simplest case is again that of the low ambient turbulence environment. In the model equations ε is used to represent the Reynolds stresses $\overline{uv}$, and can therefore be used as a sort of measure of the effective strength of the shear stresses. Mean square fluctuations (and thereby turbulence intensities) are also of interest, and are linked to the shear stresses by the effective value of the cross-correlation coefficient c. Given an estimate for c, ε can be used to give a measure of Reynolds stresses, normal stresses and turbulence intensities.

Tennekes and Lumley (7) suggest that the cross-correlation coefficient is often close to 0.4 in free shear flows. The results of Chevray (8) enable an estimate to be made of the correlation for the case of an axisymmetric wake behind a spheroid without recirculation – a similar wake flow to that under consideration here. Typically, values for c calculated from his data vary from about 0.33 (at x/D = 3, r/R = 0.2) to 0.39 (at x/D = 18, r/R = 0.4). A value of 0.4 seems a reasonable estimate, therefore, in the mid to far wake.

The stresses can now be estimated from the numerical wake model as follows:

$$-\overline{uv} = \varepsilon(U_o-U_c)/b = k_2(U_o-U_c)^2 \qquad \ldots (12)$$

$$\overline{u^2} = \varepsilon(U_o-U_c)/0.4b = (k_2/0.4)(U_o-U_c)^2 \qquad \ldots (13)$$

For low turbulence flow the numerical wake model predicts $(U_o-U_c)^2 \propto x^{-2/3}$ in the far wake. Equation (13) then agrees with the 'classical' result for the decay of turbulence intensity in an axisymmetric wake in a low turbulence environment (Uberoi and Freymuth, (9); Hwang and Baldwin (10)).

It must be remembered that the eddy viscosity used in the wake model is in some sense an average value over the wake cross-section; any turbulence estimates made using the model will, therefore, be representative of a particular downstream wake section, not a particular point in the wake. The variation in turbulence properties across the wake cannot be estimated from the model as presented here.

A comparison between wind tunnel turbulence measurements made on the centreline of an axisymmetric simulator wake at a thrust coefficient of 0.62, and calculations of wake turbulence using the numerical wake model to predict the wake decay and then estimating turbulence using equations (12) and (13), is shown in Table 1. The agreement is reasonable given the approximate nature of the calculation, the limited accuracy of the measured data, and the comparison of centreline values with profile-averaged values.

Table 1 Comparison of measured and calculated wake turbulence in low ambient turbulence

Thrust coefficient: 0.62

Downstream Distance, x/D	$\sqrt{u^2}/U_o$ Calculated	$\sqrt{u^2}/U_o$ Measured on wake c/L
6	0.080	8%
8	0.064	7%
10	0.055	7%
15	0.042	6%
20	0.035	5%

4.2 Length scale of wake turbulence

Biringen and Abdol-Hamid (11) quote the Kolmogorov-Prandtl equation for Reynolds stresses in free shear flows, and they use it for jets in particular. The equation is:

$$-\overline{uv} = 0.5 \, K_T^{\frac{1}{2}} \, L_T \, \partial U/\partial y \qquad \ldots (14)$$

where L_T is the length scale of the turbulence and K_T is the turbulence kinetic energy. Replacing $-\overline{uv}$ by $k_2(U_o-U_c)^2$, $K_T^{\frac{1}{2}}$ by $(1/\sqrt{3})(k_2/0.4)^{\frac{1}{2}}(U_o-U_c)$ and $\partial U/\partial y$ by $(U_o-U_c)/b$, we obtain an approximate equation for the turbulence lengthscale:

$$L_T = (4.8 \, k_2)^{\frac{1}{2}} b \qquad \ldots (15)$$

Using $k_2 = 0.015$ gives $L_T/b = 0.27$, which is obviously a reasonable result.

4.3 Inclusion of ambient turbulence

The calculation of turbulence properties in the case where ambient turbulence is not negligible is subject to considerable uncertainty. This arises because the calculation of turbulence properties is even more subject to uncertainties in the interaction of the wake flow with the ambient turbulence than is the calculation of gross features of the velocity field. The axisymmetric wake in a non-turbulent environment is fairly well understood in terms of the distribution of shear stress and the magnitudes of the terms in the turbulence transport equation; the same cannot be said of the wake in a turbulent environment.

However, the work that has been done on

wind turbine wakes, principally in wind tunnel studies, indicates (Wilmshurst et al., (12)) that for many purposes ambient turbulence and wake generated turbulence can be "added" in terms of their energy content, i.e., that the turbulence σ_r introduced by the rotor is independent of the ambient level σ_o, and that the total wake intensity σ_w is given by:

$$\sigma_w^2 = \sigma_r^2 + \sigma_o^2 \qquad \ldots (16)$$

σ_r can be calculated from the model as indicated in section 4.1, and hence an estimate made of σ_w in the turbulent flow case. Although a more rigorous consideration of this latter case will require the spectral distribution of turbulent energy to be considered, this simplified technique may allow an estimate of the total turbulent energy to be made. One comparison between experiment and calculation is shown in Table 2, and the agreement is satisfactory.

Table 2   Comparison of measured and calculated wake turbulence in boundary layer flow

Thrust coefficient: 0.62
Hub height turbulence intensity: 11%

Downstream Distance, x/D	$\sqrt{\overline{u^2}}/U_o$ Calculated	$\sqrt{\overline{u^2}}/U_o$ Measured at Z_H
6	0.14	13%
8	0.13	-
10	0.12	12%
15	0.12	11%
20	0.11	11%

5   CONCLUSIONS

The paper has demonstrated a form of the ambient turbulence eddy viscosity contribution which appears to give very good agreement with available experimental wake data up to the largest of wind turbine sizes. A dependence on the scale of the turbine (relative to the atmospheric boundary layer) is an automatic result, as is a dependence of the wake velocity field on atmospheric stability. Both these factors were known to be possible influences on the wake decay, but little was known previously about the magnitude and form of the effects. Good agreement has been demonstrated with experimental data without the need to introduce any extra constants.

Methods of estimating some turbulence properties using the wake model have been put forward. It is clear that only approximate estimates will be possible without further development of the wake model, and the turbulence treatment in particular. Nevertheless semi-quantitative agreement has been demonstrated both for the case of a wake in a low turbulence environment and for the case of a wake in an atmospheric boundary layer.

The numerical model described in this paper can, therefore, be used with some confidence for the calculation of wake velocity fields over a range of turbine sizes and atmospheric conditions. The model may also be of limited use for estimating wake turbulence.

6   ACKNOWLEDGEMENT

This work was performed at the Central Electricity Research Laboratories and is published by permission of the Central Electricity Generating Board.

REFERENCES

(1) AINSLIE, J.F. Development of an eddy viscosity model for wind turbine wakes, Proc. 7th BWEA Wind Energy Conf., Oxford, 1985, MEP

(2) AINSLIE, J.F. A wake interaction model for calculating cluster efficiencies, Proc. 6th BWEA Wind Energy Conf., Reading, 1984, CUP

(3) PLATE, E. (ed) 'Engineering Meteorology', 1982, Elsevier

(4) VERMEULEN, P.E.J., BUILTJES, P., DEKKER, J. and BUEREN, G.L. An experimental study of the wake behind a full scale vertical axis wind turbine, TNO report 79-06118, 1979

(5) BAKER, R.W. and WALKER, S.N. Wake velocity deficit measurements at the Goodnoe Hills MOD-2 site, Bonneville Power Admin. report BPA 84-15, 1985

(6) BAKER, R.W., WALKER, S.N. and KATEN, P.C. Wake studies at the Flowind vertical axis wind turbine site, Bonneville Power Admin. report BPA 83-10, 1984

(7) TENNEKES, H. and LUMLEY, J.L. 'A First Course in Turbulence', 1972, MIT Press

(8) CHEVRAY, R. The turbulent wake of a body of revolution, Trans. ASME, 1968, series D, p. 275-284

(9) UBEROI, M.S. and FREYMUTH, P. Turbulent energy balance spectra of the axisymmetric wake, Phys. Fluids, 1970, 13, 2205-2210

(10) HWANG, N.H.C. and BALDWIN, L.V. Decay of turbulence in axisymmetric wakes, ASME Jnl. Basic Engng, 1966, 261-269

(11) BIRINGEN, S. and ABDOL-HAMID, K. A turbulent model for free-shear flows, AIAA Jnl., 1985, 23, 1629-1631

(12) WILMSHURST, S., METHERELL, A.J.F., WILSON, D.M.A., MILBORROW, D.J. and ROSS, J.N. Wind turbine rotor performance in the high thrust region, Proc. 6th BWEA Wind Energy Conf., Reading, 1984, CUP

APPENDIX

ADDITIONAL NOTATION

d velocity deficit $(1-U_C/U_H)$ at a point (x,r) in the wake

d_o centreline velocity deficit at downstream distance x

$\hat{d}$ centreline velocity deficit as measured in field experiment

$f(r)$ wake velocity deficit profile

$p(r)$ probability of measuring at radius r in the wake

σ_θ standard deviation of wind direction fluctuations

r_o standard deviation of measurement point fluctuations

T averaging time for velocity measurements in the field

Appendix 1

A simplified correction for wake meandering effects

The wake model as described in the main part of this paper predicts the wake velocity field assuming the externally imposed windfield is stationary with time. In fact this will never be perfectly true for a wind turbine in the atmospheric boundary layer, although the approximation becomes reasonable in stable conditions. In turbulent (i.e. non-stable) atmospheric conditions the wake will meander relative to an observer fixed relative to the ground (or turbine) due to fluctuations in wind direction. This will result in the centreline velocity deficit measured by a fixed observer, $\hat{d}$, being less than the stationary value because the measurement point sweeps over a region of the wake profile during the averaging period of the measurement. Because wake measurements relate the velocity at a point in the wake to the incident, or free-stream, windspeed it is necessary to average measurements over a time T of the order of the air travel time from the turbine to the measurement point (i.e. $T \sim x/U$).

A simplified expression to relate the magnitude of this wake meandering effect to the variability in the wind direction can be deduced as follows. The velocity deficit profile in the wake can be approximated by

$$f(r) = \exp(-3.56(r/b)^2) \qquad \ldots (17)$$

beyond about three diameters downstream of the turbine (Ainslie, (1)). If the wind vector varies in direction such that a sampling point fixed relative to the ground and situated on the mean wake centreline actually samples across the wake with a probability given by

$$p(r) = (1/\sqrt{(2\pi)}r_o)\exp(-r^2/2r_o^2) \qquad \ldots (18)$$

the measured centreline deficit will be given by

$$\hat{d} = d_0 \int_{-\infty}^{+\infty} f(r)\ p(r)\ dr$$

$$= d_0(1+7.12(r_o/b)^2)^{-\frac{1}{2}} \qquad \ldots (19)$$

The value of the standard deviation r_o can be approximately related to the standard deviation of wind direction fluctuations σ_θ, measured over T, by

$$r_o = x\ \sigma_\theta \qquad \ldots (20)$$

assuming that σ_θ is dominated by wind fluctuations of timescale not much less than T; if higher frequency fluctuations dominate σ_θ the behaviour may be more like a random walk, varying as $\sqrt{x}$, although high frequency fluctuations will tend not to produce bodily fluctuation of the wake as a whole.

Equations (19) and (20) together yield the required expression relating measured and stationary centreline deficits:

$$\hat{d} = d_0\left[1+7.12(\sigma_\theta x/b)^2\right]^{-\frac{1}{2}} \qquad \ldots (21)$$

This simplified analysis now allows comparisons to be made between the wake model as presented in this paper and field data measured in circumstances where wake meandering is significant. One set of comparisons is shown in Table 3. Predicted values of $\hat{d}$, the measured centreline deficit, are compared with measured data for both the case of the 17 m VAWT (6) and the MOD-2 turbine (5) in non-stable flow conditions. In these cases the measured values of σ_θ were 10° and 8° respectively, and the values of $K_M/U_H D$ were estimated using equation (8). The agreement is very satisfying, especially considering the magnitude of the correction and the simple nature of this preliminary analysis of wake meandering. The differences between predicted average deficits and measured average deficit values are well within the range of uncertainty of the field measurements.

Table 3 Comparison of measured and predicted centreline velocity deficits for field experiments in non-stable conditions

Downstream Dist. x/d	True c/line deficit d_0	Predicted average c/line deficit $\hat{d}$	Measured average c/line deficit
17 m VAWT			
3	0.51	0.32	0.30
5	0.30	0.15	0.18
7	0.17	0.08	0.10
9	0.12	0.05	0.05
MOD-2			
3	0.59	0.44	0.38
5	0.38	0.23	0.24
7	0.23	0.13	0.10
9	0.17	0.09	-

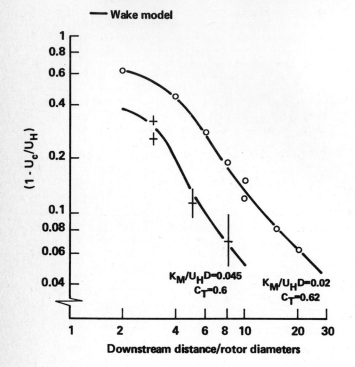

o CERL axisymmetric simulator data,
wind tunnel boundary layer flow,
$C_T = 0.62$

+ 5 m VAWT wake data (Vermeulen et al.
(4))

— Wake model

$K_M/U_H D = 0.045$
$C_T = 0.6$

$K_M/U_H D = 0.02$
$C_T = 0.62$

Fig 1   Comparison of wake model with experimental data:
small scale, neutral conditions

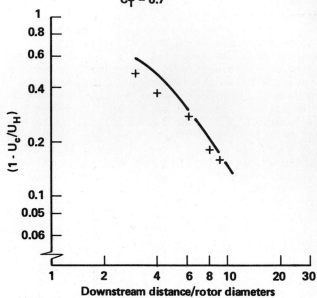

+ 17 m VAWT wake data, $C_T = 0.7$
(Baker et al. (6))

— Wake model, $K_M/U_H D = 0.02$,
$C_T = 0.7$

Fig 2   Comparison of wake model with experimental data:
17m VAWT in stable conditions

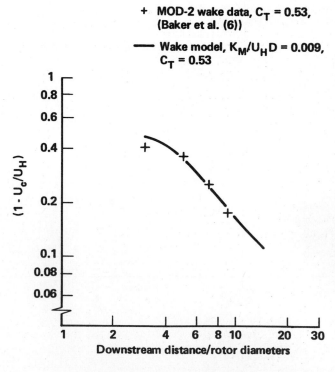

+ MOD-2 wake data, $C_T = 0.53$,
(Baker et al. (6))

— Wake model, $K_M/U_H D = 0.009$,
$C_T = 0.53$

Fig 3   Comparison of wake model with experimental data:
MOD-2 wind turbine in stable conditions

Energy yield and turbulence properties of a 320-turbine offshore windfarm

J F AINSLIE and **D J MILBORROW**
Central Electricity Research Laboratories, Leatherhead, Surrey

SYNOPSIS The energy yield and turbulence properties of a hypothetical offshore cluster of 320 wind turbines have been estimated. The calculations formed part of Phase II of the Taywood Engineering Ltd./CEGB Offshore Wind Energy Assessment. To provide an accurate estimate of the energy losses in the 320-machine offshore cluster, detailed calculations were undertaken using a wake interaction model developed at CERL, taking into account the site wind velocity distribution and the performance characteristics of the proposed machines. A square cluster was used, with the machines on a square grid.

The calculations include the effect of the wind turbine power-windspeed characteristic and the site windspeed probability distribution on the cluster energy efficiency. This, more exact, energy efficiency is higher than would be calculated from the constant performance coefficient models used previously.

The energy efficiency of the basic cluster was calculated to be 84%. By comparing results from wind tunnel tests with theoretical studies it was concluded that the tolerance on the estimate was small, although the lack of full scale wake interaction data does not allow the accuracy to be confirmed. Increasing the spacing to 15 diameters increased the efficiency to 94%.

Reference was made to wind tunnel measurements in a cluster to provide estimates of turbulence levels at the last row of the cluster for different wind speeds. These turbulence intensities relative to the local velocity ranged from 14% when the velocity incident on the cluster was 27 m/s, to 22% at 11 m/s (the incident intensity was taken as 10%). However the corresponding velocity ratios (local/incident) changed from 0.94 to 0.77 and it is interesting to note that the turbulence levels referred to the free stream were almost independent of windspeed at around 14%.

1 INTRODUCTION

The calculation of energy losses due to interactive effects in clusters of more than 100 machines has generally been based on a straightforward application of wind tunnel data. These data have generally been obtained using passive simulators and hence a number of simplifications are inherent in the approach. In particular the influence of wind turbine power and thrust characteristics on the interactive losses tends to be disregarded. A number of cluster configurations and spacings were examined as part of the UK Phase I studies (Simpson, (1)) and cluster efficiencies in the range 58-91% were quoted for a 196 machine cluster at spacings of 7 to 10 diameters. This work and other subsequent studies led to the conclusion that a 10 diameter spacing was probably the minimum acceptable for clusters of 200 or more machines.

The calculations reported here are part of a more detailed resource and cost study (Burton and Roberts, (2)) and more precise estimates of interactive losses have been made, with 10 diameters chosen as the reference spacing to contain energy losses and fluctuating blade loads within reasonable bounds (Milborrow, (3)). Calculations were also carried out using wider spacings as part of an economic analysis. The use of a 10 diameter spacing enabled 320 100 m machines to be sited within an area selected in

the offshore resource study (2) close to the Norfolk coast. The cluster rated output was fixed as near as possible to 2,000 MW.

The need to provide maintenance facilities originally led to the provision of an access corridor to a central base, but this criterion was later relaxed in the resource study, enabling a square cluster shape to be used. Cluster performance has been examined for both configurations, which are shown in Fig. 1. The study has also looked at the effect of displacements in wind turbine positions (due to local variations in sea-bed geology and topography, for instance) on the interactive effects.

A more rigorous approach to the calculation of cluster performance has been used in this study than in previous studies. The annual energy losses due to interactive effects have been calculated taking into account the site windspeed distribution, including angular variations, and the wind turbine power and thrust characteristics. The power-windspeed characteristic for the 100 m diameter 2 blade turbine design is shown in Fig. 2.

Energy yields were calculated for the following cases:

  (i) A cluster with 10 diameters spacing, the basic cluster;

  (ii) A cluster with 15 diameters spacing;

(iii) A cluster based on a 10 diameter spacing but with an access corridor to the central cluster base;

(iv) A perturbed cluster with the turbines being slightly displaced from their ideal positions.

It was necessary to make some simplifications but none are believed to introduce serious errors. In particular, variations of thrust coefficient within the array were taken into account, and detailed rotor performance calculations were not undertaken - the rotor power was taken to be a function of the mean incident windspeed only.

Turbulence intensities have also been calculated for the cluster at 10-diameter spacing. The following conditions were selected as being likely to give rise to the most severe wind fluctuations in the array, affecting both blade loads and power quality. It was reasoned that turbulent fluctuations in the array were unlikely to cause problems at low windspeeds but might do so in the vicinity of the rated windspeed (when the thrust reaches a maximum). The first test case was therefore selected to cover the region just above rated windspeed. The second test case simply represents the highest wind velocity at which all the wind turbines operate.

(a) wind speed incident on the last row in the array equal to 15 m/s (roughly corresponding to rated windspeed)

(b) wind speed incident on the first row equal to the cut-out speed of 27 m/s.

Periodic fluctuating blade loads due to the passage of the rotor blades through the non-axisymmetric velocity profiles generated by the wakes were not separately examined. Although these can cause significant fluctuations at closer spacings (Milborrow, (3)), it was reasoned that the effects would be much less pronounced at 10 diameters and, in any case, the maximum amplitudes only occur over a narrow range of incident wind directions.

2 METHODS OF CALCULATION

2.1 Calculation of the velocity field

The limitations of current methods are such that no single prediction technique can be relied on to provide all the required information for a cluster of this size. In particular, the need to calculate results for fixed speed wind turbines means that wind tunnel test data is not directly applicable. Consequently the estimates draw on data from the following sources:

(i) Wind tunnel simulations of wind turbine clusters. These have made use of passive simulators and of rotors, and detailed measurements have been made within clusters using a laser doppler anemometer (Milborrow, (4); Ross and Ainslie, (5)). They provide a rapid method of estimating cluster efficiency and turbulence intensities but the characteristics of fixed-speed machines cannot readily be simulated. A

20 row cluster was beyond the scope of wind tunnel tests but previous work had shown that equilibrium levels of velocity and turbulence were established quickly (certainly by 10 rows); extrapolation is, therefore, relatively straightforward.

(ii) Surface boundary layer theory. This can provide a useful cross-check on equilibrium levels of velocity and turbulence, but no account can be taken of spatial variations within a cluster, or of changes in wind direction.

(iii) Mathematical modelling techniques for clusters. These use linear superposition to "add" interacting wakes, and performance characteristics to predict the power output. They cope less well with the build-up of turbulence within the array. A mathematical model developed at CERL (Ainslie, (6).) formed the basis of the cluster efficiency calculations performed here.

2.2 Estimates of wind turbine output

Having calculated the velocity field in a cluster using the techniques described above, it is possible to estimate the power output of individual machines in the cluster. There are three methods which have been used to relate the incident wind to the generated power:-

(i) The "energy integration" technique. The velocities at a number of points over the rotor disc are cubed, averaged and divided by the cube of the windspeed incident on the array. This is taken to be the ratio of the power of a rotor in the array to the power generated by a rotor in the undisturbed wind. This will only be correct for wind turbines operating at constant performance coefficient and it was considered that a better technique was required.

(ii) The point values of velocity may, alternatively, be averaged to define a mean windspeed incident on the wind turbine; reference to the wind turbine power curve then gives the rotor power.

(iii) The velocity field incident on the wind turbine, defined by the point values, should ideally be used in conjunction with standard aerodynamic performance prediction techniques to calculate the rotor power. However, it is time consuming to undertake this for 320 wind turbines over a full range of wind directions.

The second and third techniques have both been applied, with only small modifications, to two interacting machines at 5D spacing (Milborrow, (3)). The results indicated that, over the full range of windspeeds, method (ii) gave acceptable results. It has therefore been used in these calculations.

3 CALCULATION OF THE CLUSTER EFFICIENCY

The cluster efficiency was calculated taking into account:-

(i) the interactions between the wakes and the downstream wind turbines;

(ii) the rotor performance characteristic;

(iii) the windspeed probability distribution for the site;

(iv) the site wind rose, although this was found to have only a marginal influence;

(v) variations in rotor thrust associated with changes in windspeed incident on the cluster.

The following were not taken into account explicitly in the calculations, but checks were made to ensure no significant errors were introduced:-

(i) Variations in thrust coefficient due to changes in windspeed within the array. Calculations were performed for three different thrust coefficients, and the effect on the power efficiency was small [changing the thrust coefficient from 0.3 to 0.9 varied the power efficiency by 10%]. An average thrust coefficient for all the machines in the cluster was therefore used for each windspeed band. The thrust coefficient was estimated from the windspeed within the cluster and the wind turbine – thrust coefficient curve.

(ii) The detailed rotor blade geometry of the wind turbines. As noted above the machine was simply characterised by power-windspeed and thrust-windspeed curves. The study by Milborrow (3) indicates that this is satisfactory.

(iii) The build-up of turbulence within the cluster. No allowance was made for the build-up of turbulence within the cluster, and the incident ambient turbulence intensity was assumed to apply throughout the cluster. Although the turbulence intensity measured relative to the local mean velocity can be considerably enhanced within the cluster, measurements from wind tunnel tests indicate that the turbulent velocity incident on a wind turbine remains little changed within the cluster. Neglecting the cumulative effect of turbulence production within the cluster is a simplification but previous studies using the cluster model (Ainslie (6)) have shown reasonable agreement with experimental data despite this simplification in all but the most extreme circumstances. The enhanced turbulence levels in each wind turbine wake due to increased shear in the wake region is, of course, included in the model through the wake mixing equations.

(iv) Wind shear. Corrections for wind shear, average yaw error and tower shadow effects were included in the power characteristic used.

The CERL wake interaction cluster model (Ainslie (6)) was the basis for the cluster efficiency calculations. The windspeed range from cut-in to cut-out was split into 7 bands as shown in Table 1 and the contribution to the annual energy yield from winds in each band was found. Calculations were made both for the original cluster geometry including an access corridor and for the geometry without access corridor which was finally chosen in the resource study.

3.1 Cluster without access corridor

The calculation of cluster efficiency was undertaken as shown in the flow diagram (Fig. 3):-

(i) The CERL wake interaction model was used to calculate the mean velocity incident on each wind turbine in the cluster for 0° incident wind direction (wind perpendicular to one face of the cluster) for each of the 7 windspeed bands.

(ii) The corresponding power for each wind turbine was then determined from the mean velocity using the wind turbine power-windspeed characteristic. The average power per machine in the cluster could then be calculated for each of the 7 windspeed bands.

(iii) Steps (i) and (ii) were repeated using incident wind directions of 10°, 20°, 25°, 30°, 35°, 40° and 45°. Symmetry of the square cluster meant that only angles between 0° and 45° need be considered.

The cluster energy efficiency over the range 0° to 45° is shown in Fig. 4. This has been averaged over all windspeeds, taking the site windspeed duration curve into account.

(iv) From this calculated variation of energy efficiency as a function of incident wind direction the energy efficiency averaged over all wind directions could be calculated. This was done both for an isotropic wind direction distribution and for the actual site wind-rose (probabilities were available in 30° sectors). For the latter calculation the cluster was aligned with the right hand edge of the area notionally selected for the cluster i.e. at approximately -30° to North (2).

(v) Multiplying the average power by the hours per year in each wind-rose sector and windspeed band enabled the average energy yield per machine to be calculated; comparison with the energy yield for an isolated machine at the same site gave the cluster efficiency.

The isotropic cluster efficiency and the wind-rose weighted efficiency were both calculated to be 84%, as shown in Table 1. The high symmetry in the cluster and the coarseness of the wind-rose data meant that the introduction of a non-isotropic wind-rose had little effect.

Table 1 Calculation of energy efficiency for the cluster (10 diameters spacing)

Windspeed (Range (m/s)	Hours Per Year	Power From Isolated Machine (MW)	Thrust Coefficient	Average Power per m/c in cluster (MW)	Annual Energy Yield Per m/c (MWh)
7-9	1508	0.69	0.9	0.41	617
9-11	1289	1.67	0.85	1.22	1572
11-13	955	3.05	0.75	2.34	2235
13-15	640	4.67	0.65	3.88	2482
15-17.2	412	6.0	0.44	5.80	2389
17.2-19.4	202	6.0	0.29	6.0	1212
19.4-27	150	6.0	0.18	6.0	900
					11407

Annual Energy yield for an isolated m/c at this site = 13,644 MWHrs

=> Energy efficiency = 11407/13644 = 84%

3.2 Cluster with access corridor

Initial calculations were made for a cluster
with an access corridor. These yielded an
estimate for the cluster energy efficiency of
38%. A simpler alternative technique was also
used: the cube root of the isotropic cluster
efficiency was used to scale the incident
windspeed, giving a 'typical' windspeed
within the cluster; the equivalent power output
was then read off from the power curve. This
method gave an energy efficiency of 87%,
demonstrating that this very much simpler
technique can be used for initial estimates of
energy efficiency for clusters of fixed speed
turbines.

3.3 Cluster performance at an increased spacing of 15 diameters

The cluster energy yield calculations were
repeated for the cluster without access
corridor at a spacing of 15 diameters. The
isotropic cluster energy efficiency was
calculated to be 94%, thus giving 12% more
energy than for a spacing of 10 diameters. The
cluster efficiency curve as a function of wind
direction is shown in Fig. 4.

3.4 Effect of displacement of turbine positions

It is of interest to investigate the effects of
small displacements in the turbine locations on
the cluster performance. Displacements might
occur in practice due to sea-bed conditions, or
they might be deliberately introduced in order
to reduce the variation of power output with
incident wind direction. The effect of small
displacements was investigated by translating
the turbines in alternate rows and columns by 1
diameter along the direction of the row or column.
The resulting cluster efficiency curve is shown
in Fig. 5. The results show that the annual
energy yield remains unaltered by the
displacements but that the variation of cluster
output with incident wind direction was
considerably smoothed. The ratio of maximum to
minimum energy efficiency was 1.2 as opposed to
1.4 for the regular array geometry. However, it
should be noted that natural variations in wind
direction over the area of the cluster would
probably reduce variations in output with wind
direction.

4 CLUSTER TURBULENCE INTENSITIES

The calculation of turbulence intensities draws
on wind tunnel measurements in clusters of
simulators (Ross and Ainslie, (5)) and surface
boundary layer equations (Milborrow, (4)). The
use of two techniques was necessary because
neither, alone, could provide the required
information. Wind tunnel measurements have not
been made at 10D spacing with the wind turbines
lined up with the wind direction, but are
available at 7D spacing, in line. The boundary
layer equations, on the other hand, enable
estimates to be made for different spacings
(by varying the momentum extraction per unit
area) but can only give a measure of the average
turbulence intensity over all wind directions.
(The application of surface boundary layer
theory assumes that the drag produced by the
turbines is spread uniformly over the surface
of the sea or land).

Estimates were made for three specific
operating conditions with the wind direction
lined up with the machine rows; this produces
the greatest energy loss and increase of array
turbulence. The aim in each case was to
quantify the turbulence intensity at the "last"
row of machines. The quoted values are simply
r.m.s. levels of fluctuation about the mean,
normalised by the mean, over a period, nominally,
of 10 minutes.

The cases examined were chosen so as to
reflect a range of operating conditions:-

(1) Incident windspeed, 11 m/s.

(2) Windspeed incident on 20th row - 15 m/s
     (near rated windspeed)

(3) Windspeed incident on 1st row - 27 m/s
     (the cut-out windspeed).

It was reasoned that either case (2) or
case (3) would produce the most severe loading
conditions. Case (1) was chosen to give an
estimate at a high thrust coefficient (wake
velocity deficits and turbulence intensities
both increase with thrust coefficient).

4.1 Details of calculation for case 1

With 10 diameters spacing, application of the boundary layer theory in a simplified form (5) for wind turbines of thrust coefficient 0.8 (corresponding to a wind velocity of 11 m/s for the wind turbine under consideration) yielded an estimate for turbulence intensity of 16% at 73.5 m (the height of the rotor centre-line above mean sea level). The estimate was made by calculating an effective roughness length for the new boundary layer profile and using ESDU correlations to find the corresponding turbulence intensity. Referred to the undisturbed incident wind velocity this corresponds to an intensity of 13%.

Wind tunnel turbulence measurements in turbulent approach flow for the in-line case were reported for 7 diameters spacing by Ross and Ainslie (5). The value measured was 19% based on the incident wind. The corresponding prediction from boundary layer theory was 15%. This gives a measure of the increased intensity when the rotors line up with the wind. The in-line intensity for 10 diameters may now be inferred in either of two ways; the available data are shown in Table 2.

(i) Increasing the spacing from 7D to 10D decreased the 'average' level of turbulence by 2%; if the in-line measurement for 7D spacing (19%) is reduced by the same amount an estimate of 17% is derived, based on the free stream windspeed.

(ii) At 7D spacing, the 'in-line' measurements were 4% higher than the average values; if the difference is similar at 10D spacing an estimate of 17% is obtained for the "in-line" case, identical with that derived above.

To derive an intensity based on the local velocity an estimate of the ratio of the velocity at the last row to the freestream velocity was required. Wind tunnel measurements yielded an estimate of 0.75 for the final velocity ratio, the cluster model 0.79; taking the mean of these two values yielded an estimate of 22% for the turbulence intensity based on the local velocity.

Table 2 Measured, calculated and inferred turbulence intensities

	Turbulence Intensity Relative to Freestream Windspeed	
	7D Spacing	10D Spacing
Wind tunnel measurements - in line case	19%	(17%)
Boundary layer calculations - average wind direction	15%	13%

A final check was needed to establish whether the fall of velocity through the array produced increased thrust coefficients, tending to raise the turbulence still further. However, the ultimate velocity, estimated at 75% of the initial velocity, was 8.2 m/s and the corresponding thrust coefficient 0.9. This increase was not sufficient to alter the turbulence estimate. It should be noted that the turbulence intensity not only varies with wind direction but also varies radially across the turbine disc. Experimental data (e.g. Ross and Ainslie, (5)) indicates that these lateral and vertical variations are relatively small at 10 diameters.

4.2 Summary of results for other cases

The procedure used to derive estimates of the turbulence intensities for cases 2 and 3 followed that outlined above and led to estimates of turbulence intensities deep within the cluster which are summarised in Table 3. It should be noted that the high intensity of 22% is associated with a reduced local velocity. The levels of turbulence build up fairly quickly and are generally established by the 6th row.

Table 3 Turbulence intensities within the 320 machine cluster

Case	Incident Velocity (V)	Ultimate Velocity	Intensity Within Cluster, Relative to Local Windspeed
1	11 m/s	0.77 V	22
2	17.2 m/s	0.87 V	21
3	27 m/s	0.94 V	14

4.3 Blade loadings due to non-axisymmetric velocity profiles

It should be noted that the increased turbulence level in a wind turbine array is not the only cause of increased blade loadings. The most severe loadings may in fact occur when the wind direction is a few degrees off the axis of the rows of wind turbines, since the blades then pass through non-axisymmetric mean velocity fields. A theoretical analysis of the interactions between the Nibe turbines (Milborrow, 1983) showed that the most severe blade loadings due to lateral shear occur when the wind is 5° off the common axis. These machines are, however, only 5 diameters apart and hence the fluctuations in an array with 10 diameters spacing will be significantly less than quoted for Nibe (+26%, -42%). The wake centre line deficits expected at 10 diameters are of order 10% and hence the corresponding blade load fluctuations will be comparable with those due to wind shear.

5 CLUSTER ENERGY YIELD

The annual energy yield per machine in the cluster was obtained by multiplying the energy yield from an isolated wind turbine at the site (estimated as 13.7 GWhr p.a.) by the cluster efficiency (84%), the efficiency of electrical transmission to the shore (97%) and the estimated annual availability (92%). The resulting average machine energy yield is 10 GWhr p.a. and the total cluster energy yield is 3270 GWhr p.a.

6 CONCLUSIONS

The cluster energy efficiency, i.e. the annual energy output per machine as a fraction of the annual energy output for an isolated wind turbine at the same site, is calculated as 87% for the cluster with an access corridor and 84% for the cluster without an access corridor. Current knowledge implies that these figures should be correct to ±3%, but little full scale data are available to corroborate the calculations and wind tunnel data. The wind probability rose does not significantly affect these figures.

The calculated cluster energy efficiency for the cluster without access corridor increased to 94% when the spacing was increased to 15 diameters. The effect of displacements in turbine locations of up to 1 diameter was investigated for the cluster geometry without access corridor at a nominal spacing of 10 diameters; the overall energy efficiency remained unaltered, but the variation in cluster power output with incident wind direction was smoothed considerably, the ratio of maximum to minimum energy efficiency being reduced from 1.4 to 1.2.

The turbulence intensity values calculated in 3 worst-case situations are 22%, 21% and 14% (compared with about 10% in the approaching airflow). These figures are based on the local wind velocity. It has been noted that rotation of a blade through a wake may give rise to additional high cyclic blade loads independent of the turbulence intensity.

7 ACKNOWLEDGEMENTS

This work formed part of a collaborative study led by Taywood Engineering Limited and funded by the Department of Energy. The work was carried out at the Central Electricity Research Laboratories and this paper is published with the permission of the Central Electricity Generating Board.

REFERENCES

(1) SIMPSON, P.B. (ed.), 1979, Report on assessment of offshore siting of wind turbine generators, Taywood Engineering Ltd.

(2) BURTON, A.L. and ROBERTS, S.C., The outline design and costing of 100 m diameter wind turbines in an offshore area, Proc. 7th BWEA Wind Energy Conference, Oxford, 1985, MEP.

(3) MILBORROW, D.J., Wake and cluster research - past, present and future, Proc. IEE-A, 1983, 130, 9, 566-574.

(4) MILBORROW, D.J., The performance of arrays of wind turbines, Jnl. Ind. Aerodynamics, 1980, 5, 403-30.

(5) ROSS, J.N. and AINSLIE, J.F., Wake measurements in clusters of wind turbines using laser doppler anemometry. Proc. 3rd BWEA wind energy conf. Cranfield, 1981, Multi-Science, London.

(6) AINSLIE, J.F., A wake interaction model for calculating cluster efficiencies, Proc. 6th BWEA Wind Energy Conference, Reading, 1984, CUP.

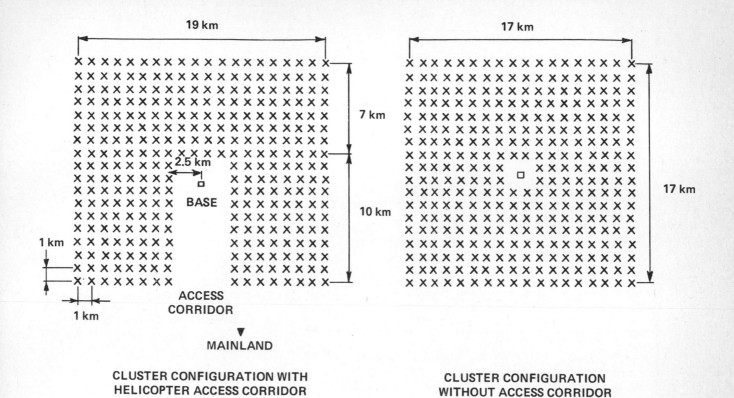

CLUSTER CONFIGURATION WITH
HELICOPTER ACCESS CORRIDOR

CLUSTER CONFIGURATION
WITHOUT ACCESS CORRIDOR

Fig 1 Cluster configurations

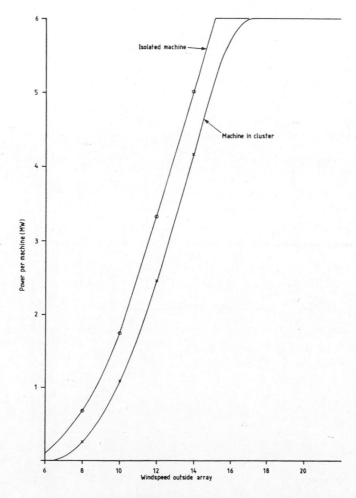

Fig 2 Wind turbine and cluster power—windspeed
characteristics

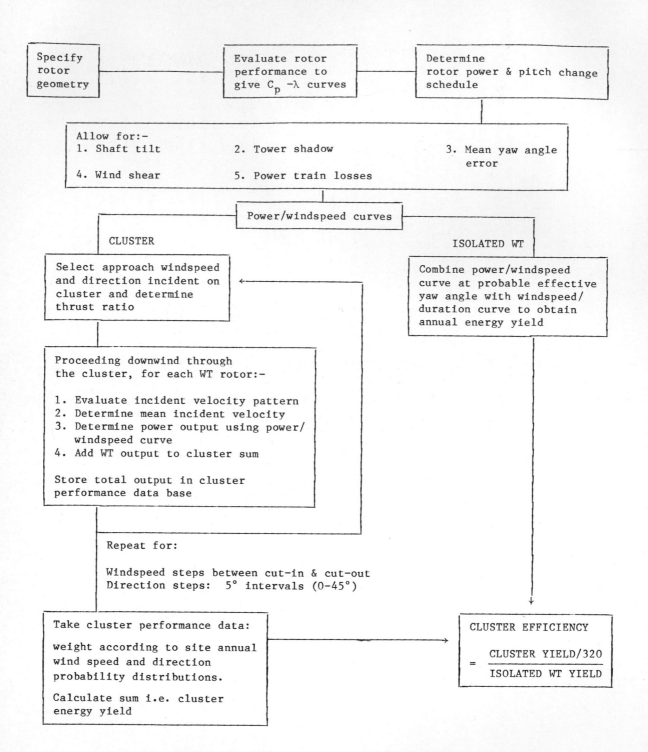

Fig 3   Flow diagram for cluster efficiency calculation

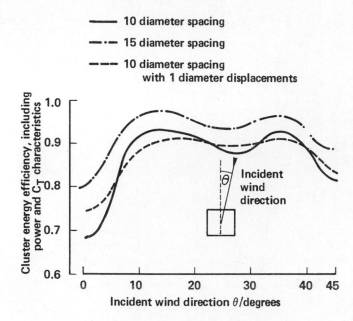

Fig 4   Cluster energy efficiency curves

Design concepts of composite materials and their applications for wind turbines

J C RIDDELL
J C Riddell and Associates

SYNOPSIS: Although Composite Materials are widely used, the methods used for composite component design have been developed from those used for plywood. The performance of composite laminates in applications where they are subjected to high stresses for continous periods is little known. New work at the German Aerospace Research Establishment has shown that well designed composites can withstand more than 10 reversals before failure occurs. This result, taken in association with their other properties, suggest that Composite Components can give superior service in Wind Turbines.

1. INTRODUCTION.

1.1. Composite Materials are used widely for Wind Turbine Rotor Blades. The materials - glass, carbon, polyamide fibres with PEEK, epoxy, polyester and other resins are however quite new in Engineering experience. This paper reviews the developments in Composite Structures in recent years and the design methods that are generally used to predict performance in service.

1.2. The principle of a composite material where the load carrying elements are held in close proximity one to another by a protective adhesive is not new. The ancient Egyptians made bricks of straw held together by Nile Mud baked in the sun. Here the mud was the binder and provided compressive strength while the straw carried the tensile and shear loads in the brick.

1.3. Wood is a Composite material where the tensile strength accross the grain is less than along the grain. Compressive failure can occur as "Compression Shakes" where the wood fibres collapse in compression over a short distance. A condition indicated by a discontinuity in the line of the wood fibres.

1.4. Wood has further characteristics that are similar to Composite materials. The grain in not always straight: tensile load on such a piece may be taken in shear with reduced tensile strength. Wood is hydroscopic and engineering properties must define the water content of the piece to which they relate.

1.5. To overcome these difficulties strength values quoted are the average of a number of samples tested. High duty pieces are usually made from a number of laminations chosen for their straight grain, glued together under pressure where their properties can be assured. Epoxy or other resin system seal the formed piece. The final component is coated with glassfibre, or with a paint finish to give weather protection. Such methods of manufacture assure that the finished component will remain stable in shaped for the course of its service life, and will be less attractive to those termites and other insects who eat wood.

1.6. As an example of the stressing values used in Wooden Aircraft Construction, Table 1.1 comes from a design office instruction in the 1950's.

SECTION 2: Properties of Composites.

2.1 The Design procedures for Composite Components have their origins in those used for wood. However much engineering experience is based on metals where the materials is considered to be homogeneous and stiff. The design stress levels are often decided by deflections allowed in the finished machine, and not by the maximum loads that the structure can reasonably carry.

2.2 Glass fibres, although immensley strong, are elastic, and the Youngs Modulus is about one third that of steel. Glass reinforced structures may undergo considerable deflections before ultimate failure occurs, that is the fibres may fracture or they break away from the resin. This property of flexibility brings allows machines using glassfibre structures to be damage tolerant, with considerable savings in running costs for boats, sailplanes, light aircraft, industrial equipment and now Wind Turbines.

2.3 The flexibility of Glass reinforced structures allows a reduction in the mechanical complexity of the structure for by careful design and use of mass counterbalance, desired deflection of the blade under varying loads can be achieved. In this way a blade can unload under gusts thus reducing the power surge in a wind turbine.

2.4 There comes a time when Glass fibre reinforcement is no longer sufficient for the deflections are too large for the efficient performance of the blade. Stiffer reinforcements are called for. Carbon and Kevlar reinforcement are available. So the laminate can be designed to have different strength and stiffness properties in different planes to reflect the loads imposed. It is quite common to use carbon fibre for the Spar Cap reinforcement on a high aspect ratio wing or blade while glassfibre is used to carry the torsion loads at right angles to the spar.

2.5   It is often desirable to use high aspect turbine blades running at a high rotational speed, the in plane bending stresses at the root, when the blade is in the horizontal position, may limit the blade section dimensions and therefore the aspect ratio that can be attained with known materials. This is a real constraint on theorectical design.

2.6.   The tensile strength of reinforcement fibres is not always matched by equivalent compressive strength. This has been a disadvantage in the use of Kevlar and recent work at Leeds University has been successful in increasing the compressive strength of Aramide fibres. A common method of overcoming this difficulty is to back the fibres with honeycomb or foam reinforcement.

2.7   There are many forms of reinforcement and it should be appreciated that the choice of the density of the foam and its thickness have an influence on the performance of the composite panel. This relationship was explored at Cranfield some years ago (Ref 6) and it was shown that poor selection of foam density and thickness could produce a panel whose flexural behavior was inferior to a glass beam of equivalent weight. However a well designed beam could carry 80% more load in bending than a solid glass piece of the same weight. This combination of Foam and Glass/Carbon/Kevlar reinforcement is used successfully in aircraft and sailplane structures.

2.8   A comparison of the Properties of the Foam and other reinforcements available are in Table 2.2 of Appendix A.

2.9.   It should be noted that these figure for Balsa Wood are for stresses parrallel to the grain. The load carrying capacity of this material perpendicular to the grain is very much less. Balsa has proved to be a stable and effective core material. It is often used where impact absorbtion is required. More recently the lighter Foam has taken its place in sailplane construction.

2.10. The Mechanical Properties of Resin Systems should also be recorded.   Table 2.3 includes the Polyethylethylketone Resin "PEEK" developed by I.C.I. as a competitor to Epoxy Resin in high duty applications.

SECTION 3: Special Characteristics of Laminates.

3.1   Composites are designed to take advantage of the very high tensile strength of the reinforcements. Under moderate loads, the fibres are assumed to be not touching one another, but both fibres and matrix are assumed to behave as perfectly elastic structures. When subjected to low tensile stress both the matrix and the fibre reinforcement are equally strained. Thus the ratio of the stresses within the composite are in inverse ratio to the tensile modulus of the constiuent materials.

3.2   Although the stress in the fibres is much higher, the strength of the panel will depend upon the proportions of the resin and reinforcement. Practical limitations have set an upper limit of about 70% of reinforcement by weight in the total laminate for a carefully controlled unidirectional piece. The values of chopped strand - random - reinforcement is much lower at about 40%. Although widely used, a low proportion of reinforcement shows small advantage over other materials and thus the increase in use of composite materials has been in high duty applications.

3.3   The difference in mechanical properties introduces a further complication in that the behaviour of the laminate is not the same in shear as under the tensile loads. In most applications, the loads are not unidirectional and the reinforcement is built in at +/- 45 degrees. The failure of the bond between the fibre and the resin matrix can occur under two conditions of load:

1.   Matrix Shear due to a failure of the fibre Resin bond.

2.   Transverse Debonding where the resin pulls away from the fibre in a direction perpendicular to the fibre axis.

SECTION 4: Fatigue and Service Life.

4.1 It is important that the resin and the fibre reinforcement must be compatible for in service for fatigue failure where the stresses undergo many reversals, depend on compatibility. In Germany at the DFVLR Aerospace Research Institute, the fatigue behaviour of Composites has been studied by Christophe Kensche (Ref 11) in the course of research to support the design of Sailplane Wings and Wind Turbines Blades.

4.2 In the course of the design study for the DEBRA 25 100 kW Wind Turbine, the DFVLR prepared the Loading Envelope for the Blade Root to define the limits of the loads on the Wind Turbine. This was then extended to produce a Fatigue Spectrum for the design to give a clear basis for design calculation.   It has been running successfully for some time.

4.3 A further point from Mr Kensche's paper sheds more light on the nature of fatigue in composites. Mr Kensche took a sample that had not failed in fatigue when it had been subjected to a similar now of load reversals to a sample that had failed. When pulled in a tensile testing machine, he found that the sample failed in tension at a the same stress as if the sample had not been subjected to the fatigue loads at all.

4.4   This suggests that Fatigue failure in Composites is due to the failure of of the Resin fibre bond.   This form of failure is encouraged by the presence of voids in the matrix. Surface cracks in the Gelcoat can allow water to enter the laminate and breakdown the interface between resin and reinforcement.

4.5 Creep, the deformation of a laminate with time underlaod is not seen as a significant problem. If deformation takes place it is usually due to poor design where loads are applied in a direction not in line with the reinforcement. The deformation takes place in the matrix material. Hald and Kensche report (Ref12) that after fifteen days of applied initial load the bolt loads on the DEBRA 25 machine were only reduced by 3-4%

SECTION 5: Damage to Laminates.

5.1 Cumulative Damage to Laminates is always difficult to assess. Cracks can occur, local delamination can take place unseen within the laminate. The need for a simple measure of the Cumulative effect of deterioration from these and other causes has led to the development of the Linear Palmgren-Miner Rule for fatigue life evaluation.

5.2 Palmgren-Miner take as their base information for the S-N curve for the laminate. The Simple theory is that the linear accumulation of damage from the different load amplitudes can be represented as a coefficient.

$$D = \sum_{j=1}^{K} \frac{n_i}{N_i} = 1$$

where
K = Sum of Load Steps
n_i = Number of spectrum load cycles at this stress level
N_i = Number of Load cycles to failure at stress level n_i

The value of D is $0.1 < D < 10$. and depends upon:

1. Load Spectrum.
2. Load Sequence.
3. Stress Level.
4. Structural Design.
5. Material and Composite Lay Up properties.

SECTION 6: DEBRA 25 100kW WTG Blade Design Details.

6.1 Hald & Kensche (Ref 12) give a very detailed account of the design details of the blades for the three bladed DEBRA 25 10-0 kW machine, and I quote their figures here as an example of the state of the art of Composite construction for turbine blades..

6.2 Eigenfrequencies required for non-resonance reasons were flapwise 2.18 Hz, and 3.3 Hz edgewise under centrifugal forces. The centre of gravity, centre of shear, and centre of stiffness should come together with the pitch axis of the blade at all cross sections. In this way moments due to centrifugal forces, and deflections are minimised.

6.3 The Blades were made in Glass Reinforced unidirectional tapes with a foam sandwich shear web faced with glass fabric. The top and bottom surfaces are laid up in seperate moulds and there the foam stabiliser sandwich is added. The two half shells are then brought together, 50 kg of leading edge weight is added, and the two halves are joined in the mould. The mass of the blade is 335 kg which compares well with an average value of 500 kg for other machines of similar size.

6.4 Static Load tests were carried out and a minimum design safety factor of 1.5 was set. A higher value of 1.725 was used in a static test to cover the dynamic effects. In the area of attachent tests were carried of at a load factor of 2.36. This was later increased to 3.39 to give added strength to withstsand the loads of the Centaury Gust case. The static deflections and loads measured on the 11.6 metre blades were:

1. Blade Tip Deflection 1.85 m.
2. Root Transverse Load 31.4 kN.
3. Root bending moment (out of plane) 172.5kNm.
4. Root torsion moment 5.56 kNm.

6.5 The measured and calculated Eigenfrequencies of the blades are given in Table 6.5.

7. Conclusions.

7.1 This paper has reviewed recent work from Germany and elswhere that demonstrate the importance of Composite construction for Wind Turbine Rotor Blades. The careful design that has taken advantage of the experience gained in the design of similar products. In this way confidence has been built up in the performance and practicalities of durable composite components.

7.2 Fatigue testing of Glass reinforced composites has been taking place over the last eight years and it is now felt that Glass reinforced composites are quite well understood. Fatigue curves for 10^3 reversals necessary for the 20 year life of a turbine are now available. Interpolation suggests that up to 10^{24} reversals are possible before total fatigue failure.

7.3 However, the same volume of investigation has not yet taken place on Kevlar Aramide and Carbon reinforced composites. The arrival of Peek resin suggest a further advance in composite performance. We can look forward to lighter and more durable Wind Turbines in the future.

REFERENCES:

1. Marine Design Manual, Gibbs & Cox, MacGraw Hill 1960.

2. Tsai and Hahn, Introduction to Composite Materials. Technomic Publishing Company, Connecticut 1980.

3. R.M.Gill, Carbon Fibres in Composite Materials. Published on behalf of the Plastics Institute by Iliff Books, London 1972.

4. N. G. McCrum. A review of the Science of Fibre Reinforced Plasatics by Published by HMSO 1971.

5. Scott-Bader & Co, Polyester Handbook. Wellingborough, England, 1983 Edition.

6. R. Tetlow, Design in Composite Applications to Aircraft Structures. Cranfield Institute of Technology. A paper presented to the Society of Aeronautical Weight Engineers Inc in Londond, June 1973.

7.   M.J.Owen & R.G.Rose, Polyester Flexibility
     versus Fatigue Behaviour of Reinforced
     Plastics. Modern Plastics November 1970

8.   M.J.Owen and S.Morris. Fatigue Resistance
     of Carbon Fibre Reinforced Plastic.
     Department of Mechanical Engineering
     Nottingham University.

9.   M.J.Owen, Fatigue Testing of Reinforced
     Plastics. Composites, December 1970.

10. Seymor Lieblein Survey of Long Term
     Durability of Fibreglass Reinforced
     Plastic Systems, NASA CR-165320 NASA
     Lewis Research Centre, Cleveland.
     Ohio 44135. USA

11. Christophe Kensche, Fatigue of Composite
     Materials in Sailplanes & Rotor Blades
     XIX OSXTIV-Congress, Rieti. August 85.

12. H. Hald & Ch Kensche, Development & Tests
     of a Light Weight GRP Rotor Blade for
     the DFVLR 100 kW WEC. Wind Power 1985
     San Fransisco CA, August 1985.

TABLE 1.1:   Properties of Wood for High Duty
             Applications.

Property/Material	Ash (V3 Spec)	Ash (Common)	Spruce (Grade V3)
Relative Density	0.64	0.6	0.4
Tension along Grain - (MPa)	87.56	55.16	68.95
Shear - (MPa)	10.34	6.895	6.2
Compression on Short Column (MPa)	36.54	27.58	34.47
Compression - accross Grain (MPa)	8.69	5.52	5.52
Young's Modulus along grain - (GPa)	1.034	1.034	1.034

TABLE 2.3:   Mechanical Properties of Resin
             Materials.

Property / Material	Polyester	Epoxy	Peek
Relative Density	1.202	1.170	1.32
Tensile (GPa)	41.3	88.9	92
Compressive (MPa)	138.0	120.0	118
Elongation (%)	2.0	7.5	50
Elastic Mod (GPa)	0.4	3.2	3.6 at R.T

TABLE 2.1:   Mechanical Properties of Fibre
             Reinforcement Yarns.

Property.	Units	"E" Glass	"S" Glass	Kevlar 49	Graphite "HT"
Relative Density		2.55	2.5	1.44	1.75
U.T.S.	(GPa)	2.41	3.4	2.76	2.58
Specific T.S.	(GPa)	26.20	37.9	68.95	48.26
Tensile Modulus	(GPa)	68.90	82.7	124.1	220.6

TABLE 6.5:   Blade Eigenfrequencies.

Mode	Test Blade (calculated)	Test Blade (measured)	Service Blade (measured)
1. Flapwise	2.04 Hz	2.19 Hz	1.95 Hz
2. Flapwise	5.87 "	5.67 "	5.50 "
3. Edgewise	3.68 "	3.32 "	3.25 "

TABLE 2.2:   Mechanical Properties of Reinforce-
             ment Materials.

Property/ Material		Balsa Wood	Rigid PVC Foam	Rohracell Foam
Relative Density		0.10	0.09	0.07
Tensile Strength	(MPa)	9.48	1.72	2.8
Compressive "	(MPa)	5.17	1.24	1.5
Shear "	(MPa)	1.24	1.03	1.3
Mod of Elasticty	(MPa)	193.0	27.58	92.0

Towards lighter wind turbines

D J MILBORROW

Technology Planning and Research Division, Central Electricity Generating Board, London

SYNOPSIS   The "most economical" size for a wind turbine has been the subject of debate for some time. Pragmatic analyses of wind turbine costs, together with relatively simplistic scientific arguments, have tended to infer that large wind turbines are not yet cost effective. The scientific analysis has hitherto been confined to noting that rotor weight increases as the third power of the diameter and energy as the square. This paper seeks to quantify these trends in a little more detail by examining the relative influences of three primary design drivers:- flapwise bending loads, extreme gust loads and gravitational bending stresses. The effects of design options such as number of blades, design lift coefficient and blade slenderness are also explored and the implications on rotor weight of using materials such as wood, aluminium and the reinforced plastics are examined.

Although the analysis is qualitative rather than quantitative, the results have been correlated with available design data and a number of interesting trends can be detected. With most materials, the extreme gust loading has an important influence on rotor weight unless it is assumed that the blades can be feathered to minimise the high drag forces. Stresses due to flapwise bending also have a strong influence on design and gravitational bending stresses become dominant only when mild steel or alminium alloy is used for large rotors. Under these circumstances rotor weights increase with the fifth power of the diameter.

Very light rotors - down to one fifth the weight of steel rotors - can only be envisaged by the use of expensive materials such as titanium alloy or CFRP. However the analysis provides a basis for examining the trade-offs between material costs and rotor weight, which also influences the design of other components of a wind turbine.

NOTATION

c	Blade chord	m
C_D	Aerofoil drag coefficient	
C_L	Aerofoil lift coefficient	
d	Diameter of blade spar	m
h	Non-dimensional radius, r/R	
H	N.D. radius of max. stress	
F	Design windspeed	m/s
F_E	Design max. gust speed	m/s
g	Acceleration due to gravity	m/s^2
M	Bending moment	N-m
r	Radius	m
R	Radius of wind turbine	m
s	Stress	N/m^2
S_F	Fatigue limit	N/m^2
S_M	Allowable mean flapwise stress	N/m^2
S_U	Ultimate tensile stress	N/m^2
x	Non-dimensional thickness of blade, t/c	
X	Maximum N.D. thickness	
W	Blade weight	N
z	Number of blades	
Z	Section modulus for bending	m^3
λ	Tip speed/wind speed ratio	
ρ	Density of air	kg/m^3
ρ_s	Density of blade spar material	kg/m^3

1.   INTRODUCTION

A debate concerning the size of horizontal axis wind turbine likely to produce the cheapest energy generation costs has been in progress for some time. One simple argument, for example, notes that energy yields increase with the square of rotor size, weights as the cube, and hence that very large rotors are unlikely to be economic. Some recent analyses of machine costs have tended to reinforce this view (Hau, ref. 1, Spencer, ref. 2), whilst estimates of the weight of very large rotors - e.g. 150 tonnes for the 97 m MOD-5B design (Lowe and Wiesner, ref. 3) can also be interpreted as signifying that diminishing returns are realised at this scale.

Wood has recently become established as a blade material and machine sizes employing this material are increasing. It is not yet clear whether wooden rotor sizes will eventually be subject to some restraint. One important factor is what will be termed the "crossover size" beyond which gravitational bending forces dictate rotor spar dimensions and hence weight. It may readily be shown that rotor spar stresses increase with size (Milborrow, ref. 4) and it follows that spar dimensions must increase to keep stresses within acceptable limits. This, in turn, implies that rotor weights may increase with some higher power than the cube of the size. Frandsen et al (ref. 5) have suggested that the boundary between "small" and "large" is 30 m diameter but did not appear to explore the implications of different materials.

The purpose of this analysis, therefore, is to examine:-

(1) The factors which determine the weight of wind turbine rotors, and whether flapwise or gravitational bending loads dictate rotor spar geometry.

(2) Once gravitational bending becomes the design driver, how rotor weights vary with size.

(3) The effect of material properties on rotor design. The study focuses on rotor weight, partly as a guide to their cost, but also due to the effect of rotor weight on the cost of support towers, transmissions and hence the structure as a whole.

2.   METHOD OF ANALYSIS

A simplified approach has been adopted for this study. Blade loads and stresses have been calculated by working from first principles and neglecting the effects of aerofoil drag. (Since drag has little effect on flapwise bending loads - which control the design of small rotors - and torque is small compared with gravitational bending loads - which control the design of large rotors - this approach is quite justified). It is assumed that rotor loads are supported by hollow circular or elliptic spars within an aerofoil of uniform, or varying, (non-dimensional) thickness. The chord increases with decrease of radius. The geometry is thus fully comparable with that of modern rotors (Figs 1 and 2).

The analysis builds on results from a previous study (Milborrow, ref. 4) to which reference should be made for the derivation of some of the basic equations.

3.   ROTOR GEOMETRY

A simple relationship for the blade chord distribution was derived from the earlier analysis and was found to yield blade shapes similar to those used in practice:-

$$cr/R^2 = 5\pi/3z\lambda^2 \ C_L \qquad (1)$$

In practice this relationship is not used in the root region, where modifications are made so that the blade chord does not become excessive.

It is assumed that blade loads are supported by spars with minor axes equal to the thickness of the aerofoil section. Initially these spars are assumed to be circular and solid - simply to aid the development of the design equations - but the use of elliptic spars is later assumed for the development of usable numerical estimates of blade weights. Such spars weigh less and have higher edgewise moduli.

If the non-dimensional aerofoil thickness is x it follows that the diameter of the load-bearing circular spars, d (= xc), at any non-dimensional radius, h, is given by:-

$$d/R = 5\pi x/3z\lambda^2 \ C_L \ h \qquad (2)$$

4.   ROTOR DESIGN DRIVERS

The previous study showed that the principal load governing the rotor construction of "small" machines tends to be the flapwise bending moment, whereas edgewise bending tends to control the design of "large" rotors. The implications of each of these design drivers on rotor weight will be considered, in turn, before the boundaries between "large" and "small" are discussed. A third possible design driver - bending of a parked rotor at the maximum design windspeed - will also be discussed.

4.1   Flapwise Bending

The previous study derived the following expression for flapwise bending moment:-

$$M = 5\pi\rho F^2 R^3 \ (2 - 3h + h^3)/36z \qquad (3)$$

and the following expression for the surface stress:-

$$s = 0.03\rho F^2 z^2 \lambda^6 \ C_L^3 \ h^3 \ldots$$
$$(2-3h + h^3)/x^3 \qquad (4)$$

which has a maximum value at a non-dimensional radius, h, of 0.618. Setting the surface stress at this radius to some limit prescribed by the properties of the spar material enables the thickness of the aerofoil section to be defined:-

$$x = 0.14\lambda^2 \ C_L \ \sqrt[3]{z^2 \ F^2/S_M} \qquad (5)$$

The relationship between the design stress, S_M and the fatigue limit/UTS is examined later.

To determine blade and rotor weights designed on flapwise bending criteria, it is necessary, first, to define a blade profile in the root region which eliminates the need for the very large chords implicit in the basic equation (1). The equation defining surface stresses (4), can be used to show that the ratio of the maximum stress (at 0.618R) to the stress at 0.1R (taken as the root) is 52.9. If the spar diameter at 0.1R is reduced by a factor $\sqrt[3]{52.9}$, i.e. 3.75, then the stresses at the two locations will be equal. Since the use of an aerofoil section of constant non-dimensional thickness has been assumed, it follows, from (1) that the chord is reduced by a factor 3.75 also, giving a root chord:-

$$c/R = 14/z\lambda^2 \ C_L \qquad (6)$$

This chord can be maintained constant out to a radius 0.375R, where it blends with the original profile, without raising the surface stress above the maximum value.

In order to ensure that there is adequate thickness in a spar to keep flapwise stresses at an acceptable level under all operating conditions, it is necessary to relate the predicted mean design stress to some property of the spar material. The analysis of cyclic fluctuations of the flapwise loads measured on MOD-2 (Thresher, ref. 6) showed that, at the design windspeed, the amplitude of 99.9% did not exceed the magnitude of the design mean value. The probability of larger fluctuations at higher windspeeds was low. However, if a safety factor of 1.5 is put on the corresponding acceptable stress for the material it is considered that the integrity of the design would be satisfactory, at least for an approximate analysis. Fluctuations equal to the mean imply a stress ratio of zero, so that, using the standard Goodman diagram approach, the acceptable mean level (before taking the safety factor into account) is:-

$$S_M = S_U \cdot S_F / (S_U + S_F) \qquad (7)$$

4.2 Edgewise Loads

If edgewise loads control rotor design geometry, the location of the maximum stress again needs to be known. The bending moment about any radius hR, assuming a constant thickness aerofoil section with the blade weight concentrated in a solid circular spar, is given by (Milborrow, 1982)

$$M = 25\pi^3 \rho_s g \ x^2 R^4 \ (h - \log(h) - 1)/36 \ z^2 \lambda^4 \ C_L^2$$

The surface stress at any radius, $s = M/Z$:-

$$s = 24 R \rho_s g z C_L \lambda^2 h^3 \ (h - \log(h) - 1)/5x\pi$$

$$\qquad (8)$$

which has a maximum at $h = 0.546$.

Assuming gravitational loads are much larger than edgewise torque loads, the maximum stress may be related to the fatigue limit, with allowance for a safety factor, giving:

$$S_F = 0.0376 R \rho_s g z C_L \lambda^2 / x \qquad (9)$$

Rearranging this equation it is possible to define an aerofoil thickness for the rotor and hence calculate its weight. However an examination of the stress predictions from equation (8) revealed that they fell off sharply both outboard and inboard of 54.6% radius, which implied that some economy of material was possible. The possibilities are explored later.

4.3 Extreme Gust Loading

A further design loading condition imposed on a wind turbine rotor is the "extreme gust" loading on a stationary rotor. Unlike the aerodynamic and gravitational loads, it can be circumvented if the blades are feathered or the rotor yawed so that the blade edges face the wind. Whether or not these courses of action can be guaranteed is beyond the scope of this analysis and the relevant blade loadings for this case are derived as follows:-

The bending moment of an elemental portion of blade, at radius r, due to the wind load, about radius hR is:-

$$dM = 1/2 \rho F_E^2 \ C_D c(r - hR) dr$$

and hence the total moment, substituting for c from (1).

$$M = \int_{hR}^R (1/2 \rho F_E^2 \ C_D .5\pi R^2 (1 - hR/r)/3z\lambda^2 C_L) dr$$

$$\qquad (10)$$

Integrating and then writing $s = M/Z$ to determine the stress at any section gives:-

$$s = 432 \rho F_E^2 C_D z^2 \lambda^4 \ C_L^2 h^3 (1 - h + h \ \log h)/25\pi^3 \ x^3$$

$$\qquad (11)$$

which as a maximum, at $h = 0.577$, of:-

$$s = 0.0113 \rho \ F_E^2 C_D z^2 \lambda^4 C_L^2 / x^3 \qquad (12)$$

An examination of the implications of using a parallel chord section inboard of 0.375R showed that the root stress ($r/R = 0.1$) would be 1.74 times the stress given by eqn. (24). Setting this root stress equal to the ultimate tensile (or compressive, whichever is lower) bending stress (divided by safety factor) and rearranging gives the necessary aerofoil thickness:-

$$x = 0.27 \ \sqrt[3]{(\rho F_E^2 C_D C_L^2 z^2 \lambda^4 / S_U)} \qquad (13)$$

5. BLADE AND ROTOR WEIGHTS

If it is assumed that the blade weight is concentrated in the spar the total weight can be calculated by adding the weight of the tapered section, from $r/R = 1$ to $r/R = 0.375$:-

$$W = \int_{hR}^R (\pi d^2/4) \rho_s g dr$$

$$= 25\pi^3 R^3 \rho_s g x^2 (1 - h)/36 h z^2 \lambda^4 C_L^2 \qquad (14)$$

to the weight of the parallel section, inboard of 0.375R. In practice solid spars are not used and the previous explored explored the weight reductions achievable with hollow spars – factors of 2 or more are easily possible. Allowing for the weight of the aerofoil skins, fixtures and fittings, a factor of 2 is considered realistic. This yields a blade weight, for constant thickness blade sections, of:-

$$W = 1.25\pi^3 R^3 \rho_s g x^2 / z^2 \lambda^4 C_L^2 \qquad (15)$$

5.1 Variable Thickness Blades

In practice, constant thickness blades are only used for small rotors since weight savings can be made using variable thickness, without changing the maximum stresses. The scope for reductions of thickness may be assessed by calculating the ratio of the stress at any given radius to the maximum stress. The cube root of this ratio defines the allowable thickness ratio. Empirical relationships were then devised to describe the thickness variation, of the form:-

$$x_h = [1 - m \ (H - h)^2] \ X_H \qquad (16)$$

Values of the parameters in this equation varied, as shown in Table 1. The thickness distributions predicted by this equation are comparable with those used in practice. The allowable tip thickness, for example, becomes about one-third of the maximum - or 10% if the maximum is 30%.

Further weight reductions were achieved by equalising the stresses at the root and maximum stress locations in the main blade. This modification (iterative with gravity as the design driver) yielded reduced values of the coefficient in equation (15), which are shown in Table 1.

TABLE 1

Thickness Variations
and Weights

Driving Load	Flapwise, aerodynamic	Gravity	Gust
Max stress location, H	.618	.546	.577
m, eqn (15)	4	3.33	4
Ratio $X_H/x_{0.1}$	1.0	0.81	0.83
Coefficient, eqn (15)	1.20	0.82	1.00

The calculation for gravity-driven designs took into account the additional savings possible with variable thickness in the constant-chord section ($r/R = 0.1$ to 0.375).

5.2 Flapwise Loads Driving Design

It is now instructive to substitute from (5) into (15) with the appropriate coefficient, giving:-

$$W = 0.73R^3 \rho_s g (\rho F^2/zS_M)^{2/3} \qquad (17)$$

for the blade weight, or

$$W = 0.73R^3 \rho_s g z^{1/3} (\rho F^2/S_M)^{2/3} \qquad (18)$$

for the rotor weight. It follows that the weight of a three blade rotor is always greater than that of a 2-blade rotor.

5.3 Edgewise Loads Driving Design

The use of elliptic spars (ref. 4) can reduce weight and enable the edgewise modulus to be increased. The combined effects enable stresses to be reduced by factors possibly as high as 5, but if a value of 4 is used, to scale down the maximum stress and eqn (9), is substituted into the relevant version of (15), we have:-

$$W = 0.014 (\rho_s g)^3 R^5/S_F^2 \qquad (19)$$

5.4 Gust Loads Driving Design

Finally, to establish rotor weights pertinent to this design criteria, we may substitute eqn. (13) into eqn. (15) giving:-

$$W = 2.3 \rho_s g R^3 z \ F_E^2 C_D/S_U \qquad (20)$$

6. "CROSSOVER" RADII

Since gravitational bending stresses increase with size it may be inferred that the design of "small" rotors will be controlled by the need to limit stresses due to flapwise bending, whereas, with "large" rotors gravitational stresses will predominate. The boundary between "small" and "large" may be calculated by establishing the radius at which the required weights become equal. However, the loads due to aerodynamic and gravitational loads act in orthogonal planes and the principal stress plane in the blade spar will lie between them. When the two cyclic stresses are equal (assuming them to be in phase) the magnitude of the principal (cyclic) stress will be 2 times either stress. The "crossover" radius has therefore been scaled down by this factor giving:

$$R = 7S_F \sqrt[3]{(\rho F^2/zS_M)/_s g} \qquad (21)$$

7. LIMITING SLENDERNESS AND MAXIMUM SPEEDS

The predictions for rotor weights imply that these are independent of design tip speed windspeed ratio (T.S.R), and hence of blade slenderness. In practice it is desirable to use a high T.S.R. and slender blades, partly to keep down the complexity and cost of the aerofoil skins, partly to minimise rotor torque and gearbox ratios, which helps to reduce costs.

For the purpose of estimating limiting slenderness, the maximum aerofoil thickness may be set at 30%. This yields the following estimates of limiting tip speed/wind speed ratio:-

Flapwise bending criterion

$$\lambda^2 \text{ MAX} = 2 \ \sqrt[3]{(S_M/z^2 \rho F^2)}/C_L \qquad (22)$$

Edgewise bending criterion

$$\lambda^2 \text{ MAX} = 25S_F/R \rho_s gz \qquad (23)$$

Maximum gust criterion

$$\lambda^2 \text{ MAX} = (1.17S_U/\rho F_E^2 \ C_D C_L^2 z^2) \qquad (24)$$

8. RESULTS

The objective of the study was to identify influences and trends, rather than calculate precise numerical values. Blade weights, in practice, are likely to vary considerably - due to differing design procedures, methods of construction, and variations in the weight of peripheral equipment, such as control gear, mounted on the blades. The relationships developed in the foregoing analysis have been used to derive estimates of rotor weights, crossover radii and limiting design tip/speed wind speed ratios for a variety of blade materials, and these are listed in Table 1. In the case of steel, titanium alloy, GRP and CFRP, there were quite small differences between the weights derived from the "flapwise bending" and "extreme gust" criteria. As the design parameters for the latter case were quite stringent, weights appropriate to the "flapwise bending" criterion have been tabulated. For the purposes of this analysis fixed values of certain design parameters have been used, as listed below.

Design Wind Speed (F) - 9 m/s

Most of the world's large wind turbines are designed to reach their peak efficiency around 9 m/s, which roughly corresponds to the mean windspeed at hub height. For rotors designed on flapwise bending criteria, weight varies as $F^{4/3}$ and hence the effect of reductions or increases will be significant.

Design Lift Coefficient (C_L) - 0.8

Most aerofoils reach peak efficiency (C_L/C_D) near this value. Changes in C_L would have a small effect on λ MAX ($\propto 1/C_L$), for rotors designed for flapwise bending or extreme gusts.

Extreme Gust Level (F_E) - 60 m/s

There are considerable variations in this figure, which is a function of the wind turbine site. Careful selection is important since the choice can change the design driver; moreover weight varies as F_E^2 if the gust criteria drives the design.

Drag Coefficient of Parked Blades (C_D) - 1.6

Uncertainty surrounds this figure. The chosen value is simply a compromise between the "conventional" drag coefficient of an infinite aspect ratio blade at 90° incidence (2.0) and more recent estimates, derived from actual measurements (ranging down to 1.2), listed by Jamieson and Hunter, (ref. 7).

The material properties listed in Table 1 have been taken from several sources, not all of which agree. The values quoted are thought to be conservative and include a safety factor of 2 on the U.T.S. and fatigue limit, 1.5 on the allowable mean flapwise bending stress. The first line of the table quotes values for a "typical" structural steel and the corresponding weights and material costs have been used as baseline values in the discussion.

8.1 Validation of Analysis

The principal objective of this study was to develop relationships which show the effects of the principal design variables on rotor weights. Nevertheless it is instructive to check the numerical values generated by these relationships against available data. However, precise "calibration" is, of course, very difficult, partly due to the difficulty of obtaining data in the right form (the boundaries between "hub" and "blades" vary considerably) and also due to variations in design philosophy.

Table 2 compares data from actual wind turbines, or detailed design studies, with predictions from the analytical model. Not surprisingly there are variations in the degree of agreement, with the largest discrepancy implying that the assumed properties of aluminium were conservative. Finally it should be noted that the estimates of maximum permissible tip/speed wind speed ratios in Table 1 are consistent with those used in practice.

9. DISCUSSION

9.1 Small and Large Wind Turbines

The results of this analysis indicate that the crossover diameter for rotors using "ordinary" constructional steel is about 100 m. This is treble the value quoted by Frandsen et al (ref. 5), using a more empirical approach, which took less account of variations due to different materials. The present analysis indicates that "ordinary" steel and aluminium alloys are the only materials likely to lead to "crossover" sizes within current size ranges. The use of these materials for larger rotors would lead to very weighty structures as rotor mass then increases as the fifth power of diameter. The design of large rotors constructed of all other materials apart from wood appears to be controlled by flapwise loads unless a high extreme gust speed needs to be used. Although a simple design criteria for flapwise loads was particularly difficult to define, the use of alternatives is unlikely to alter the principal conclusions drawn from the study. With wood, it appears that extreme gust loads may, in some cases, control the design. However, changes in blade chord distribution may reduce the maximum stresses and enable weight savings to be made.

9.2 Two Versus Three Blades

The analysis has clearly shown why three bladed rotors - assuming an optimum design - will always be heavier than two bladed ones, irrespective of the design criteria, (eqns. 18, 19 and 20). Moreover, the maximum achievable design tip speed/wind speed ratio will always be lower with three blades (eqns. 22-24) which is why three bladed rotors run slower. A pointer to the validity of the analysis is that the maximum tip speed/wind speed ratios quoted in Table 1 are quite consistent with those used in practice. The value for CFRP (12.8) is lower than used on the Voith machine (16), however, (Hau and Windheim, ref. 8) indicating that the properties of this material have been underrated in Table 2.

9.3 Fatigue Loads on Wind Turbines

Of the three principal design drivers - extreme gust loading, gravity bending and flapwise bending - the latter involves some uncertainty, whereas the former two can be predicted with reasonable accuracy. The procedure used for establishing rotor design for flapwise loads in this analysis followed the philosophy - but not the exact detail - of most actual studies. However it is clearly important to establish:-

(a) The appropriate mean stress level

(b) The appropriate cyclic amplitude

(c) A safety factor and finally

(d) Relevant properties of the chosen material.

since over-conservative design procedures will lead to unnecessarily heavy (and expensive) rotors, whereas inadequate design could clearly have serious consequences.

9.4 Rotor and Wind Turbine Costs

Although a detailed assessment of rotor costs is beyond the scope of this analysis, the results do shed light on current trends and controversies. One reason for the recent emergence of wood as a blade material, for example, is clear. Even a 3-blade wood rotor is likely to weigh less (about 2/3) than a 2-bladed mild steel rotor. Since wood costs around 1/2 to 2/3 the cost of mild steel the advantages are obvious. The analysis shows that the design driver with wood is likely, in some instances, to be the extreme gust loading. It follows that such rotors will have a generous safety factor against turbulent flapwise loads. The last line of the tabulated weight estimates is intended to indicate the potential savings possible if blade drag, gust levels and safety factors can safely be reduced. Even a three-blade rotor then weighs only 28% of the weight of the "baseline" rotor.

GRP costs around twice the cost of mild steel and offers the promise of rotors around half the weight. Given the consequential savings elsewhere on the structure, it may have promise but high blade deflections can sometimes cause problems.

Even allowing for possible conservatism in its tabulated properties, aluminium appears to offer few advantages. Reduced weight (1/2 steel) is an attraction, but offset by higher costs (8-10 x steel).

The very high price of titanium alloys (50 x steel) seems unlikely to offset possible weight reductions, but CFRP may offer attractions. The current price appears to be upwards of 10 x steel but the achievable weight reductions approach this factor. Once again, consequential savings in the drive train and support structure might tip the balance — especially if costs and manufacturing difficulties ease.

Finally, and to a first order estimate, the cost of steel rotors is probably only slightly modified by the grade used. The high quality grades cost 2-3 times ordinary mild steel prices and offer weight reductions of similar magnitude, but they are often are more difficult to fabricate.

10. CONCLUSIONS

An analysis of the factors controlling rotor weights, and the way in which these vary as a function of material properties is capable of predicting trends rather than precise numerical values, due to uncertainties in the design processes and material properties and the difficulty of allowing for weights of aerofoil skins, control mechanisms and other vital components.

Nevertheless, an analytical technique has been developed which enables "design drivers" to be predicted, rotor weights to be assessed and the effect of design changes to be examined. Predictions from the analysis are generally in line with measured data. The following conclusions can be drawn:-

1. Large (above 80-100 m dia) mild steel and aluminium alloy rotor designs are controlled by gravitational bending stresses and their weight then increases as the fifth power of the diameter.

2. With most other blade materials the boundary between "small" and "large" is well beyond present sizes. Most rotor designs are controlled by flapwise bending or gust loads and weight increases as the cube of the size. Wooden rotors, however, may possibly be designed to "extreme gust" criteria.

3. With the exception of wood, the margin between the "extreme gust" criteria and the "flapwise bending" criteria is small. More accurate assessments of the drag coefficients of parked rotors will resolve this question.

4. The "flapwise bending" design criterion is the most complex and, of necessity, has been simplified for this analysis. In practice, however, a better understanding of the applied loadings and rotor response may enable the design criteria to be eased and rotor weights reduced.

5. In the short term, the popularity of wood seems likely to continue, due to fairly clear cost advantages; moreover wooden rotors may possibly be conservatively designed for fatigue loads, if the gust loadings are more severe. In the long term, CFRP may emerge as a potential cost-saver —for the wind turbine as a whole — due to large rotor weight savings, but the costs of the material itself would need to fall considerably.

6. Analytical bases for the differences between two and three blade rotors suggest that the latter must always be heavier (by a factor equal to $\sqrt[3]{1.5}$) and, for structural reasons, need to rotate slower.

11. REFERENCES

1. Hau, E., 1984. What is the most economical size for a large wind power plant? Proc. European Wind Energy Conf. Hamburg. H.S. Stephens & Assoc., Bedford.

2. Spencer, D., 1985. Advanced power systems division, EPRI, R&D status report. EPRI Journal, October.

3. Lowe, J.E. and Wiesner, W., 1983. Status of Boeing wind-turbine systems. IEE Proc-A, 130, 9, 531-536.

4. Milborrow, D., 1982. Performance, blade loads and size limits for horizontal axis wind turbines. Proc. 4th BWEA Wind Energy Conf. Cranfield, BHRA.

5. Frandsen, S., Madsen, P.H. and
    Hansen, J.C., 1984. What is the
    difference between a large and a
    small wind turbine? Proc. European
    Wind Energy Conf. Hamburg.
    HS Stephens and Assoc., Bedford.

6. Thresher, R.W., 1985. Wind turbine
    structural loads resulting from wind
    excitation. EPRI report AP-4089.

7. Jamieson, P. and Hunter, C., 1985.
    Analysis of data from the Howden
    300 kW wind turbine on Burgar Hill,
    Orkney. Proc. 7th BWEA Wind Energy
    Conf. Oxford. MEP Ltd.

8. Hau, E. and Windheim, R., 1982. The
    wind power programme in Germany –
    present status. Proc. 4th Intl.
    Symposium on Wind Energy Systems,
    Stockholm. BHRA, Cranfield.

9. Moser, D., 1980. Growian composite
    rotor blade; design, concept and
    testing. IEA Expert meeting,
    Stockholm. KFA Julich GmbH.

10. Boeing Engineering and Construction
    Co., 1982. Mod-2 wind turbine system
    development final report. US DOE
    report DOE/NASA/0002-2.

11. National Swedish Board for Energy
    Source Development, Undated. The
    National Swedish Wind Energy
    Program.

12. General Electric Co., 1979. Mod-1
    wind turbine generator analysis and
    design report. US DOE report
    DOE/NASA/0058-79/2.

13. Glasgow, J.C. and Miller, D.R., 1980.
    Teetered tip-controlled rotor:
    preliminary test results from Mod-0
    100 kW experimental wind turbine.
    Proc. Wind Energy Conf., Boulder,
    Colorado, AIAA.

14. Lark, R.F., 1983. Construction of low
    cost Mod-0A wood composite wind
    turbine blades. DOE/NASA report
    20320-43.

12. <u>ACKNOWLEDGEMENT</u>

      This paper is published by permission
of the Central Electricity Generating Board.

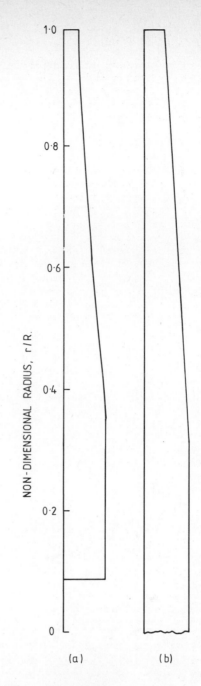

Fig 1   Blade profiles
         (a) Analytical model $C_l = 1$, $\lambda - 10$
         (b) MOD-2

Fig 2   Blade construction
         (a) Reference design
         (b) Hollow elliptic spar

TABLE 2

Rotor weights for different materials

MATERIAL	S.G.	PROPERTIES MN/m² (Safety factors incl.)			Blades	Max. T.S.R. (λ_{MAX}) and Driver	Rotor weight (Tonne) Diameter (m)				Flap/edge x-over dia.(m)	Alt. λ_{MAX}, driver and weight change
		UTS S_U	Fatigue limit S_F	Flap bend S_M			40	60	80	100		
Steel												
(ST52)	7.85	330	60	64	2	11.7 F	14.8	27.0	64.0	125.0	100	11.7G Same
(SAE6150)		930	280	267	2	14.8 F	3.0	10.3	24.5	47.8	290	–
Al Alloys	2.7	120	10	12	2	8.8 F	8.5	28.6	67.7	132.1	85	–
		240	20	24	2	9.9 F	5.4	18.0	42.6	83.2	106	–
Ti Alloy	4.42	500	300	250	2	14.8 F	1.8	5.9	14.1	27.6	560	14.0G x 1.25
CFRP	1.4	300	100	100	2	12.8 F	1.0	3.4	8.0	15.6	790	12.3G x 1.18
GRP	1.7	210	25	45	2	10.8 F	2.4	8.1	19.1	37.3	224	–
					3	9.4 F	3.0	10.2	24.1	47.0	196	9.2G x 1.11
Wood												
Spruce	0.5	30	10	10	2	6.9 G	4.3	14.6	34.6	67.6	460	(8.8F x 0.37)
Compressed, or plus expoxy	0.65	45	20	18	2	7.6 G	2.9	9.7	23.1	45.0	480	(9.4F)
					3	6.2 G	5.6	19.0	45.0	87.9		(8.2F)
(With F_E = 50, C_D = 1.2, Safety Factor 1.5 on UTS)					3	7.9 G	2.2	7.4	17.6	34.3		(8.2F)

Notes: F – Flapwise bending loads, G – Extreme gust load
The values quoted for aluminium alloys represent the approximate range
The value quoted for GRP is mid-range.

TABLE 3

Blade weights – actual and predicted

Wind Turbine	Rotor dia.(m)	Material	Values Actual	Predicted	Unit
Growian (Moser, 1980, ref. 9)	100	Steel	28	23–54	tonne
	100	GRP	19.4	18.6	tonne
	100	CFRP	9.9	7.8	tonne
MOD-5B (Lowe & Wiesner, 1983, ref. 3)	93	Steel	42	37 Max	tonne
MOD-2 (Boeing, 1982, ref. 10)	91.4	Steel	25	34	tonne
WTS3 (ref. 11)	78	GRP	14	8.9	tonne
WTS75 (ref. 11)	75	Steel	20	15	tonne
MOD-1 (General Electric, 1979 ref. 12)	60	Steel	9.1	5–12	tonne
MOD-0 (refs 13 and 14)	38	Steel	1820	2280	kg
MOD-0A (NB: F = 7.2 m/s)		Al	1067	1349	kg
		Wood	1105	930	kg

Maturation of the US wind industry

PAUL GIPE
Paul Gipe and Associates

SYNOPSIS Concentrated in California, the wind industry grew rapidly from 1981 through 1984. Growth during 1985 was tempered by the approaching expiration of the Federal energy tax credits and the resulting industry consolidation. As the industry expanded, energy generation and productivity have risen dramatically. While the average machine size in California wind power stations has nearly doubled since 1981, costs have continued to decline. The advantages of modularity and declining costs have led several studies to conclude that wind energy is one of the least cost generation options for the 1990's. However, expiration of Federal and California energy tax credits pits the maturing wind industry against conventional sources.

1. INDUSTRY CONCENTRATED IN CALIFORNIA

The California Energy Commission (CEC) reports that the 8,500 wind turbines in the state at the start of 1985 account for 95% of the U.S.'s and more than 75% of the world's commercial wind-electric capacity.

Denmark contains the world's second largest concentration of wind turbines where approximately 1,000 units have been installed. Unlike California, where clusters of turbines are being installed in wind-driven power plants or wind power stations, turbines in Denmark are used singly or in small groups for residential and light commercial applications.

Small numbers of turbines in wind power stations have also been installed in New Hampshire and Hawaii.

2. INDUSTRY CONSOLIDATING

After four years of intensive development the wind industry in California, which includes a number of foreign manufacturers, has learned what works. Operating experience and production history now dictate success in the market.

This emphasis on performance has led to the industry's increasing concentration. From the many different turbine designs and manufacturers of the early '80's have emerged less than a dozen manufacturers producing turbines in 1985.

Not suprisingly the success of the Danish wind industry in Denmark has led to the increasing use of Danish wind turbines in California. During 1984 Danish turbines captured nearly one-third of the U.S. market. Danish turbines may comprise one-half of the turbines installed in California during 1985.

Similarly, from the more than 40 developers of wind power stations in 1982 and 1983, have emerged one half dozen major players. Three firms now account for the bulk of California's generation: U.S. Windpower, Zond Systems, and Fayette Manufacturing. With 43% of the total number of turbines installed during 1984 these three developers generated 58% of the total energy produced. During the first nine months of 1985 they produced more than one half of the wind-generated electricity in California.

3. INDUSTRY RAPIDLY GROWING

The California wind industry has grown phenomenally since commercial development began in 1981. From 1982 to 1983 generating capacity in California tripled. In 1984 it increased 2-1/2 times. (See Table 1) The number of wind turbines installed doubled in both 1983 and 1984. By the end of 1985 nearly 1,000 Megawatts (MW) of capacity had been installed in California wind power stations.

Expansion has leveled off during 1985, reflecting the early stages of the industry's retrenchment as the expiration of the tax credits near. (The Federal energy credits expired at the end of 1985.) Capacity during 1985 will increase by a more modest 1-1/3 times over that installed in 1984. Though the growth in capacity has been striking, energy generation has grown even faster.

4. PRODUCTIVITY IMPROVES

Energy generation has increased much more rapidly than what would be expected

Table 1 California wind industry development*

Year	81	82	83	84**	85	Total
Installed Capacity (MW)	7	64	172	350	500	1,100
Number of Turbines	144	1,145	2,500	4,700	5,000	13,500
Energy Generated (million kWh)	NA	6	49	195	670	920
$ Invested (millions)	21	139	326	680	850	2,000
Oil Saved(bbl)***	NA	11,000	86,000	340,000	1,120,000	1,560,000

*California Energy Commission [1]
**From industry Sources: CEC, and AWEA 1985 Wind Industry Update [2]
***Primary Energy Equivalent. Estimate by Zond Systems, Inc., assumes 567 kWh/barrel

by the growth in turbine numbers and capacity alone. The industry generated nearly four times more electricity in 1984 than in 1983, and in 1985 than in 1984.

Increased production relative to the increases in capacity can result from improvements in reliability, in siting, and available wind energy. Because neither siting practices nor the available wind have changed dramatically, increased productivity is largely due to improved reliability.

Several of the principal developers, for example, report that their turbines are available for operation (the industry's measure of reliability) from 90% to 97% of the time.[3] Though the definition of availability varies widely, the reported availability reveals that most of the turbines at these wind power stations are operating most of the time wind is present.

High availability and rapidly improving productivity demonstrates that the wind industry is reaching levels of reliability and performance required by commercial generating stations.

4.1 Measures of Productivity
There are three measures of productivity in use: generation per turbine (kWh/unit), generation per unit of capacity (kWh/kW), and Capacity Factor (CF). Developers, investors, and their attorneys use generation per turbine to gauge performance because it is easily understood. For example, if investors are buying single turbines the generation per unit will clearly state how much energy they can expect. In the same way they can also easily monitor performance by comparing what the turbine did deliver with that expected.

Generation per unit of capacity is more useful to project planners where a broad measure of productivity is important not necessarily the number of specific machines. This measure is easily convertable to total expected generation once the total project capacity is known.

Capacity Factor is a related parameter in common use within the electric utility industry. It is defined as the ratio of actual generation to that of potential generation if the generator operated at 100% of capacity for the entire period.

Reliability, available wind energy, and turbine rating affect each measure in varying degrees. Turbine rating has a direct influence on generation per unit but an inverse influence on generation per kW and to CF, that is, it may increase per-turbine production while lowering CF if the rotor diameter remains constant.

The available wind energy produces a sizeable effect on all productivity measures. Increasing a site's average wind speed from 14 mph to 17 mph can boost productivity by 80% because of the cubic relationship between speed and power. Such variations in wind speed are common from one site to another in California as is year to year variations in wind speed at the same site.

In the following tables all three measures are used to illustrate the industry's maturation.

4.2 Actual Productivity
Energy generation per unit from 1982-83 to 1984-85 doubled. Generation per kW increased 63% during the same period. (See Table 2)

Energy generation per unit is increasing faster than energy output per kW because of greater generator loading per unit of swept area or rotor disc. (Higher reliability and high winds would produce an equivalent increase in both productivity measures.) For example, U.S. Windpower increased the rating on their 17-meter turbine from 50 kW to 100 kW. Similarly, the swept area on Danish turbines has increased 28% (from 15-meter diameter to 17-meter diameter), but peak generator ratings have increased 47% (from 75 kW to 100 kW). These increased loadings take advantage of California's high wind speeds relative to sites elsewhere in the U.S. and in Europe. At California sites these higher ratings improve total production per kW or increase CF.

Table 2 Productivity improvement
(California Average)

| Capacity Installed* | 1982 | 1983 | 1984 | 1985 |
Year of Generation	1983	1984	1985	1986 (proj.)
kWh generated	50 GWh	195 GWh	670 GWh	1800 GWh
Capacity	71 MW	243 MW	590 MW	1100 MW
kWh/kW	700	800	1140	1600
Capacity Factor	8%	9%	13%	19%
Turbines	1,300	3,800	8,500	13,500
kWh/unit	38,000	50,000	79,000	133,000

* At year end

The average productivity of turbines in Denmark is considerably greater than that in the U.S. (See Table 3). However, the data on productivity in Denmark represents less than 500 turbines.[4] (There are 17 times more turbines in the U.S.) Nevertheless, the productivity of Danish turbines in Denmark represents an industry nearing maturity for wind conditions there.

Table 3   Productivity of selected arrays

	1984		1985 (9 months)	
	kWh/kW	CF	kWh/kW	CF
U.S.:				
Peak array	1700	19%	2000-2400	20%-28%
Denmark:				
Average*	1600	18%	--	--
Average of				
55 kW unit	1800	21%	--	--

*From Windpower Monthly data. [4]

With more powerful wind regimes in California (average speeds from 14 mph to 22 mph) than in Denmark (10 mph to 15 mph) U.S. productivity should rise substantially above that now found there. As evidence of this, the productivity of selected arrays in the U.S. have exceeded 2,000 kWh/kW and CF's of 20%. And two of the industry's leading developers with nearly one-fourth of the total turbines installed in the U.S. produced 1,600 kWh/kW (a CF of 24%) during the first nine months of 1985. Collectively with 2,100 turbines, these two developers matched the performance of the Danish industry from four times more units.

4.3 Potential Productivity
The trend towards rising productivity will continue for several reasons. First, less than normal winds occurred throughout California from 1983 through early 1985 possibly due to the El Nino-Southern Oscillation phenomenon's influence on regional weather. Winds in California now appear to be returning to normal. Second, productivity will rise as idled turbines installed during the industry's early stages are returned to production or are replaced. Third, these early less productive turbines will become a smaller portion of the whole as the industry expands. Fourth, with the expiration of the

Federal energy tax credits growth will be curtailed, and new turbines will represent a smaller portion of total installed plant. (Productivity is often low during the first several months after a new turbine is commissioned. Once the turbine is fully operational, these start-up losses do not reoccur.)

Capacity Factors of 30% or more are possible. Southern California Edison projects a CF of 30% for the mature technology. [5] One array of 300 turbines in Southern California has achieved CF's of 24% during the first nine months even though the available wind energy was substantially less than normal. In Northern California one array of over 1,000 turbines achieved CF's of 26% during the first nine months of 1985.

5.   INCREASING SIZE
Greater productivity as well as increasing unit size are also improving wind energy's cost-effectiveness. Greater productivity increases the generation per unit of swept area, or per unit of capacity. Increasing size, on the other hand, increases total production without an equivalent increase in cost owing to economies of scale. (See Table 4.)

Even though the number of turbines installed during 1984 doubled, generating capacity more than doubled, indicating that average machine size and average capacity per turbine is increasing in California wind power stations.

Intermediate size wind turbines, for example, predominate. The bulk of the turbines installed during 1984 fall within the 15- to 17-meter size class (65%). All Danish turbines installed during 1984 were 15 to 16 meters in diameter. One of the largest domestic manufacturers, U.S. Windpower, has used 17-meter turbines solely since development began in 1981.

During 1985 developers installed a growing proportion of 17-meter turbines, possibly up to 50% of the marker. Nearly 85% of the Danish turbines installed during 1985 were nominally 17 meters in diameter with a 100 kW rating.

The complementary trends of improved productivity and increasing unit size are making wind energy more attractive. So are declining costs.

Table 4 Increasing unit size

	81	82	83	84	85
Avg.Capacity(kW)	49*	56*	69*	78*	80-100**
Rotor Diameter					
-10 m (33 ft)	-	-	-	9%	
10-14 m(33-48 ft)	-	5%	42%	25%	
15-17 m(50-56 ft)	-	48%	55%	65%	
18-25 m(59-82 ft)	-	2%	3%	3%	
26+ m(85 ft)	-	-	-	-	

*California Energy Commission [1]
** Estimate by Zond Systems Inc.

6. COSTS DECLINE

Independent energy producers are installing wind power stations at a cost from $1,500 to $2,000/kW of capacity ($1.5 billion to $2 billion per 1,000 MW) The U.S. Congressional Office of Technology Assessment (OTA) estimates that the typical wind power station in the 1990's will cost from $900 to $1,200/kW ($0.9 billion to $1.2 billion/1,000 MW). [5] This, in fact, may be high. Some developers are projecting capacity costs in this range for the late 1980's. (See Table 5.)

Table 5 Average installed capacity cost*
(1985 $/kW)

1981	1982	1983	1984	1990's**
3,100	2,200	1,900	1,860	900-1,200

*California Energy Commission [1]
**Office of Technology Assessment, U.S. Congress [5]

With 1,000 MW of capacity installed at an average cost of $1.9 billion, the industry has achieved an effective per kW cost of $15,000 to $16,000/kW (12% to 13% CF) based on actual operating experience. Yet at least one-fourth of the industry this year delivered at an effective cost of $9,500/kW (20% CF), and $6,300/kW is reasonably achievable (30% CF) from turbines already in place during coming years with normal winds. Newer turbines will deliver even lower effective costs as productivity continues to improve and costs per unit of capacity decline.

Nuclear power plants being completed in the U.S. today range in cost from $3 billion to over $5 billion per 1,000 MW of capacity. For example, the recently completed 800 MW Shoreham plant on Long Island cost $4.2 billion ($5,200 per kW); Consumer Power's soon to be be completed 800 MW plant near Midland, Michigan will cost $4 billion, and Niagara Mohawk's 1,100 MW Nine Mile Point plant in New York will cost $5.1 billion ($4,700 per kW).

Nationwide, nuclear plants operate at a capacity factor of 55% or roughly twice that of a wind power station. [6] The effective cost (adjusted for CF) of new nuclear capacity from the more expensive plants is from $5,500 to $9,500 per kW.

At least one-fourth of the wind industry is currently competitive with new nuclear capacity in the U.S. from wind turbines that are already in place including the more expensive units installed during the first few years of development. Using turbines installed in 1984 at $1,860/kW at a reasonably attainable 20% CF, wind capacity is less expensive at $9,000/kW effective cost than the more costly nuclear plants in the U.S, and even more so as the capacity factor rises to 30%: $6,200.

This comparison of effective capacity costs between nuclear power plants and wind power stations assumes that fuel and decommissioning costs are equivalent. They are not. Decommisioning the damaged Unit 2 reactor at Three Mile Island has cost $500 million, and it will cost another $500 million to complete ($1,100/kW).

Though individual turbines at wind power stations may need replacement after 20 to 30 years there is no reason for abandoning the site. New, more cost-effective turbines can be added, the infrastructure (e.g. the power transmission system) upgraded if necessary, and the plant returned to use. This is possible at a wind power station because there is no accumulation of dangerous wastes as in the case of a nuclear plant, or the exhaustion of the fuel source as in the case of a fossil fuel-fired plant.

More importantly, while the cost of both new nuclear and coal capacity is increasing, that of wind power stations is declining.

7. FUTURE COSTS

Within five years, 1981 to 1985, the U.S. wind industry has nearly reached commercial maturity when compared to new nuclear capacity, a technology that has been under development since 1948. The wind industry has made rapid strides by almost any standard, yet it is instructive to examine technology development and see how wind energy fares and what that holds for the future of the industry.

7.1 Technology Development
Energy technologies normally take 15 to 30 years before they move from the research, development, and demonstration stage to their first commercial units. The time required depends upon the technology's complexity, the market it will serve, and any outside incentives

designed to shorten the process. Nuclear power, because of federal involvement, reached the commercial stage in half the time it took for heat pumps to reach the market (15 years vs 30 years).

Additional time, often 15 or more years, is necessary before the technology contributes significantly to the market. For example, nuclear power in the U.S. achieved commercial status in the mid '60's, but only began to noticeably affect utility generation in the mid to late '70's.

Wind turbine development has taken a much shorter path. Work began in the early '70's and by the early '80's the industry had grown beyond the demonstration stage. By the mid '80's the industry has already made a sizeable contribution to utility capacity in California with the equivalent of one large nuclear power plant.

The most directly comparable example of such rapid development is the personal computer industry. During 1980 sales of personal computers totaled $1 billion. Sales doubled during 1981 and again during 1982. By 1985 sales of hardware alone are expected to total $15 billion. [7] Investment in the wind industry, in contrast, will near $1 billion this year after three years of doubling capacity annually.

The rapid growth of both industries in part reflects their lower per unit costs and their modularity in comparison to existing technology. Personal computers cost a fraction of the cost of a mainframe, and a business can buy as many as they need, when they need them, instead of buying one large mainframe that may be larger than necessary.

7.2 Modularity

The modularity of wind technology has contributed significantly to the industry's rapid growth. Modularity is also a powerful lure to further industry expansion because it offers substantial cost savings to utilities.

Wind turbines individually cost a fraction of the cost of a conventional power plant. And independent energy producers or the utilities themselves can install as many of them as they need, when they need them. This modularity has driven the technology at a faster pace than conventional energy development because wind technology has a much shorter lead time. Modularity and short lead time result in quick adoption of technological refinements.

In an era of uncertain growth in electricity demand, modularity not only offers utility planners flexibility but cost savings as well. When growth is uncertain the addition of a large central station may commit too much of the utility's resources for too long. The

utility then runs the risk that some of the capacity may stand idle once completed. The utility must both absorb the financing costs until the plant enters the rate base as well as any plant costs determined by the utility commission as not contributing to used and useful capacity. Instead of spending a decade and committing large amounts of capital to construct one plant, developers can bring wind power stations on line within one year.

Modeling studies by OTA indicate that the "cash flow benefits of (small-scale, short lead-time generating) plants in the short term could be considerable". [5] Los Alamos National Laboratory is more specfic. Los Alamos finds that "utilities could afford to pay as much as four times more in overnight construction costs for 5 year lead-time plants than for 15 year lead-time plants." For example, if a conventional plant cost $1,000/kW, a utility could afford up to $4,000/kW for a plant with only a 5 year lead time. [8] Thus, even where new wind capacity is more expensive than conventional base-load plants, the benefits of modularity make wind energy more attractive.

Findings such as these and that of several recent studies on the costs of electricity from new plants provide an encouraging view of wind energy's future.

7.3 Future Costs

The CEC estimates that by 1990 wind energy will rival conventional hydro as the low-cost energy resource for future utility generating capacity in California. The CEC calculates that wind energy will cost less than all other energy forms including nuclear, coal, oil, and natural gas. Wind energy will cost half that of nuclear and coal according to the CEC. (Coal-fired plants in the state would have to meet extremely strict and expensive emissions requirements.) [9] (See Table 6.)

Worldwatch Institute estimates that, nationwide, wind-generated electricity will cost almost half that of nuclear energy in 1990, and that wind energy will even be competitive with coal. Like the CEC, Worldwatch finds wind energy competitive today with nuclear energy from the more expensive plants. [10]

In the most recent analysis yet, OTA found that wind energy ". . . shows the lowest cost among the new generation technologies" and that the "cost of wind technology is significantly lower than the other non-base load technologies". According to OTA, "wind also has the potential for competing with the base-load technologies", and "could be profitable at buy-back rates above 6 cents/kilowatt-hour (kWh)." "If wind power achieves its most optimistic capital cost and capacity factor ranges, wind could require only a 4 cents/kWh rate." [5]

Table 6
Estimated cost of electricity
from new plants

Energy Source	WW 1983	WW 1990	CEC 1990	OTA 1990
(cents per kilowatt-hour)				
Nuclear Power	10-12	4-16	5-15	-
Coal	5-7	7-9	8	6
Small Hydropower	8-10	10-12	2-10	-
Cogeneration	4-6	4-6	-	-
Biomass	8-15	7-10	6	-
Wind Power	12-20**	6-10	3-5	7
Photovoltaics	50-100	10-20	7-10	21
Energy Efficiency	1-2	3-5	-	-

*Costs expressed in 1982 dollars except
 for OTA, which expressed costs in 1983
 dollars, CEC in 1984 dollars.
**Before tax cost
WW--Worldwatch Institute. New capacity
   in U.S.
CEC--California Energy Commission. New
   capacity instate only
OTA--Office of Technology Assessment,
   U.S. Congress

For comparison, Niagara Mohawk's
soon to be completed Nine Mile Point
plant will generate electricity at 18
cents/kWh. [11]. Pacific Gas & Electric
Co. (PG&E) estimates that the 1985 cost
for a generic in-state coal plant is 11.8
cents/kWh and that for a natural gas-
fired combined cycle plant is 18.1
cents/kWh. Assuming a CF of 33% and an
installed cost of $1,250/kW (a reasonably
attainable near term cost for wind
capacity), PG&E calculates that wind
energy costs 10.6 cents/kWh in current
1985 dollars. In constant 1985 dollars,
i.e. without accounting for inflation,
the costs are 6.7 cents/kWh, slightly
less than PG&E's 1985 avoided cost of 7
cents/kWh. [12]

8. MARKETPLACE EQUITY

The tax credits have worked successfully.
They have pushed a promising technology
towards commercial maturity at a pace
much faster than that recorded by conventional energy technologies. Ironically,
just as the industry nears maturity
growth will be curtailed by expiration of
the credits.

The prospect that wind power
stations will improve their
competitiveness against entrenched technology and become more cost-effective is
good. But wind energy remains at a disadvantage in the U.S. because an
unfettered market does not exist.

Wind energy must compete in the post
tax-credit market against conventional
fuels long subsidized by the Federal
government. Incentives to the production
and consumption of oil, natural gas,
nuclear, coal and large-scale hydro
continue unabated. As Washington pundits
note there has never been a "level
playing field".

In the last 60 years the Federal
government has invested enormous sums in
energy development. The $175 million
spent on energy tax credits for wind
power stations from 1981 to 1984 is
negligable in comparison. The Federal
government, in contrast, has spent $385
million on wind turbine development yet
has erected only 19 MW of capacity since
1973. [13]

Table 7 1984 Federal energy susbidies*
         ($ billion)

oil	8.6	nuclear electric	15.8
natural gas	4.6	fusion	0.6
coal	3.4	hydro	2.6
syn. fuel	0.6	renewables**	1.7
fossil elec.	7.2	conservation	0.9
		Total	46.0

* H. Richard Heede; "A Preliminary
Assessment of Federal Energy subsidies in
FY 1984". [14]
** Includes wind.

The Rocky Mountain Insitute
estimates total Federal energy subsidies
including tax expenditures, agency
outlays, and loan costs reached $46
billion during 1984 alone. (See Table 7.)
[13] Subsidies to the wind industry for
the 10% investment tax credit (available
to any other industry as well), 15%
energy tax credit and depreciation
deductions during 1984 amount to less
than 1% of total Federal energy
subsidies. (See Table 8.)

Table 8 1984 Federal cost public-private
         sector wind energy development
         ($ billion)

R & D Programs*	0.026
Energy Tax Credits (15%)**	0.102
Investment Tax Credits (10%)**	0.068
Depreciation (21% at 50% bracket)**	0.071
Total	0.267

*Alternative Energy Institute, WTSU. [12]
**1984 private investment of $680 million

9. ENVIRONMENTAL PERSPECTIVES

Should the energy tax credits be
extended, the continued exponential
growth of the wind industry in California
may be tempered by increasing conflicts
over land use. In a sense these conflicts
are in themselves a measure of an
industry that is fast moving from the
category of an "alternative" to one of a
"conventional" source of energy.

With the scheduled expiration of the
energy tax credits and with installed
capacity of nearly 1,000 MW ,now is an
opportune time to examine some of the

environmental costs as well as benefits.

9.1 Costs

Though often considered benign relative to conventional energy sources, wind turbines have raised concerns about aesthetics, noise, and land use impacts. Questions of aesthetic and noise impacts are comparable to those surrounding conventional energy generation. However, the land use requirements for wind power stations are unique to this technology.

Wind power stations use land. This is the most serious environmental concern. But wind power stations use less land than commonly thought and the trend is towards, as Buckminster Fuller would say, "doing more with less".

Designers array turbines in geometric patterns to minimize the interference from upwind turbines on those downwind. These patterns and the spacing between turbines depends upon the terrain and the prevailing winds. In areas where the winds are omnidirectional, designers space the turbines equidistant from one another, e.g. 8 to 10 rotor diameters apart. Where winds are unidirectional or bidirectional as they are in California's mountain passes, the turbines are tightly packed in rows across the prevailing wind, e.g. 1.5 to 3 rotor diameters apart, with greater spacing between rows, e.g. 8 to 10 rotor diameters.

The Oregon Department of Energy estimates that wind power stations require from 15 Acres/MW (A/MW) in linear arrays for more unidirectional winds to 80 A/MW in widely spaced arrays for omni-directional winds. [5] The U.S. Geological Survey in examining existing wind power stations estimates that the turbines occupy 7-10 A/MW in linear arrays on flat land (0.7 to 1.0 A/100 kW turbine) to 20 A/MW on steep terrain. [14] In practice one major developer has found that linear arrays in steep terrain occupy from 15 A/MW to 20 A/MW depending upon project boundaries and how much land within the boundaries was suitable for wind development.

With 1,000 MW of installed capacity, California wind power stations today occupy from 23 to 31 square miles. This land requirement, about one township (36 square miles) per 1,000 MW, contrasts sharply with prior studies that assumed more dispersed arrays. For example, one projection of a 1,400 MW array calculated that wind turbines would occupy 150 square miles, 3.5 to 4.6 times more land per MW than in use today with existing technology. [15] For perspective, consider that the land occupied by wind turbines in California comprises less than 7% of Edwards Air Force Base, the California landing site for the Space Shuttle.

As experience has grown with operating turbines on rolling or hilly terrain, designers have found that they can pack the turbines closer together without seriously compromising performance. Because land prices have risen as well as the difficulty of developing new parcels, there is a strong incentive to minimize land use or find multiple uses for the land not actually occupied by the turbines.

An innovative concept called a "stepped array" (the turbines are clustered vertically on towers of varying heights as well as horizontally in parallel rows) may use even less land per MW than at present. This concept also opens existing wind power stations, which are located at prime wind sites, for expansion vertically. By installing larger, more cost-effective turbines on taller towers than those already in place, project managers can upgrade the capacity and production of the wind power station with a modest cost penalty for the taller towers.

The CEC estimates that 13,000 MW of potential capacity exists within the state and has set a goal of developing 4,000 MW by the year 2000. [16] With present technology this would require from 90 to 120 square miles of land or from 3 to 4 townships, still less than one-fourth the area of Edwards Air Force Base (470 square miles), and less than 0.02% of the state's land area.

The CEC's goals also includes the equivalent production of 25 million barrels of oil per year by the year 2000. This would be difficult for existing technology without the addition of another 1,000 MW in generating capacity to the 4,000 MW goal. Using energy production as the criteria, the CEC's year 2000 goal would use from 115 to 150 square miles of land, still significantly less than that occupied by Edwards Air Force Base.

Though the land required for wind development is much less than that first thought, even less land is actually disturbed by the construction or operation of the wind turbines. Development and continued use disturb only a small portion of the land occupied by the wind power station. The percentage of land affected depends upon the terrain, the size and number of roads, mounting pads, buildings, and other structures as well as what is chosen as the project boundaries.

Wind projects disturb more land in steep terrain than on level ground. Grading for roads and mounting pads on hillsides cut the slope on the uphill side and push the soil down slope. These cut-and-fill slopes extend the disturbance beyond the immediate vicinity of the road or pad. On level terrain grading disturbs less soil because there are no cut-and-fill slopes.

Project boundaries in most cases do not directly determine the soil disturbed. (Infrequently, a legal boundary will require an unnecessarily circuitous road pattern.) But they strongly influence the total amount of land occupied by the wind power station. As the total land area increases for a given amount of turbines, the percentage of soil disturbed decreases. For example, if the boundary circumscribes a group of ridge-top turbines the percentage of soil disturbance is greater than if the boundary encompasses both the ridge-top turbines and the lower slopes without turbines.

Examining two projects on steep terrain that occupy 15 A/MW to 20 A/MW one developer found that the soil disturbed for road and pad construction was from 7% to 10% of the total project area. Even assuming that upslope and downslope disturbance doubles the land area affected, only 15% to 20% of the project's soil was disturbed.

The rest remains in its natural state, or can be used for other purposes. In the Altamont Pass, open grazing co-exists with wind turbines. On the level ground of one project in the Tehachapi Mountains, dry-land farming is being considered.

9.3 Benefits

Wind energy produces no waste heat, requires no cooling water, and generates no emissions nor toxic wastes.

About two-thirds of the fuel consumed in fossil-fired and nuclear power plants is lost as waste heat and is dissipated into the environment. Because they generate electricity directly, wind power stations do not need cooling water or cooling towers, nor do they contribute to thermal pollution of lakes and streams.

Wind power stations produce no toxic wastes other than those associated with heavy machinery. Each wind turbine uses a transmission. Every 2 to 5 years maintenance crews will replace the transmission fluid. Like motor oil, this fluid can be reprocessed rather than being disposed.

Because wind turbines generate electricity directly, they do not emit air pollutants. (Construction does produce fugitive dust. This temporary effect is reduced by watering the roads.)

California utilities burn primarily oil and natural gas. These fuels are both low in ash, the precursor of particulate matter (smoke), and low in sulfur content, the precursor of sulfur oxides. In Table 9 the emissions from the average of two power plants, one in Northern and one in Southern California that burn both fuels have been used to calculate emission offsets. Because the industry is concentrated in California, it primarily offsets oil- and gas-fired generation. [17]

Oil and gas-fired plants emit considerably less pollutants than Midwestern power plants burning high-sulfur coal. In light of the concern about acid rain, it is helpful to examine the emissions offset if the wind industry was located in the Midwest.

Two plants were used for comparison. One in Illinois emits 5.2 pounds of sulfur oxides/million Btu. Another, in Indiana scrubs the stack gases, reducing sulfur oxide emissions to 2.9 pounds/million Btu. [18] (See Table 9.)

There are Midwestern plants that emit less pollutants than those used in the example. Some plants emit considerably more sulfur oxides, e.g. one Missouri plant spews out 10 pounds/million Btu. Some meet the emission standard of 1.2 pounds/million Btu. Most do not.

Table 9   Emission offset from 1985 estimated generation*(lbs)

	California Oil/NG-fired	Midwestern Coal-fired
Sulfur Oxides	2,040,000	31,200,000
With scrubber		17,400,000
Nitrogen Oxides	1,260,000	9,600,000
Particulates	288,000	600,000

*Derived from several sources. [19]

The Natural Resource Defense Council (NRDC) surveyed 539 units that are collectively responsible for 65% of the sulfur emissions in the U.S. Two-fifths are in violation of the Clean Air Act according to NRDC, the remaining three-fifths were allowed exemptions. All exceed the emission standard for sulfur oxides. [20]

Wind power stations can permit greater industrial expansion in the regions where they are located. By generating electricity without emissions other plants, factories, or enterprises that do produce emissions can be built without reducing the region's air quality. Or wind power stations can be used to improve the quality of life by contributing to improved air quality where further population and industrial growth is undesired.

TRENDS

The wind industry will continue to consolidate as the larger firms begin to offer operations and maintenance services to the smaller developers who are unable to adequately support their own staffs. With improving productivity and declining costs utilities will take a more active role in developing wind power stations either directly or through unregulated subsidiaries. Individual turbines as well as the wind power stations of which they are a part will continue to increase in size to gain economies of scale. With the expiration of the tax credits the industry will turn towards conventional project financing to raise development capital in contrast to raising capital from thousands of individual investors.

REFERENCES

1. California Energy Commission, Wind Program Staff; "Summary of Large Scale California Wind Projects'; Sacramento, California; November, 1984.
American Wind Energy Association; "Wind Industry Update"; by analysis, Review, and Critique; Washington, D.C. March, 1985

2. Batham, Michael; California Energy Commission; Personal Communication; January, 1985.

3. Lynette, R.; "Wind Power Stations: 1984 Experience Assessment'; Research Project 1996-2, Interim Report for Electric Power Research Institute; R. Lynette and Associates, Bellevue, Washington; January, 1985.

4. Sander, Niels; "2.9 Million kWh From 429 WECS in December"; Windpower Monthly Vol. I, No. 2; Vrinners Hoved, Denmark; February, 1985.

5. Office of Technology Assessment, U.S. Congress; "New Electric Power Technologies: Problems and Prospects for the 1990's"; Washington, D.C.; July, 1985.

6. Lovins, Amory; "Approximate Short-Run Marginal Capital Cost of a Nuclear Power System"; Rocky Mountain Institute; Old Snowmass, Colorado; 1, 1984

7. Rosen, Benjamin; "Calm Before the Storm"; Lotus; October, 1985

8. Sutherland, R.J.; Drake, R.H.; "The Future Market for Electric Generating Capacity: A Summary of Findings"; Los Alamos National Laboratory; December, 1984.

9. Ringer, Mike; "Relative Cost of Electricity Production"; California Energy Commission, Technology Assessments Project Office; Sacramento, California; July, 1984..

10. Flavin, Christopher; "Nuclear Power:The Market Test"; Worldwatch Institute Paper S-17; Washington, D.C.; December, 1983.

11. Cook, James; "Nuclear Follies"; Forbes; February, 1985.

12. Pepper, Janis; "Wind Farm Economics From A Utility's Perspective"; Pacific Gas & Electric Company; San Francisco California; August, 1985.

13. Nelson, Vaughn; personal communication; Alternative Energy Institute, WTSU; Canyon, Texas; March, 1985.

14. Heede, R.R.; "A Preliminary Assessment of Federal Energy Subsidies in FY 1984"; Rocky Mountain Institute; Old Snowmass, Colorado; June, 1985.

15. Wilshire, H.; "Harvesting the Wind in California;' draft; U.S. Geological Survey; unpublished; Menlo Park, California; September, 1985.

16. White, S.; "Towers Multiply, and Environment is Gone With the Wind"; Los Angeles Times; November, 1984.

17. California Energy Commission; "Wind Energy: Investing in Our Energy Future'; Revised Edition; Sacramento, California; March, 1984.

18. Batham, Michael; Average of selected Northern and Southern California dual-fuel plants, personal communication; California Energy Commission; Sacramento, California; February, 1985.

19. McCormick, John; Selected midwestern coal-fired plant emission of SOx, and PM from coal with 2.9% sulfur, 11% ash, and with 99% PM removal, personal Communication; Environmental Policy Institute; Washington, D.C.; February, 1985. NOx emissions: Environmental Protection Agency; "Compilation of Air Pollution Emission Factors", AP-42, Supplement 13; Environmental Protection Agency; Washington, D.C.

20. Natural Resources Defense Council; "Tall Stacks: A Decade of Illegal Use, a Decade of Damage Downwind"; Washington, D.C.; March, 1985.

Wind energy standards development in North America

VALERIE A CHAPMAN
Thermax Corporation, Burlington, Vermont, USA

SYNOPSIS - An up-to-date review is given of wind energy standards development in North America, along with references to parallel standards programs elsewhere.

1. UNITED STATES

The American Wind Energy Association started a working committee for standards development six years ago, originally chaired by Michael Bergey. Under the current chairmanship of Ward Slager, the group's first project to draft a Performance Testing standard was completed in 1985.

The publication is AWEA Standard 1.1 - 1985 - 'Standard Performance Testing of Wind Energy Conversion Systems', available from American Wind Energy Association, 1516 King Street, Alexandria, Virginia 22314, U.S.A.

This document has five main sections:
- (a) Definitions and units of measure.
- (b) The field testing procedure used for determining the power production of a machine, as well as for instrumentation, data collection and possible corrections to the data.
- (c) Noise level tests.
- (d) Standard presentation of performance ratings and results, and
- (e) A format for equipment certification.

A second project, to publish a guide to terminology, was completed in 1985 under the chairmanship of Hugh Curran. This publication is AWEA Standard 5.1 - 1985 - 'Wind Energy Conversion Systems Terminology'.

Several other standards are still in preparation at AWEA - these are:
- (a) Structural Design Criteria.
- (b) Safety.
- (c) Electrical interconnection, and
- (d) Siting.

ASME (The American Society of Mechanical Engineers) has also been working on certain wind energy standards, which may possibly be proposed for publication by ASTM, along with the AWEA standards.

2. CANADA

The Canadian standards development effort was started in mid-1981, one year later than in the United States, but has proceeded somewhat faster, for two reasons:

First, the Canadian program has been conducted from the start through the professional facilities of the Canadian Standards Association in Toronto, which is the official Canadian certification body (parallel to the Underwriters Laboratories in the United States), and second, the Canadian standards program has been financially underwritten until 1985 by the National Research Council, and now by the Department of Energy, Mines and Resources.

The Canadian standards program has so far produced a Safety, Design and Operation Criteria standard, a Performance standard, and a Utility Interconnection standard. The latter two standards have been approved for publication but as of this moment are still at the printers.

The Safety standard has six main sections:
- (a) Definitions and general requirements.
- (b) Environmental considerations.
- (c) Design considerations.
- (d) Component design guidelines, including rotor, support structure and electrical system
- (e) Marking, and
- (f) Field and shop tests and reporting.

The Performance standard has five main sections:
- (a) Definitions and references.
- (b) Test machine and field test methods.
- (c) Data gathering and analysis procedures.
- (d) Noise test methods, and
- (e) Performance reporting.

The interconnection standard consists of six main parts:
- (a) Definitions and general requirements.
- (b) Owner and user requirements.
- (c) Utility requirements.
- (d) Operating requirements, including line outages, stability and synchronization.
- (e) Equipment requirements.
- (f) Performance requirements.

Still in preparation in Canada are standards on Siting and Installation, as well as a Terminology document.

It is also proposed to adapt an existing standard on photovoltaic power conditioning to the requirements of the wind industry.

Finally, when the standards are completed, the group will then go to work on a certification program.

Copies of the Canadian standards are available from Canadian Standards Association, 178 Rexdale Blvd., Rexdale, Ontario M9W 1R3. Catalog numbers of the Canadian standards are:

    F416 – Safety
    F417 – Performance
    F418 – Interconnection
    F428 – Siting
    F381 – Power conditioning.

3. INTERNATIONAL COORDINATION

The two parallel standards development groups in the United States and Canada have worked closely together since the beginning - with cross-membership on committees, and have the advantage of a common language. Further, since Canada is a bilingual country, the resultant standards will be available in French as well as English. The final results of the North American efforts are expected to be a complete set of fully compatible wind energy standards, each of which also reflects the special national requirements of its country of origin.

With respect to the overall progress of international standards development, reference is made to the very well-documented review prepared by Beurskens, Frandsen, Lundsager and Pedersen for the IEA workshop held in Denmark in June 1985.

Our industry is very fortunate to be developing worldwide standards in relative harmony, right now, before it is too late - as has happened in certain other rapidly emerging technologies, such as computers and video recorders, where everyone ended up fighting over mutually incompatible standards.

The end customer always suffers (and eventually pays for) such shortsighted human folly.

Of course it is true that wind industry standards are being developed simultaneously in several places, in several languages, and are being forced to accommodate the unfortunate electrical industry - whose standards are so totally uncoordinated that the US electric kettle cannot be plugged into the UK outlet, but if it could plug in it would explode!

What is heartening to see, however, is the willingness of the wind standards writers the world over to mutually consult and communicate with each other before casting their work in concrete, and to be sensitive to the need for universal and compatible world standards.

Our company has been actively involved in the technical development of the standards, as well as helping to support the programs.

We are hoping that the eventual result is going to be a set of universal, fully coordinated international standards, so that the entire wind energy industry will be speaking with <u>one</u> voice.

The economic prospects for wind energy on the UK supply grid – a probabilistic analysis

M J GRUBB
Energy Research Group, Cavendish Laboratory, Cambridge

SYNOPSIS. This paper presents a detailed analysis of the economic prospects for wind energy as a supply source on the British electricity system. Results indicate that installed capacities of 10-40GW are likely to be profitable, bringing a national economic benefit of several hundred million pounds a year. The conclusions appear robust to a range of assumptions, and indicate the importance of analysing in detail both the resource and the impact which wind energy has on the structure of electricity supply.

1. INTRODUCTION

Over the last five years wind energy has developed to a point at which a number of major electricity utilities now treat it as a serious contender for future supplies. With a world market running into hundreds of millions of dollars, wind energy costs have become clearer and our knowledge of the wind energy resource has increased. Along with these developments, a number of studies have investigated the role which wind energy can play on grids as structured at present.[1-2]

Electricity utility planning necessarily focuses on time periods of decades, this being the period over new plant will be operating. When considering developments such as the Fast Breeder reactor, even longer periods are commonly considered. Generally such investigations have not included wind energy, for a number of reasons including the methodological problems of including such intermittent sources in planning models.

In a previous paper[3] the author discussed these difficulties and presented an assessment method which could include many of the factors important in the long-term integration of wind energy. In this paper the methods are developed further, to a state which appears adequate for investigating the optimal integration of intermittent sources on thermal-based supply systems. A detailed analysis of the wind energy resource in Britain is then presented. The results from this are used as input to a study of the economic potential for wind energy on the British electricity grid, using techniques of probabilistic analysis developed at Cambridge and previously applied to nuclear power.[4]

2. DEVELOPMENT OF ANALYTICAL METHODS

2.1 Modelling basis

The analytical technique developed centres upon two functions which carry information about the state of the electricity grid. Details are given in Grubb[3,5-6] and only a summary is presented here.

The load distribution function describes the probability distribution of load on the system. The output from intermittent sources can be included by suitable manipulations to produce a reduced load function which represents the net load to be met by dispatchable plant on the system. Correlation among the intermittent inputs is accounted for using an orthogonalisation procedure. The dispatchable units are then included iteratively, taking into account their own forced outages. In this way a measure of both the energy supplied by each thermal unit and the reliability of the generation system is obtained.

This information is supplemented by a transition frequency function which describes the frequency with which the load crosses a given power level. Again, this can be combined with the transition functions of intermittent sources to give a reduced transition function. This gives information on the cycling loads to be met by the dispatchable plant on the system – the frequency with which they must be part loaded or taken offline. Such actions incur costs, and one uncertainty has always been the degree to which the fuel savings from intermittent sources would be offset by increased losses in thermal plant operation.

2.2 System limit costing

With intermittent sources on the grid, it is not possible to use planning methods based on a fixed planning margin of reserve capacity. Also, two factors make it undesirable to plan on the basis of a fixed reliability criterion.

Firstly, it is impossible to construct a 100% reliable grid, and the optimal reliability can only be determined by a comparison of the consumer costs of supply failures with the national costs of increasing system reliability. Since the latter changes with grid structure it is best to include outage costs explicitly in planning models if possible. Secondly, the electricity supply system is not a rigid structure which fails when load exceeds the rated capacity available. Minor excursions can be dealt with by slight changes in

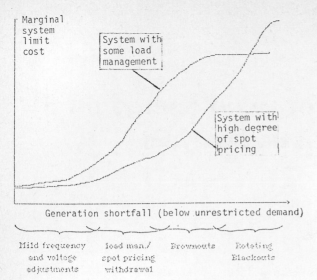

Fig 1   System limit costs

grid frequency and voltage, which increase station output and reduce demand.   Larger excursions can be curtailed using various forms of load management.

Some loads, such as process heating and refrigeration, can be interrupted at slight cost. Arranging separate circuits with special interruptible tariffs is one of the simplest forms of load management, and it can bring significant benefits to both consumer and the utility. About 1.6GW of industrial consumption currently is on interruptible tariffs.[7] Such options may be extended to domestic consumers, if 'intelligent' electricity meters are adopted. These can manage a number of separate circuits, one of which could be for interruptible loads.   Several such meters are currently undergoing trial.[8]

Utilising load management extends the amount of power shortfall which can be accomodated before blackouts occur. There are clearly costs incurred, in terms of lower revenue to the grid or disutility to the consumer. These would increase with the degree of shortfall. The likely form of cost increase is shown in figure 1(a).

Extension of these ideas leads to various forms of spot pricing, in which a partial market exists in electricity, with a short-term price reflecting the current cost of production and consumers adjusting demand accordingly. There are severe obstacles to basing supply entirely upon spot pricing, but a mixed system is quite possible and perhaps likely because of the greater economic efficiency it offers. As unrestricted demand rises above available capacity, the associated costs to the nation rise as in 1(b).

The method developed includes such costs explicitly in place of a fixed reliability constraint. The actual figures involved are highly uncertain, though some indication can be gained from estimates of feasible load management capacities and work on the costs of electricity supply failures. For modelling purposes it is assumed that the costs/kwh of demand above nominal available capacity increase linearly with the degree of shortfall. The associated costs are termed 'system limit costs' to emphasise the fact that they encompass several complex issues surrounding the behaviour of electricity systems when generating at the limit of available capacity.

2.3 Seasonal division and maintenance scheduling

Both electricity demand and wind energy are highly seasonally dependent, and for this reason all the input functions are defined in each of four seasons. Operating costs (including system limit costs) are calculated for each, and summed.

Major baseload plant require a significant amount of scheduled maintenance and at present every effort is made to carry this out during the summer months. The ability to schedule some maintenance is of relevance for the integration of intermittent sources, and in the model a simple algorithm is used to allocate maintenance between seasons. Using the difference between peak and average availabilities as an indication of the amount of schedulable maintenance, the capacity of plant type i withdrawn for maintenance in season s is defined to be inversely proportional to the mean reduced load (demand net of intermittent input) in that season, i.e.

$$\text{Capacity withdrawn} = m(P/P_s)(r_i-\alpha_i)/r_i \times 100\%$$

where P is the annual average reduced load, P_s is that during season s, r_i is the peak availability and α_i is the annual average availability. m is then a normalisation constant given by:

$$m = (T/P) / \sum_s (T_s/P_s)$$

where T = 365 days and T_s is the season length.

The ability to schedule maintenance is thus partial, in an attempt to reflect the real life constraints which prevent full flexibility.

3. TREATMENT OF WIND RESOURCES & COSTS IN BRITAIN.

To analyse the prospects for WECS (Wind Energy Conversion Systems) we require information about the output characteristics of installed arrays, and the associated costs as a function of installed capacity. This section summarises the methods used to determine wind resources; full details are given in Grubb.[9]

3.1 Wind power output functions

First it is necessary to divide the overall resource base - assumed to be mainland Britain - into regions. In this way the effects of geographical diversity and regional differences in the resource can be captured. Six regions were chosen:

    1 - North Scotland
    2 - South Scotland
    3 - North England
    4 - Wales & West Midlands
    5 - Midlands & South East
    6 - South West England

To form the wind power output functions, hourly time series data was obtained for a number of sites within each region. Hourly smoothing may be taken to approximate the effects of dispersal over several hundred km[2]. The wind speed was converted to power using the standardised conversion characteristic shown in figure 2 with power coefficient 0.4 at the rated speed. The resulting series were summed over the sites within

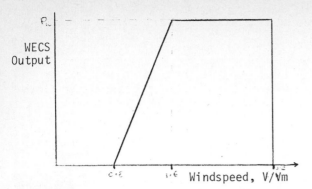

Fig 2   Generalised WECS conversion characteristic

the region and normalised. From these time series, the distribution and transition functions for wind output in each region were derived, together with the inter-regional correlational statistics. The process is summarised in figure 3.

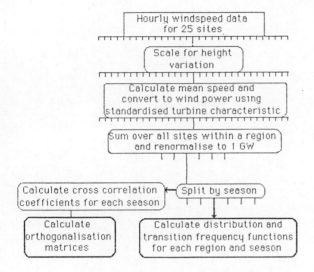

Fig 3   Derivation of wind power output functions

3.2 Mean speed resource functions

In estimating the costs of wind energy, the first step is to establish the profile of wind resources - how much is available at given site characteristics. In deriving this it is essential to take account of the effects of terrain on the available wind sites.

Data was obtained from the ETSU study R17[10] which investigated the effects of terrain on the wind field over a 100x100km area of S-W Scotland. This was matched with results from Moore's upper air study[11] which describes the 'flat land' mean speed to be expected at all points in the UK. By dividing the region into terrain types, the comparison gives an indication of the 'divergence' which different types of terrain impose on the mean wind field, i.e. the expected distribution of site mean speeds resulting from the effect of terrain on a given flat land mean speed. Four terrain types were chosen, and the resulting divergence functions are shown in figure 4.

Figure 5 shows the way in which this information was applied. The six regions were divided into sectors according to the flat land mean speeds expected - usually an inner and outer band for each coast, and one inland sector. By

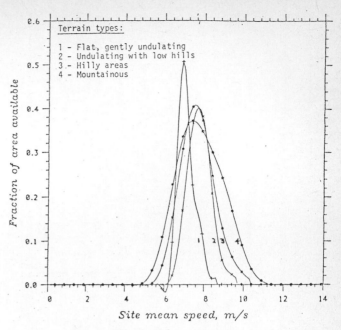

Fig 4   Effect of terrain type on distribution of site mean speeds — plot of divergence functions for flat-land mean 7.0 m/s

estimating the amount of each terrain type in each sector, applying the divergence functions to the associated flat land mean speed, and summing over terrain types and sectors, a wind resource profile can be built up for each region.

First order siting restrictions were included at this stage. These include a ban on sites within 300m of dwellings or roads, and no siting on lakes, military zones, amenity areas (e.g. golf courses, race tracks), quarries etc. On such grounds, the IEA European Wind Resource Study[12] estimated that 40% of land area within the UK would be excluded, and ETSU estimated a figure of 72% for the R17 study area. A value of 60% first-order exclusion was used. Also, the effects of WECS interference were included at this stage, with an array efficiency taken as 90% averaged over all WECS.

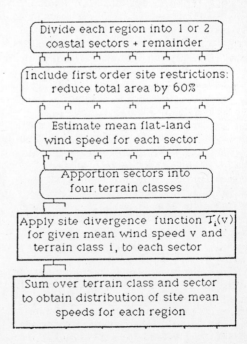

Fig 5   Derivation of mean speed resource functions

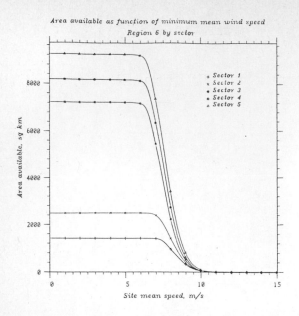
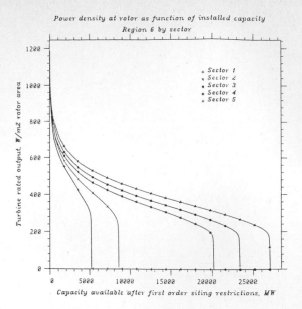

Area available as function of minimum mean wind speed
Region 6 by sector

Power density at rotor as function of installed capacity
Region 6 by sector

Fig 6   Resource functions for region 6 (South-West)

From this analysis charts of mean speed vs area are obtained. Assuming the turbine characteristics given before, this leads to resource functions described in terms of the rated energy density vs. installed capacity, on the assumption that the best available sites would be used first. The resource functions for the South-West region are shown in figure 6.

Summing over the six regions gives an estimate for the technical British wind resource (approximately equivalent to the "technical potential" as defined in ETSU R20[13]) of 215GW. At 27% capacity factor this is equivalent to ≈510TWh/yr, twice British electricity demand in 1984.

3.3 Wind cost analysis

Wind energy costs vary between sites, but not in a simple manner. If WECS are rated according to site characteristics, it would be incorrect to assume a constant cost/m^2 rotor area because some components (e.g. generator, gearbox and drive shaft) are directly capacity-related. Such costs form a significant part of the total, and in addition there are some changes in structural costs due to the greater load borne at higher windspeed sites.

The solution adopted was to break down WECS capital costs into two components, one power-related c_p/kw, the other rotor-related c_r/m^2. Given the resource function $R(p)$ in terms of the marginal rated energy density at the rotor, and assuming that the best available sites are utilised first, the total costs of an installed capacity K can be calculated as:

$$\text{Cost} = c_p K + \int_0^K c_r / R(p) \, dp$$

Operation and maintenance costs can be simply added as an annual fixed cost; they are specified by region on the grounds that maintenance in remoter areas is likely to be more expensive. Transmission losses and turbine availability effects are included by reducing the delivered output from a given installed capacity accordingly. Again, these are defined by region.

3.4 Overall model structure

With these developments the structure of the model used to calculate total supply costs is complete. It is summarised in the flow diagram of figure 7 which also shows the data requirements and principal outputs. The main costing algorithms execute in about 0.08 CPU seconds. This is fast enough for the model to be used as a subroutine for determining optimal grid structure; this is defined as the capacity mix which gives minimum yearly costs, including system limit costs and capital annuitised over the machine lifetime.

4. A PROBABILISTIC STUDY OF ELECTRICITY FUTURES INCLUDING WIND ENERGY.

4.1 Nature of analysis

Any attempt to investigate the long-term economics of supply sources is bedevilled by the uncertainty which surrounds any future projections. The need for such projections cannot be avoided, however. One technique which can help to increase the useful information available from studies is probabilistic analysis.[4] The remainder of this paper describes a probabilistic study of the economics of supply technologies on the CEGB grid, including wind energy.

The aim of the method is to make future uncertainty explicit, by describing the important uncertain variables as probability distributions of possible values. For example, coal station capital costs are described by a triangular probability distribution with minimum £850/kw, mode £1050/kw and maximum £1400/kw. In this study, all important uncertain variables are described by triangular probability distributions.

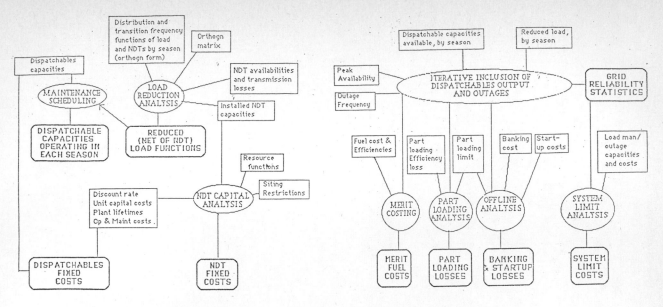

NDT = Non-Dispatchable Technology (e.g. wind energy)

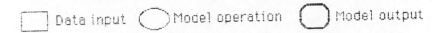
Data input   Model operation   Model output

Fig 7   Operations chart for probabilistic electricity grid analysis model

4.2 Data assumptions and results

The analysis proceeds as shown in figure 8. For each model run, one value is selected from each of the uncertain distributions to form a complete sampled set of input assumptions. The gross yearly benefit from WECS was chosen as a measure of the importance of wind energy technology, this being defined as the difference between minimum grid costs without wind energy and the minimum costs with it.

The calculations can be repeated for a series of sampled sets of the uncertain variables. The stratified technique of Latin Hypercube Sampling[14] was used to ensure full coverage of the ranges defined, in as few runs as possible. For the main analysis, 16 uncertain variables were defined and 35 sampling runs made.

Full data inputs are listed in Appendix 1. The following may be noted:

(i) The input costs (in £1985) and characteristics were estimated as mean values over the period 2010-2030. They are based on a review of current values and the forces acting for change,[15] but inevitably they reflect subjective judgements of the author.

(ii) Five dispatchable plant types were included:

   FBR
   Thermal nuclear (PWR or AGR)
   Coal 1 (Main coal units)
   Coal 2 (Smaller units for frequent cycling)
   Gas Turbines, or other peaking plant

(iii) Dispatchables capital costs are not dissimilar from current values, but the fuel costs of both fossil and thermal nuclear plant are assumed to rise.

(iv) Data on thermal part loading efficiencies is available from several sources, and French experience of part-loading PWRs suggests that there are no special problems associated with nuclear plants.[16] Offline costs are dominated by the stresses involved in repeated cycling, and are estimated from data on the assumed cycling lifetimes of the major CEGB oil plant.[17] The model was set to offload coal units in preference to part loading, keeping them banked at one-hour standby. Nuclear units are part-loaded if possible.

(v) Wind energy capital costs are based on an analysis of four recent machines and a few detailed engineering and weight-based projections. The breakdown of power-related to rotor-related costs was taken as 35%:65% for current machines. For a site with hub-

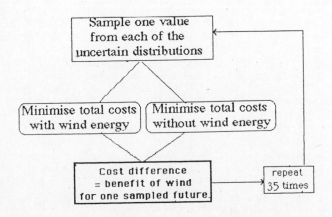

Fig 8   Structure of probabilistic analysis

159

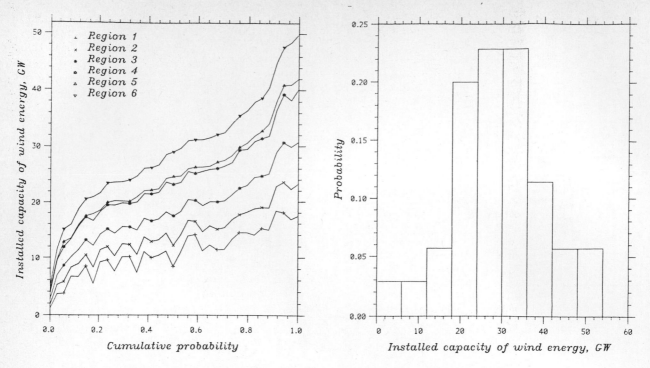

Fig 9   Distribution of installed wind energy capacities over 35 sampled futures

height mean speed of 8m/s, the ranges chosen correspond to total costs with minimum £550/kw, £275/m^2 ; mode £765/kw, £385/m^2; maximum £1000/kw, £500/m^2. The latter approximates the costs of recent machines.

Results are shown in figures 9 and 10. The mean yearly benefit from wind technology is £845m/yr (undiscounted £1985) and the mean installed WECS capacity is 28GW. These compare with total grid costs mostly in the region 9000-14000 £m/yr, and thermal capacities of 35-90GW.

5. DISCUSSION AND CONCLUSIONS

5.1 Accommodation of wind energy on a thermal grid

Figure 11 shows how the grid structure changes under the impact of wind energy for the central values used in the probabilistic analysis. The following observations can be made From such plots.

(i)   There is a capital shift from baseload plant to peaking plant. This arises from the impact of wind energy on the net load profile - more energy has to be supplied by stations operating for relatively small durations in

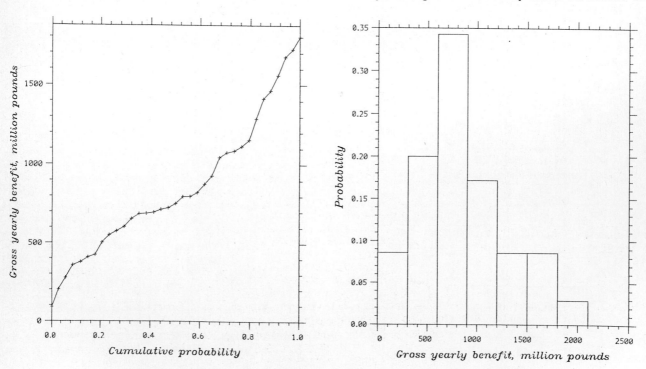

Fig 10   Distribution of yearly benefit from wind energy over 35 sampled futures

the year. The effect is dependent on the fact that wind energy is distributed, and hence has relatively few periods with no wind energy. When the model was run with only one region available (North Scotland, without siting restrictions), less wind capacity was built and the capacity shift was less marked.

(ii) There is a reduction in total dispatchable capacity of about 7%, and a slight increase in system limit costs. This is due to three effects. Because of the increased steepness of the load duration profile around the peak, the same reliability is achieved for lower dispatchables capacity, and the marginal costs of increasing system reliability in energy terms is greater - hence more load management is utilised. In essence these effects quantify the value of the capacity credit of the wind energy. Also, the reduction reflects the greater peak availability of peaking plant: in terms of meeting grid reliability requirements, peaking plants and baseload plants are not equivalent.

These features become more marked with higher wind capacities. In futures with particularly favourable cost conditions for wind, the capacity division was approximately 2:1:1 between wind energy, nuclear baseload and cycling/peaking units, and the reduction in total dispatchables capacity was ≈15%.

(iii) The combined effects of capacity credit and capital substitution are of great importance even at high wind penetrations. On systems dominated by nuclear power, capital savings typically accounted for 50-60% of the value of distributed wind power. These results are broadly in agreement with those of Diesendorf and Martin.[18-19]

(iv) Cycling costs, under the assumptions used, remain fairly small, with a shift from part loading to offline costs as some baseload nuclear capacity is forced offline. This reflects the fact that wind energy is not an exceptionally variable source - fluctuations from well distributed machines occur on the timescale of major weather patterns, i.e. several days. At high wind penetrations, part loading and offline costs form about 7% of the total. The utilisation of tidal energy, with a 12-hour cycle, would impose more severe cycling costs.[20]

(v) The amount of wind energy which it is economic to accommodate varies with the cost conditions. Over the 35 cases examined, the installed capacity of wind energy was commonly 60-110% of mean demand (average 90%), with wind accounting for 10-35% (average 26%) of the delivered energy.

5.2 Important Uncertainties

One of the outputs available from the analysis is the ranked correlation of the uncertain inputs with output values. These are not sensitivities in the normal sense, but a measure of the relative importance of the input uncertainties in determining the result. Seven uncertainties were significant (90% confidence level) in determining the net benefit:

Installed capacities for run MODAL

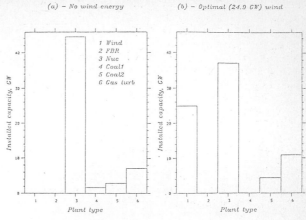

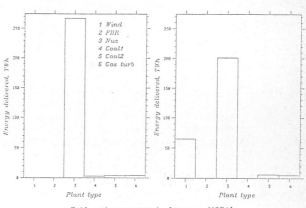
Delivered energy by plant type, run MODAL

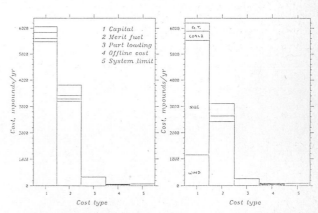
Grid cost components for run MODAL

Fig 11 Accommodation of wind energy for modal projection

Nuclear capital costs
Wind energy siting restrictions
Coal fuel cost
Dispatchables forced outage rate
WECS rotor-related capital costs
Nuclear fuel costs
System limit costs

The prime place of nuclear capital costs emphasises the fact that nuclear dominates thermal supplies in most futures, and that in such circumstances the capital savings from wind energy are of great importance. The importance of siting restrictions requires little elaboration, nor does the impact of coal fuel costs; coal is always the dominant cycling fuel and occasionally takes the baseload supply. The dispatchables forced outage rate features because of its role in determining the capacity value of dispatchable units in comparison with that of wind energy. The relatively low place of WECS capital costs reflects the input judgement that the uncertainties in future WECS costs, though they appear large, are in fact less than those of nuclear plant. Nuclear fuel costs are also significant, while system limit costs also affect the result because they are largely responsible for determing the value of the capacity credit, and are regarded as highly uncertain.

5.3 Distribution of wind resources

From figure 9 the distribution of wind energy between regions is apparent. Averaged over all futures, the mean division is:

Region	Mean installed capacity	
	(GW)	(%)
N. Scotland	10.50	36
S. Scotland	3.07	11
N. England	4.16	15
Wales & West Midl.	5.23	19
Midl. & SE England	0.95	3.4
SW England	4.36	15
Total	28.27	100

The dominance of Northern Scotland, despite the transmission losses, lower availability and maintenance penalties, is striking. This is partly due to the relatively large number of sites with high mean windspeeds. However, in all model runs the rated energy density of the marginal site used was lower in this region than any other. On diversity grounds alone the opposite would be expected. This observation indicates that an important factor is the greater reliability of the winds in Northern Scotland, resulting both in more energy output per kw and a significantly greater capacity value.

Equally clear is the relative paucity of the wind resource in the South-East and Midlands, despite the fact that geographically this is the largest of the regions.

Conclusions

In view of the complexity of this work and the number of new features which it contains, it would be premature to draw firm conclusions regarding the economic potential for wind energy. Nevertheless the following may be stated with some confidence:

(i) The economic potential for onshore wind energy in Britain appears to be greater than is often supposed, and considerable economic benefits may arise from wind utilisation.

(ii) Intermittent sources cannot be analysed by simple methods. The issues of capacity credit, capital displacement and the inhomogeneity of the resource are all of great importance. Studies of new energy sources which are based on 'average' characteristics, and which do not take account of capital effects, may be highly misleading.

(iii) The electricity grid of the future is likely to be well diversified. Cheap baseload supply from nuclear power remains important but most of the futures projected involve considerable capacities of wind energy, cycling plant and peaking units in addition.

(iv) It is known that as more wind energy is utilised, the marginal value to the grid is reduced and the marginal costs of supply increase. Both factors appear to be important in determining the optimal capacity.

(v) Serious analyses of electricity futures should no longer ignore the potential contribution from wind energy.

Acknowledgements

I should like to thank all those who have helped me in this work, in particular: Dr Nigel Evans (CERG), for his consistent support and invariably wise advise; Drs Page, Taylor and Birch of ETSU, who arranged for me to be sent data from the ETSU R17 study of wind resources and who have offered many helpful comments; Dr David Moore (CEGB) and Dr Jean Palutikoff (UEA) for arranging access to the CEGB long-term windspeed data records; Dr Bob Lowe (formerly Open University) for access to hourly windspeed data records; and Dr Jim Halliday (ERSU, Rutherford) for his extensive help and encouragement in the early stages of the work. Thanks are also due to SERC and the University of Cambridge for their financial support.

REFERENCES

1 Halliday J., Lipman N., Bossanyi E.A and Musgrove P.E.: "Studies of wind energy integration for the UK national grid", Wind Workshop VI, Minneapolis, USA (June 1983).

2 Bossanyi E.: "Use of a grid simulation model for long-term analysis of wind energy integration", Wind Engineering Vol.7 No.4 (1983).

3 Grubb M.J.: "Assessing the economics of wind energy at high penetrations: a statistical method for reliability and operation considerations", Wind Energy Conversion (7th BWEA Conference), Ed A.D.Garrard, Oxford (1985).

4 Evans, Nigel: "The Sizewell Decision: A Sensitivity Analysis", Energy Economics Vol.6, No.1 (Jan 1984).

5 Grubb M.J.: "Probabilistic Electricity Grid Analysis (PEGA) Model, Part 1: Derivation and Manipulation of Effective Load Functions.", Cambridge Energy Research Group Report ERG-84/63 (1984).

6 Grubb, M.J.: "Probabilistic Electricity Grid Assessment (PEGA) Model, Part II: costing the grid", Cambridge Energy Research Group Report ERG-85/16 (1985).

7 Allen J.C.J. and Murray R.D.: "Electricity Load Management in England and Wales", Fourth IEE International Conference on Energy Options, London (April 1984).

8 Plumpton A.: "Probing load management's new ground rules", Electrical Review Vol 212 pp.24-25 (May 1983).

9 Grubb M.J.: "Analysis of onshore wind energy resources in Britain", Cambridge Energy Research Group Report ERG 86/2, (Feb 1986).

10 Newton K. and S. Burch: "Estimation of the UK wind energy resource using computer modelling techniques and map data: a pilot study", ETSU R17 (1983).

11 Moore D.J.: "10 to 100m winds calculated from 900mb wind data", 4th BWEA wind energy conference. BHRA fluid engineering, Cranfield, UK (1982).

12 Soan B.S. and M.G. Mytton: "Prospects for harnessing wind energy for electricity generation in the UK", Solar Energy R&D in the European Community, Series G. Vol 1. (1983 (cited in Newton, ETSU R20)).

13 Newton K.: "The UK wind energy resource", ETSU R20 (September 1983).

14 McKay M.D., Conover, Whiteman: "Report on the application of statistical techniques to the analysis of computer codes", Report LA-NUREG-6526-MS, LASL (Aug 1976).

15 Grubb M.J.: "A discussion of cost and performance assumptions for electricity supply in the long term", Cambridge Energy Research Group internal paper (March 1986).

16 Dehon F.: "Fuel operating experience under power cycling", Nuclear Engineering International Vol.30 No.374 (September 1985).

17 Beatt R.J.I. et al: "Two-shift operation of 500MW boiler/turbine generating units". Proc Instn Mech Engr Vol 197A (October 1983).

18 Diesendorf M., Martin B. and Carlin J.: "The economic value of wind power in electricity grids", Proc. Int. Coll. on Wind Energy, Brighton, U.K. (BWEA) (August 1981).

19 Diesendorf M. and Martin B.: "Integration of wind power into Australian electricity grids without storage: a computer simulation", Wind Engineering, Vol.4 No.4 p.211-223 (1980).

20 Bossanyi E.: "Wind and tidal integration into an electricity network", Proceedings of the Fourth BWEA Wind Energy Conference (BHRA Fluid Engineering) (1982).

Appendix 1:   Summary of data inputs for long term supply analysis

I. ECONOMETRIC AND ELECTRICITY DEMAND DATA

0.05	DISCOUNT RATE
<0.9, 1.4, 2.1>	ELECTRICITY DEMAND: ESCALATION FROM 1971-79 CONDITIONS
<0.4, 0.8, 1.1>	ELECTRICITY DEMAND: CHANGE IN STANDARD DEVIATION FROM 1971-79 CONDITIONS
<1.5, 7.5, 25 >	INCREMENTAL COST OF ELECTRICITY SUPPLY SHORTFALL, p/kwh/%msd shortfall

II. DISPATCHABLE UNITS DATA

	A FBR	B PWR/AGR	C COAL1	D COAL2	E GAS TURBINE	
	<200,800,1800>					FBR plutonium core cost, £/kw
1	B1*<0.65,1.0,1.25>	<800,1200,2400>	<1400,3200,6500>	C1	7000	Fuel cost, £/TJ
2	9.5	10.0	9.8	10.5	13.0	Heat rate, MJ/kwh
3	B3*<1.05,1.25,1.5>	<1100,1550,2200>	<850,1050,1400>	C3*1.10	400	Capital cost, £/kw
4	A3*0.1	B3*0.1	0.0	0.0	0.0	Decomissioning costs, £/kw
5	30	35	40	40	40	Lifetime, years
6	A3*0.014	B3*0.014	C3*0.015	D3*0.015	E3*0.015	O&M costs, £/kw/yr
7	800	600	600	200	100	Unit capacity, kw
8	98	<0.55,0.70,0.82>	<0.64,0.72,0.82>	C8	0.95	Average Availability
9		<0.25,0.50,0.75> x (1 - average availability)				Forced outage rate
10	2000	2000	2000	2000	2000	Average time before failure, hours
11	A1*A2*0.8	B1*B2*0.8	C1*C2*0.8	D1*D2*0.4	E1*E2*0.1	Cold start-up cost, £/kw/start
12	12	12	12	6	2	Cooling time constant, hours
13	A3*0.00003	B3*0.00003	C3*0.00003	D3*0.00001	E3*0.00001	Turbine start-up costs, £/kw/start
14	A1*A2*0.065	B1*B2*0.065	C1*C2*0.065	D1*D2*0.08	E1*E2*0.1	Banking cost, £/kw/hr
15	72	72	24	5	1	Minimum time for switch-off, hours
16	0.3	0.3	0.3	0.3	1.0	Part loading limit
17	0.13	0.13	0.13	0.10	0.10	Part loading efficiency loss parameter

III. WIND ENERGY

Power-related costs:	<220,280,320> '/kw
Rotor-related costs:	<170,250,340> '/sq metre
Conversion characteristics:	Cutin speed 0.8Vm increasing linearly to
	Rated speed 1.6Vm with power coefficient 0.4
	Cutout speed 3.2Vm
Turbine lifetime:	30 years
Array efficiency:	90%

First order siting restrictions: 60 % of area excluded
Second order siting restrictions: <20,60,100> % further reduction in
available area

Characteristics by region:

	1-North Scotland	2-South Scotland	3-North England	4-Wales & West Mid	5-South East	6-South West
O&M (%capital/yr)	1.5	1.2	1.0	1.0	0.8	0.8
Availability	0.88	0.90	0.91	0.91	0.93	0.93
Relative transmission efficiency	0.96	0.98	0.99	0.99	1.01	1.01

<x,y,z> Indicates a triangular distribution of uncertain values, with min x, mode y and max z.

Wind energy feasibility study for Ascension Island

P T BARRY
BBC Transmitter Capital Projects Dept
R JOHNSON
ERA Technology Ltd (now with British Maritime Technology, Applied Fluid Mechanics Division)

SYNOPSIS
A feasibility study has been undertaken on the economic prospects for a wind turbine generator project on Ascension Island. The programme was carried out in three main stages, computer windfield modelling, collection of wind data, technical and financial appraisal.

The study was based on two commercially proven designs from UK Manufacturers of medium size machines. (Ratings in the range of 200 to 330 kW).

A payback period of 5 to 7 years appears possible at the best site.

1. INTRODUCTION

1.1 Background

Ascension is a small, barren, volcanic island with no agriculture, commerce or industry. Because of its location between continents it is useful for radio and undersea cable communications networks and as a refuelling stop for sea and air routes.

The BBC operate a diesel-fuelled power station which supplies the short-wave Transmitting Station and the overhead distribution system.

A study was commissioned into the feasibility of wind generation. The following aspects were considered when formulating the study objectives:

- There are virtually constant winds on Ascension.
- The main load on the system is the BBC Transmitters and these cause rapid changes in power demand.
- There is no significant and/or convenient storage load available.

It was decided to concentrate attention on commercially proven machines of 200 or 330 kW rating produced in the UK and to establish whether they could meet the technical requirements for connection to the system.

1.2 Geography of Ascension Island

Ascension Island is situated 8° South and 14° West (figure 1) and has an area of about 87 square kilometres. At its latitude of 8° South and its location in the mid-South Atlantic, the island's climate and wind characteristics are relatively straightforward compared with the temperate mid-latitude climate of the British Isles. The winds on Ascension Island are governed by two principal meteorological features; the semi-permanent South Atlantic anticyclone and the tropical convective or Hadley cell. The resultant effect of these features, combined with the remoteness from a significant land mass, is a climate which exhibits very little seasonal or even monthly variation and in which the wind conditions are dominated by the ESE Trade Wind. The constancy of the climate, in particular the wind, allows straightforward and reliable estimates to be made of annual wind energy potential. Existing data from the island suggest that site wind measurements over a period of only several days would be adequate for site evaluation purposes.

2. APPLICATION OF NOABL WIND MODEL

The numerical windfield model NOABL was applied prior to the wind measurement survey to identify a number of high windspeed sites. This model is becoming fairly well established in the wind energy R&D sector.

A contour map of Ascension Island was digitized

3. WIND MEASUREMENTS

3.1 Mean Windspeed

The wind measurements were undertaken during a 2-week site visit from 31 October to 15 November 1985. The wind data of primary interest is the mean windspeed. The constancy of the winds on the Island means that a detailed record of wind direction is not essential. The anemometry comprised 6 cup counter anemometers mounted on 10 metre telescopic towers. The counters were read at regular intervals and average windspeeds calculated in the conventional way. Figure 3 shows the location of the measurement sites. An anemometer was installed near to the power station for the duration of the visit to provide a fixed reference. The remaining 5 anemometers were deployed at more than one site. In addition to the cup anemometers, a TALA kite was deployed to assess, albeit on a visual basis, qualitative turbulence characteristics. The cup anemometer data was referenced to annual wind conditions using data supplied by the RAF Meteorological Station at Wideawake Airfield.

At two of the sites, North East Bay and Lady Hill, a data logger was used to record the diurnal variation in the form of 15 min. av. windspeeds for periods of up to 24 hours.

The windspeeds measured at each site and those measured over the same period at the power station are shown in Table 1. The site windspeeds have been scaled using the power station and the Wideawake reference data to estimate the annual mean windspeeds for each site.

3.2 Turbulence Observations

The observed physical behaviour of the TALA kite suggested that the Cowsheds, Weather Gardens, Paddock and North East Cottage sites would be unsuitable for wind turbine operation. The smoothest flow was observed at the Dark Slope Crater site.

3.3 Comparison of Measurements with NOABL Prediction

The comparisons with NOABL predictions are shown in Table 2. In general, the predictions are reasonable and justify the decision to use the model. The only major discrepancies are for Lady Hill and Paddock where NOABL produced respectively under-and-over estimates. These discrepancies can be attributed almost entirely to model spatial resolution limitations imposed by the contour to grid conversion.

4. VIABLE SITES

The final candidate wind turbine sites are Lady Hill and Dark Slope Crater.

Lady Hill is near the centre of the island and was formed in recent geological times by volcanic activity. It is composed of non-cohesive volcanic ash and pumice which is reasonably dense and stable. The summit ridge rises some 150m above its surrounding base to a height of 329m above sea level. The windward facing slope is very steep. There is only room for one WTG at this site and considerable soil investigation would be required before a suitable foundation could be designed and costed.

The ridge at Dark Slope is about 150m above sea level and faces the wind coming from the sea across the land to the south of the airstrip. The ridge is composed of hard volcanic rock and foundation costs will be much cheaper than for Lady Hill. However, it is not close to an existing overhead line so the connection costs will be higher. There may be space for more than one machine.

5. OPERATION IN PARALLEL WITH THE POWER STATON

The main reason for establishing and operating the Electricity Utility run by the BBC is to supply the short wave broadcast transmitters at the Atlantic Relay Station. These Transmitters have a high utilization and operate throughout the year broadcasting to South America and to Africa.

The system total load is largely determined by the broadcast schedule as can be seen from figure 4. For simplicity, some of the transient step load changes due to transmitters being switched on or off have been omitted from the diagram.

When broadcasting, the instantaneous load follows the transmitted programme content and therefore fluctuates rapidly at syllabic or tonic frequencies, producing an infinite variety of load change patterns. Figure 5 shows the effect of normal programme and of the GMT pips.

To limit voltage and frequency transients on the electricity supply it is necessary to select diesel alternator sets with added inertia and suitable governing systems. It is anticipated that only diesel generation will continue to provide the necessary firm capacity. In these circumstances a limited contribution from wind energy could be used to effect fuel savings provided the wind machine operating characteristics were compatible with the diesel-alternators.

The study found that wind speed fluctuations would not cause system instability and that either induction or synchronous generators up to 330 kW would be suitable. However, if the wind energy contribution were increased to, say 750 kW then a synchronous machine would be necessary to limit voltage disturbance. A control regime would also be necessary in this case to ensure that a minimum of two diesel sets were running when the likelihood of WTG cut-out was high.

A wind machine would generate at 415v which would be locally transformed to 11 kV for connection to the overhead line system. Radio telemetry or mains signalling would provide control and monitoring facilities in the Power Station Control Room.

6. POWER OUTPUT ESTIMATION

Examination of the 15 minute average windspeed data and comparison with annual mean windspeed indicated that a power output estimate could be based solely on the annual mean windspeed. It is emphasised that this simple method of calculating power output is valid only because the winds on Ascension are highly consistent, with no significant departures from annual mean conditions.

The power output estimation assumes 100% availability for simplicity. Whilst this is probably optimistic, it is not unreasonable in view of manufacturers now claiming availabilities in excess of 95%. Appropriate adjustments to the calculations can be easily implemented if availabilities different from 100% are important. The annual power output estimates for the two sites are as follows:

Lady Hill (KwH)   1.88 x 10^6 330 kw rating
 per annum)   1.18 x 10^6 200 kw rating

Dark Slope Crater 0.876 x 10^6 330 kw rating
                  0.526 x 10^6 200 kw rating

7. FINANCIAL ANALYSIS

The analysis was carried out using discounted cash flow techniques. Estimated costs were obtained from suppliers for a complete installation including shipping, foundations, road access and electrical connection.

It was assumed that all the power available from the wind machine would be used to reduce the operating costs of the Power Station. The maximum value of wind generated units was assumed to be equivalent to the sum of the fuel and operational costs elements of the 1985 Tariff which had been applied to consumers on the distribution system. The following comparison was obtained:

Lady Hill   330 kw   200 kw

Payback Period   3 years   4 years
Internal rate of return   36%   31%

Dark Slope Crater

Payback Period   8 years   9 years
Internal rate of return   15%   12%

The minimum value of wind generated units was assumed to be equivalent to the fuel cost element of the Tariff only. This gave a 5 year payback and an IRR of 24% for the 330 kW machine at Lady Hill.

8. CONCLUSIONS

Of the potential sites examined, only Lady Hill offers sufficiently strong winds to enable a wind turbine generator to return a payback period of 5 years or less at 1985 prices. On the basis of acceptable foundations being feasible, this site represents an attractive proposition.

If the effect of recent reductions of fuel prices were to be applied then the payback period would be extended. However, the study has shown that the wind generation option is worth pursuing. A further analysis can be carried out when results from a soil investigation are available and when the value of wind-generated units have been re-assessed in line with the current trend in fuel costs.

TABLE 1 :   Speed up — ratios relative to winds at the power station

Site	Measured mean windspeed (ms^{-1})	Corresponding mean windspeed at the power station (ms^{-1})	Speed up ratio = Site / Power station
Transmitter Station	6.3	6.7	0.94
Dark Slope Crater	8.2	6.5	1.26
North East Cottage	7.1	5.5	1.3
Ladies Loo	5.3	5.8	0.91
North East Bay	6.4	5.5	1.16
Weather Gardens	9.0	6.2	1.45
Cowsheds	9.3	6.0	1.55
Lady Hill	10.1	6.0	1.68
Below Weather Gardens	9.0	6.4	1.41
Hospital Hill	4.8	5.0	0.96

TABLE 2 : Summary of mean wind data

Site	$\bar{U}_{site}$ (ms⁻¹)	$\bar{U}_{site}/\bar{U}_{PS}$ Measured	$\bar{U}_{site}/\bar{U}_{PS}$ Predicted	$\bar{U}_{site}/\bar{U}_H$
Transmitter Station	5.8	0.94	1	1
Dark Slope Crater	7.8	1.26	1.14	1.47
North East Cottage	8.1	1.3	1.42	1.78
Ladies Loo	5.7	0.91	1.14	1
North East Bay	7.2	1.16	1.14	1.06
Paddocks	7.7	1.25	1.67	1.83
Cowshed	9.6	1.55	1.67	1.83
Lady Hill	10.4	1.68	1.21	1.65
Weather Gardens	9.0	1.45	1.42	1.82
Hospital Hill	6.0	0.96	0.85	1.62
Airfield	7.9	1.28	1.15	1.29

Notation:

$\bar{U}_{site}$ = annual mean windspeed at site at 10 m agl

$\bar{U}_{PS}$ = annual mean windspeed at the power station at 10 m agl

$\bar{U}_H$ = annual mean windspeed in undisturbed boundary layer at same height above sea level

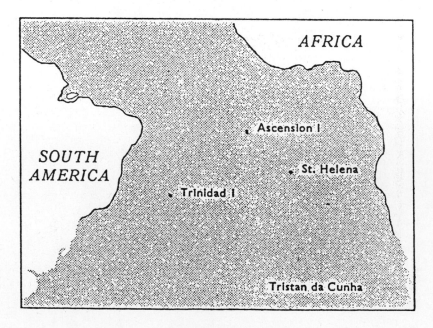

Fig 1   Location of Ascension Island

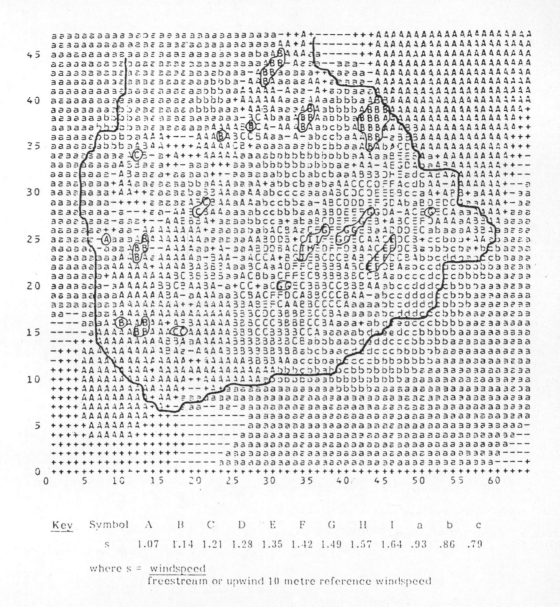

Key Symbol A   B   C   D   E   F   G   H   I   a   b   c

s   1.07 1.14 1.21 1.28 1.35 1.42 1.49 1.57 1.64 .93 .86 .79

where s = windspeed / freestream or upwind 10 metre reference windspeed

Fig 2   NOABL prediction of the 10 metres windfield for
         wind direction South East

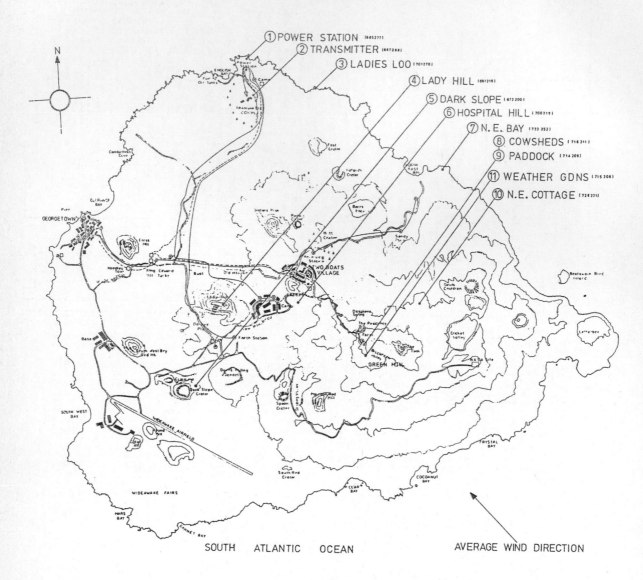

① POWER STATION (685277)
② TRANSMITTER (687266)
③ LADIES LOO (701270)
④ LADY HILL (681219)
⑤ DARK SLOPE (672200)
⑥ HOSPITAL HILL (700219)
⑦ N.E. BAY (733252)
⑧ COWSHEDS (716211)
⑨ PADDOCK (714209)
⑪ WEATHER GDNS (715208)
⑩ N.E. COTTAGE (728221)

SOUTH ATLANTIC OCEAN   AVERAGE WIND DIRECTION

Fig 3

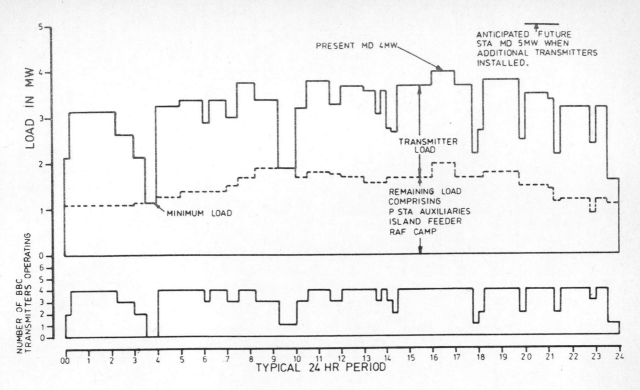

Fig 4

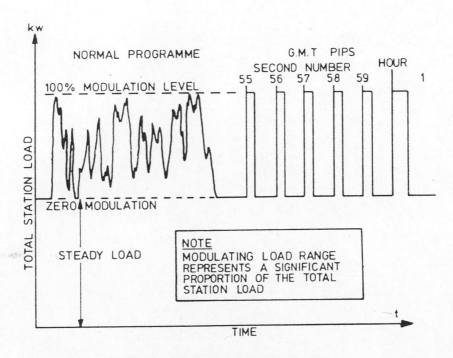

Fig 5   Theoretical instantaneous load

A 55kW wind energy conversion system for a holiday chalet complex in the Shetland Isles

G S SALUJA
School of Surveying, Robert Gordon's Institute of Technology, Aberdeen
P ROBERTSON and **R S STEWART**
Energy Design, Aberdeen

SYNOPSIS   This paper describes the selection, installation, commissioning and initial operational experience of a Vestas 55/11 grid connected wind energy conversion system (WECS) at Scalloway, in the Shetland Isles. Installed in August 1985, as part of the European Economic Community Energy Demonstration Projects Scheme, it consists of the first Vestas aerogenerator in operation in the United Kingdom and its application to supply electricity to a tourist site is unique. The daily recorded data for the first six months of the WECS is analysed and the results of the wind energy production and utilisation are presented. In addition, operational problems encountered during this period are highlighted.

1 INTRODUCTION

In 1983 Energy Design were commissioned by Easterhoull Chalets Ltd, Scalloway, Shetland Isles to investigate the feasibility of installing a medium wind energy conversion system to generate heat and power for the tourist chalet site. The unique aspect of the project is the application of WECS to isolated and island tourist communities with particular reference to the electricity usage patterns, system integration and load matching problems when compared with mainland residential communities. As far as can be ascertained, this wind energy application has not been demonstrated, and as a result this project has been awarded a grant under the EEC Energy Demonstration Projects Scheme 1983 round of applications (1).

2 SITE DESCRIPTION

The tourist site is situated 8 kilometres East of Scalloway, Shetland Isles, 2 kilometres from the sea inlet, East Voe of Scalloway. A location plan of the site is illustrated in Fig 1, and a detailed site layout showing the position of the aerogenerator is illustrated in Fig 2.

The aerogenerator is located on the site at an altitude of 90-100 metres above sea level, and there is no screening or obstruction to the prevailing south/south-westerly wind. The surrounding ground is composed of rough pasture and moorland, consisting of a topsoil layer of 300-900mm peat on a rock substrata. The layout of the site is illustrated in Fig 2 and consists of eleven 3-person chalets with washroom and laundry facilities being provided in a seperate building. The client's house is located within the site. A summary of the site accomodation is as follows:

    11 three person chalets, 425 sq.m. total area
    1 four person house, 90 sq.m. total area
    1 washroom/laundry block, 15 sq.m. area.

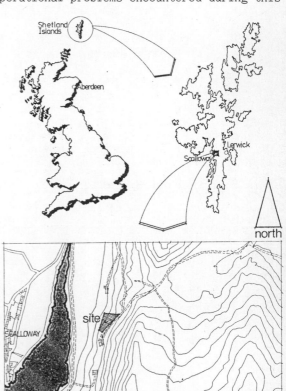

Fig 1   Location plan of site

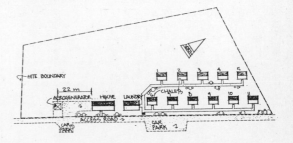

Fig 2   Site layout of aerogenerator installation

3 CLIMATE

No detail information was available on local wind conditions, as a result, the data recorded at the Lerwick Meteorological Station was utilised to determine the potential power generation from the aerogenerator. The predicted wind energy profile for the site is illustrated in Fig 3.

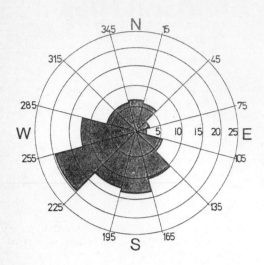

Fig 3   Wind energy profile for site

4 ENERGY REQUIREMENTS

The monthly electricity load for the site, based on the records of 1982 is given in Table 1. The breakdown of the space and water heating within the total electricity consumption is estimated.

Table 1 Energy Consumption 1982

Month	Space/Water Heating (kWh)	Cooking/ Lighting (kWh)	Total Energy (kWh)
January	10,430	2,120	12,550
February	9,570	2,080	11,650
March	9,780	1,990	11,770
April	8,210	1,810	10,020
May	6,620	1,590	8,210
June	4,690	1,340	6,030
July	3,160	1,060	4,220
August	3,160	1,340	4,500
September	4,150	1,590	5,740
October	6,310	1,810	8,120
November	7,820	1,990	9,180
December	9,300	2,080	11,380
Total	83,200	20,800	104,000

Electric oil-filled radiator heating and electric radiant panel ceiling heating units are used for the space heating of the chalets, water heating is provided by electric immersion heater units. All heating is operated by on-peak electricity. The electricity costs for 1982 was £5,335 based on a rate of 5.68p/kWh.

5 AEROGENERATOR SELECTION

Based on a client's budget of £35,000 and an annual energy production equivalent to the site energy requirement of approximately 100,000kWh, a review of the wind energy conversion systems marketed within the United Kingdom highlighted that only two machines were suitable to meet these requirements:
  (a) Polenko WPS 11 A40
  (b) Vestas 55/11

After assessment, a Vestas 55/11 was chosen; this machine has two asynchronous generators, 11kW and 55kW, driven by a 15 metre diameter, three bladed horizontal axis system with fixed pitch blades constructed of polyester reinforced with glassfibre. The turbine has active yawing via a gear mechanism and motor controlled by a wind vane, the turbine is mounted on a 22 metre lattice tower and is the first machine of its kind to be installed in the United Kingdom.

The predicted electricity generation from the Vestas 55/11, based on the available wind data is illustrated in Fig 4.

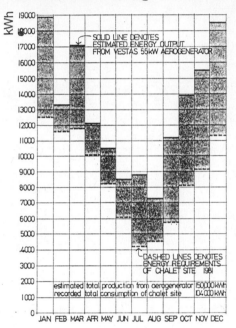

Fig 4   Predicted energy generation from Vestas 55/11

6 DESCRIPTION OF INSTALLATION PROCEDURE

Full planning permission was granted by Shetland Islands Council on November 1984 to permit erection of the aerogenerator. Prior to this date, Energy Design found it necessary to lodge three amended planning applications due to a degree of dubiety arising from the positioning of the aerogenerator. This problem appears to be prevalent in isolated countryside areas where ownership of land is often disputed. The planning permission was also further delayed by concern by the Planning Department over electromagnetic interference and noise emission levels eminating from the machine. These concerns were allayed by guarantees given by the manufacturers, Vestas. The final series of delays before installation in August 1985 was the unexpected rerouting of the electricity supply cable and construction of a new transformer for the aerogenerator supply. This supply was provided in June 1985. Foundations for the aerogenerator were laid in late July by local contractors and the machine

and tower shipped to Scalloway was erected
without difficulties over a period of two days.

7 DESCRIPTION OF MONITORING SYSTEM

A schematic diagram of the monitoring system is
illustrated in Fig 5. The data acquisition
system consists of the recording at hourly
intervals, wind speed, electricity generation,
import supply from grid, export excess
generation capacity to grid and reactive power
requirement. The monitoring system is
presently being installed.

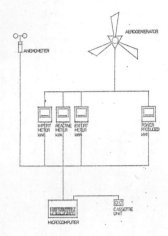

Fig 5   Schematic diagram of monitoring system

8 SYSTEM PERFORMANCE

The system was commissioned on 19 August 1985
and the following daily data has been recorded
by the client since 27 August 1985; Table 2
summarises the operational experience to
28 February 1986.

Recorded daily data parameters are as follows:
  –energy generated by WECS (A)
  –operating hours of WECS
  –energy imported from the grid (B)
  –energy exported to the grid (C)
  –energy consumed by the site (D)
This data is used to evaluate the energy
balance for the site as follows:

$$C = A + B - D \tag{1}$$

The system performance since commissioning,
highlighted that less energy was generated by
the aerogenerator than theorrectically possible
due to a number of operational problems.
The six month energy production for the
aerogenerator is 67,000kWh, as against a
predicted annual performance of 110,000kWh.
This figure compares unfavourably against the
manufacturer's performance estimate of
140,000kWh.

The operational problems encountered to
date can be categorised into three areas:
  (i) machine malfunction;
  (ii) control malfunction;
  (iii) power failure.
The aerogenerator malfunction was caused by the
yawing mechanism and associated control system;
failure to date has occurred on three seperate
occassions, with the gear mechanism being
finally replaced in February 1986. Power
failure, a common occurrence on the Shetland
Isles, has been highlighted on a number of
occassions during the past six months.

Analysis of the WECS output for the six
month period outlined in Table 2 and Fig 6
indicates the highest generated output to be
achieved in November.

Table 2 Summary of Operational Experience of the 55kW Vestas Aerogenerator at Easterhoull Chalets Ltd,
        in Scalloway, Shetland Isles, UK.

| | PERIOD | | | | | | | |
	Aug 85 (1)	Sept 85	Oct 85	Nov 85	Dec 85	Jan 86	Feb 86	Aug 27 1985 to Feb 28 1986
NO OF FULL DAYS OF MACHINE SHUT-DOWN	0	0	7	0	0	0	7	14
NO OF DAYS OF PARTIAL SHUTDOWN (2)	0	3	3	3	3	5	not recorded	17+
PERCENTAGE OF PERIOD MACHINE PRODUCING POWER	63	62	80	68	60	61	56	67
ENERGY PRODUCED BY THE AEROGENERATOR	1901	7867	11777	16027	12609	15008	3605	68794
AVERAGE POWER DURING OPERATIONAL PERIOD (3)	15.8	10.9	20.5	22.3	17.0	20.2	7.2	16.7
MAX. AVERAGE POWER IN ANY DAY	27.5	48.6	48.8	65.0	52.3	55.0	31.5	65.0
ENERGY CONSUMPTION	1016	7918	8946 (6730) (4)	13833	11440	11781	11320	66254
ENERGY IMPORTED FROM GRID	521	4631	3462 (2352) (4)	8000	6343	6092	9015	38064
PERCENTAGE UTILISATION FROM MACHINE (5)	26.0	41.8	37.2	36.4	40.4	37.9	63.9	41.0
ENERGY EXPORTED TO GRID	1406	4522	7500	9320	7637	9301	1093	40779

NOTES

(1) Machine commissioned on 19th August 1985. Monitored data available for last five
    days of month.

(2) Inoperational period approximately two hours or more per day.

(3) Calculations do not include the period when the aerogenerator was not in operation
    for complete number of days.

(4) Figures in brackets correspond to the operational periods

(5) Percentage utilisation=((Energy consumed – energy imported) / energy produced))/100

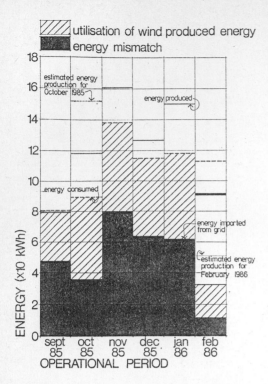

Fig 6   Energy balance of Easterhoull site

Further analysis of the prevailing wind speed regime during this period is given below.

Table 3 Analysis of Site Wind Speed Regime

Monthly Mean of Daily Hourly Mean Wind Speed (m/s)	Standard deviation of Daily Mean Wind Speed (m/s)
September	6.4	2.5
October	7.1	2.6
November	8.0	4.1
December	7.5	3.1
January	8.6	3.2

These figures highlight anomalies in system performance during January and February as a result of yawing gear malfunctions which significantly reduced performance. In addition, a relatively constant site energy consumption is upset by unexplained peaks during November; this problem is presently under investigation. In all months, apart from February, production exceeded consumption and import, presenting a significant energy mismatch and a relatively low utilisation factor of 40%.

In order to analyse the performance of the system in detail, the daily performance figures for December 1985 have been plotted in Fig 7.

Fig 7   Daily energy balance at Easterhoull site for December 1985

From this graphical analysis, eight possible and six distinct performance senarios have been identified:
1. import=0; export=production-consumption
2. export=0; import=consumption-production
3. production=0;consumption=import
4. production>consumption>import
5. consumption>import>production
6. consumption>production>import
7. import=0;export=0;production=export
8. consumption=0;production=export

From this analysis, the load matching of energy consumption to energy generation can be assessed and methods of improving the low utilisation factor put forward. One method which will be studied will be the conversion of the existing direct heating system to storage heater and bulk hot water storage tanks. This study awaits the commissioning of the monitoring system to provide the necessary detailed hourly performance data.

9 CONCLUSIONS

Given the considerable time taken to recieve grant funding and achieve the statutory planning and electricity authority approvals, the construction and installation of the aerogenerator has proceeded without major delay or problem. The operation of the aerogenerator has in general been trouble free, although the yawing gear mechanism has been of some concern, and has now been replaced. The performance of the machine is estimated to be 20% below manufacturers predicted performance, however, it is difficult to accurately judge the reasons for the reduction in performance until detailed monitoring figures become available later this month.

In conclusion, the system is operating successfully and no major problems have been experienced and the aerogenerator, once monitored, should provide a valuable source of operational information and research.

ACKNOWLEDGEMENTS

The authors acknowledge the financial support of the European Economic Community Energy Demonstration Scheme, Science and Engineering Research Council, and Shetland Islands Council for the aerogenerator installation, monitoring and research studies and greatly appreciate the assistance given by the client, Mr A Williamson.

REFERENCES

(1)   Internal Report, EEC Support for Energy Demonstration Projects, Application Ref.No.096/83/UK, 1983.

Dynamic analysis of wind turbines for fatigue life prediction

A D GARRAD and **U HASSAN**
Garrad Hassan & Partners, 10 Northampton Square, London

SYNOPSIS   Historically   the   dynamic analysis of wind turbines has   tended   to produce results of rather academic interest.   This paper shows that a realistic wind input may be combined with an aeroelastic model of the rotor system and the material characteristics of the blades to produce fatigue damage estimates.   The paper describes the model, discusses its strengths and weaknesses and then gives details   of   a   brief   analysis   that investigates the   affects   of   turbulence intensity and yaw misalignment on fatigue damage rate.

1.   INTRODUCTION

The design procedures used   in wind energy   are   in   their   infancy.   Wind turbines   are   unlike other   engineering structures   in   that they   are   designed both to be wind sensitive and to be   in windy places.   As the designs evolve the structures,   both blades and towers, have   become more slender and thus   more active   dynamically.   During   a   wind turbine's design life of, perhaps, 20 or 30   years   it   may   well   perform 10^9 rotations.   Wind   turbines   are   in addition   then   subject   to: natural excitation   from the wind to which they readily   respond,   high   cycle   fatigue loads and possibly very high extreme loads   or   marginally   less   severe   loads that   appear   as low cycle fatigue cases. Until   relatively   recently most   of   the design calculations   were forced to omit, or   include   in a very approximate way, the impact of the wind turbulence on the loading and response experienced by wind turbines.   The purpose of this paper is to   show that it is possible to   provide an   analysis   that   includes   all   the salient aspects of the realistic loads and   furthermore to process them into a form   in which they will be of use to   a designer.   This final step is   of   great importance.   Much   effort   has   been expended on the dynamic analysis of wind turbines over the past five years or so, but   the results from the analyses have been   of   little   use   to   the   design engineers,   who   have   no   means   of incorporating a power spectral density or   cross correlation of bending moment into their detailed design calculations. The aim of the present analysis is   to overcome this difficulty and provide not only useful,   but also useable, results.

2.   DETAILS OF THE MODEL

The flatwise loads on a horizontal axis wind   turbine   blade   are   due   to   both deterministic   and stochastic sources   in approximately   equal   proportions.   In order   to   evaluate   the   fatigue   damage both sources must therefore be included. The   present   model deals with   the   two sources separately.   A block diagram of the program is shown in Figure 1.

The   deterministic   solution   is performed entirely in the time domain. The aerodynamic model is based on strip theory.   The   structural model   is   modal and   derived   from a finite   element representation.   The   forcing comprises all the standard deterministic sources: tower   shadow,   shaft   tilt,   yaw misalignment and wind shear.   This part of the procedure is therefore fairly standard.   The   results are cyclic time histories   that   are   conveniently represented   in terms of Fourier series. In   principle there is no limit   to   the complexity of the structural model used for   the deterministic phase since the solution   is   performed   by   numerical integration   in the time domain.   There are   more constraints to the stochastic solution which requires a linear set of differential   equations   with   constant coefficients.   For convenience the same structural   model is used for both parts of the analysis.   This constraint means that   the   rotor may be modelled with reasonable detail but the tower can only be   represented   by a single lumped mass on a spring that is free to move in   the fore-aft direction.

The stochastic phase of the model is conducted in the correlation time domain or,   when Fourier transformed,   in the frequency   domain.   The   choice   of   this type of solution rather than a time domain, simulation, approach is an important one since it has repercussions at later stages.

The use of spectral methods for the dynamic analysis of stochastically excited structures is fairly standard and common practice in wind engineering. It has firm foundations in many branches of engineering. Furthermore, the use of the von Karman correlation function, as the starting point for the description of the turbulence is a common choice. There are, however, a range of alternative time domain methods. Some of these lie very much in the realms of black art, based on some empirical representation of a gust - the shape, size and duration of the gust varies depending on the author! Others, which may truly be called simulation, start from a point very similar to our own analysis and generate "runs of wind", [1]. These techniques are very expensive to use and do not lend themselves to parametric studies. However, they are, unlike the present spectral technique, able to deal with non-linear events. In short, the advantages of spectral methods are: firm foundations in reality, economy of use, ease of validation and flexibility for parametric use. Their disadvantages are the need for linear systems and their requirement to treat the stochastic load quite separately from the deterministic - the latter feature is particularly problematic when using the results to compute fatigue damage.

Once the stochastic wind input has been derived it must be turned into loads via an aerodynamic transfer function computed from perturbation of the steady state solution used in the deterministic phase. This function can accommodate stalled and unstalled blades in the sense that it allows positive and negative thrust derivatives. The results from the stalled parts may, however, be rather unreliable since the stalled flow itself will have frequency characteristics that are not linearly related to those of the ambient flow. The transformation of the wind input into loads could therefore be a very non-linear process.

The combination of the spectral loads and the modal model of the structure is relatively straightforward, as is its solution. The only complicating issue is the fact that the aerodynamics couple all the modes together making the mathematical manipulation rather clumsy. The output from the stochastic phase are power spectral densities of loads.

3. PROCESSING RESULTS

For a particular load case defined by wind speed, turbulence intensity, yaw angle etc. two sets of results are thus obtained: a Fourier series and a power spectral density for each blade station. In that form the results are of little interest to the designer. They require further processing. The most useful parameters are statistical moments. The moments may be used to calculate various characteristic quantities for the signals. The simplest is the variance which is the zero moment. A combination of the zero and second moment gives a frequency parameter.

Other moments and combinations of moments are useful but less easy to appreciate in physical terms. Suffice to say that the moments may be used to give a description, albeit an incomplete description, of the processes considered. Functions of these moment are used to give a rapid means of evaluating the properties of the blade loads. Such an approach gives rise to a closed form solution for fatigue damage rate, [2].

The calculation of the fatigue damage rates using this technique requires various assumptions. A representative S - N curve is required for the material; Miners Law is invoked; the statistics for the stochastic response are assured to be Gaussian and the combined stochastic and deterministic response is assumed to behave in a similar way to the combined response of a narrow band system.

An alternative approach is to invert the Fourier series and power spectral densities into the time domain, add them together and treat them as a complete, characteristic, time history. This method has some attractions since it allows rather more sophisticated fatigue evaluation. Computationally it is clumsy and will not be adopted here.

4. EXAMPLE RESULTS

There is little more to be said about the modelling of the wind. It is, however, worth giving some thought to the next step in the stochastic chain - the aerodynamic transfer function. Figure 2 shows three typical sets of power spectral density of loads along a blade at different wind speeds. At the lowest speed the blade is unstalled and at the highest it is nearly unstable in the sense that the thrust derivative is only just positive. At low wind speeds the high aerodynamic activity both gathers load and prevents dynamic response, whereas, at high wind speeds, the stalled blade cannot collect the load but responds very strongly at resonance. Figure 2 demonstrates very clearly the important part that the aerodynamics plays in the response.

Figures 3, 4 and 5 show the standard deviation of blade flapwise bending moment normalised against the mean value at rated wind speed as a function of radius for some typical load cases.

Figure 3 shows the deterministic loads produced by wind shear and tower shadow but in the absence of any yaw misalignment. Under these conditions the curves are essentially straight lines parallel to the x-axis indicating that the relative importance of the static and dynamic loads is independent of radial position. Figure 4 shows the same plot for the conditions of Figure 3 but including 10° of yaw.

Figure 5 shows a similar plot giving the corresponding values for the stochastic loads. There is not a great deal of variation with radius at low wind speeds, but the changes grow more marked as the wind speed increases. The difference in shape between this graph and the deterministic graphs is due to the difference in energy distribution between the two sources.

The most striking feature of this set of Figures is the similar value of the dynamic loads obtained from the two sources. For this machine it seems that the turbulence induced loads at a low, 12%, turbulence intensity site are comparable to the deterministic values. Neither source can therefore be ignored for rigid hub machines. This point is made still clearer by reference to Figure 6 in which the standard deviations from the two sources near the blade root are plotted against wind speed.

Based on the data plotted in these Figures and other statistical parameters derived from the blade response spectra and Fourier series fatigue damage estimates have been made. These estimates are presented in Figure 7. They are plotted in normalised form. The actual damage rates are extremely sensitive to material data which, for wood, which is the material used here, is still fairly unreliable.

Figure 7(a) shows the damage rate as a function of mean wind speed. Note that the higher wind speeds produce damage rates of about four orders of magnitude

greater than the low speeds. Fortunately very little time is spent operating in the high wind speed region. This operational consideration can be accounted for by weighting the damage rates by the probability of their occurence. Figure 7(b) shows the probability density function describing the wind distribution, which may be used as the weighting function. The points in Figure 7(c) are determined by taking the product of those in Figures 7(a) and 7(b). The total damage is therefore calculated by integrating the function of which the few points in Figure 7c form a part. Although the high wind speed operation still emerges as the most damaging its relative importance is now more clearly demonstrated.

5. CONCLUSIONS

This paper has outlined a mathematical model that computes all the salient loads experienced by a wind turbine. The model has been used to obtain illustrative results showing the relative importance of the deterministic and stochastic loads in terms of fatigue life.

It should be a relatively straightforward task to validate the load prediction phases of the program. The interpretation of these loads in terms of fatigue is rather more complicated. It is considered that this stage in the analysis is rather uncertain, but nevertheless required in order to make the results accessible to design engineers.

REFERENCES

1. M B Anderson, S J R Powles and E A Bossanyi. The response of a vertical axis wind turbine to fluctuating aerodynamic loads. Proc. 7th BWEA Conference 1985, Oxford UK, Ed. A D Garrad

2. P H Madsen et al. Dynamics and Fatigue Damage of Wind Turbine Rotors. Riso-R-512, July 1984.

PROGRAM STRUCTURE

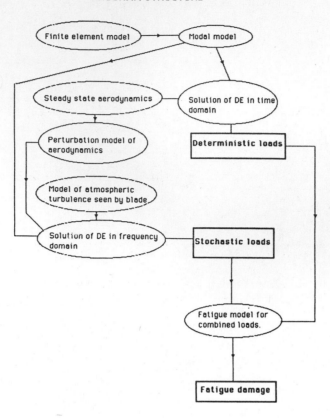

Fig 1   Block diagram of the computer code

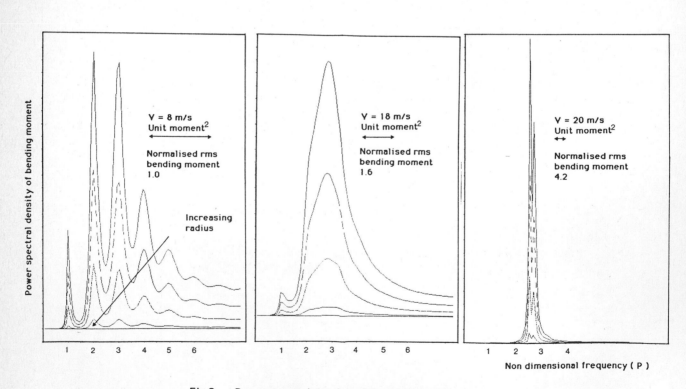

Fig 2   Power spectral density of flapwise bending moment
distribution as a function of wind speed

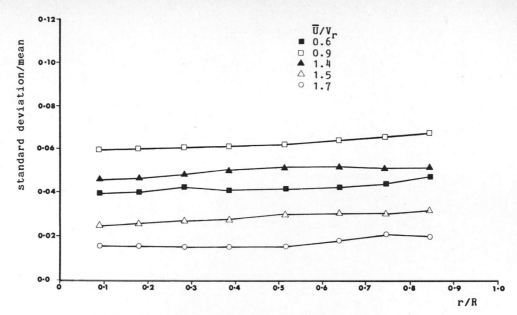

Fig 3   Normalised standard deviation of bending moment
as a function of radius — deterministic with no yaw

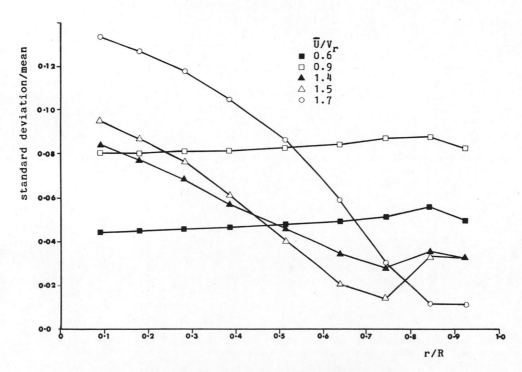

Fig 4   Normalised standard deviation of bending moment
as a function of radius — deterministic with 10° yaw

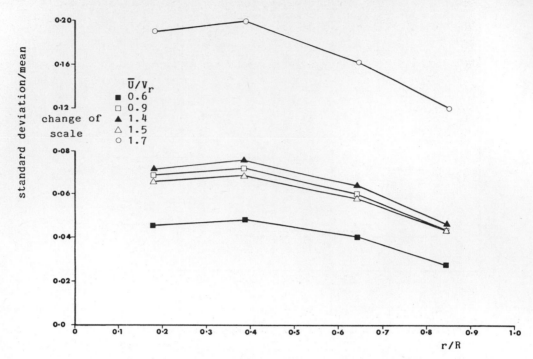

Fig 5   Normalised standard deviation of bending moment
as a function of radius — stochastic with 12 per cent
turbulence intensity

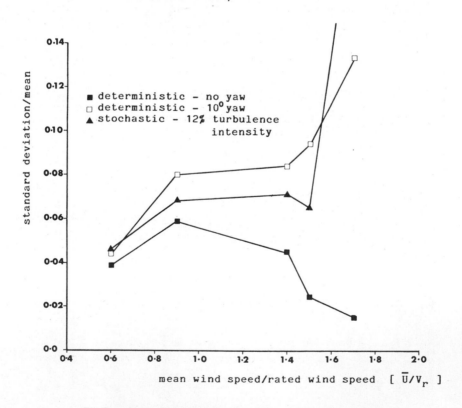

Fig 6   Normalised standard deviation of bending moment
as a function of wind speed

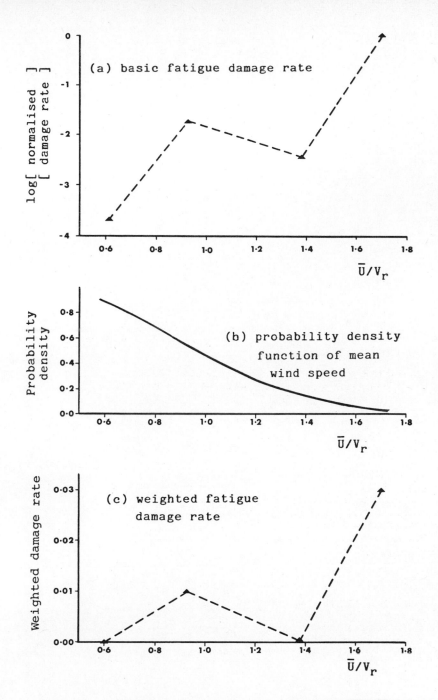

Fig 7   Predicted fatigue damage rate as a function of wind speed

Blade dynamics and yawing activity of a HAWT in a sheared wind

G T ATKINSON, S M B WILMSHURST and **D M A WILSON**
Cavendish Laboratory, Cambridge

1. ABSTRACT

This paper reports measurements in the natural wind of variations in coning angle for a free-yawing downwind machine with independently hinged blades. The experiment is significant because of the light it casts on the way yawing activity and blade dynamics are interrelated.

A simple theory is outlined by which the variation of yaw error with tip speed ratio may be predicted for machines with independently hinged blades.

Measurements of the forces on the Cambridge MkIV machine provide further tests of the proposed theory.

2. INTRODUCTION

The turbine, described in Ref.1, has a 5m diameter rotor with its axis in the horizontal plane 10m above ground level. The rotor axis intersects the veer axis. The top part of the tower is fitted with a fairing to reduce tower shadow.

The veer axis is 41cm from the hub and the blade hinge 15cm from the rotation axis.

Vertical wind shear was measured using two anemometers supported on the same mast 6.5m upwind of the tower. The time of passage of the wind from the anemometers to the turbine is about a second, which is the timescale of variations in shear, so it is necessary in analysis to allow for this delay, which changes with windspeed. It has to be assumed that the turbulence is simply convected downstream without developing. A positive shear is taken to mean that the wind increases with height, so that the average windshear is positive.

Inspection of the variation in cone angle (e.g. Fig.1) shows it to be close to simple harmonic with slowly modulated amplitude. Given this assumption, the amplitude can be simply calculated from the r.m.s. variation of cone angle over the preceding rotational period.

Measurements of the total force on the rotor were made using two pairs of strain gauges mounted near the base of the tower.

3. RESULTS AND DISCUSSION

3.1 The Effect of Yaw upon Blade Dynamics

Figs. 2 and 3 show that, roughly speaking, the amplitude of cone angle variations increased with

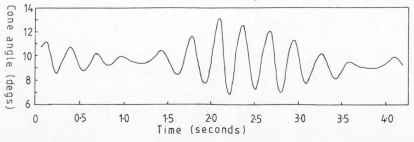

Fig 1

the magnitude of the wind shear, as is to be expected. However, the wind shear corresponding to minimum flapping amplitude is not in general zero. In fact, Fig. 4 suggests that the shear for minimum amplitude varied systematically with the running conditions, particularly with tipspeed ratio. This is the effect of an average yaw error, changing with tipspeed ratio, on blade dynamics. Rasmussen and Pedersen reported a related effect (Ref.1), namely a difference in the magnitude of cyclic blade loads on rotors with yaw error +/- 30 degrees due to an average wind shear.

3.2 Timescales of Variation in Yaw Error and Blade Flapping

A typical value for the instantaneous vertical windshear is 2m/s in a wind of average speed 8m/s. The eddies responsible for these velocity differences between points separated by 5m are clearly a maximum of a few tens of metres in size, passing in a few seconds. This is confirmed by direct observation of shear and consequent cone angle excitation (Fig.1). However, the out-of-plane flapping driven by such shears is subject to heavy aerodynamic damping: changes in the

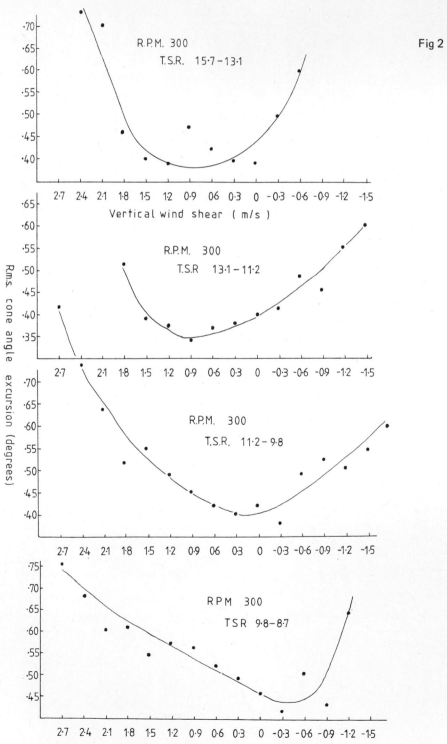

Fig 2

188

response after an abrupt change in shear are complete in a fraction of a revolution, that is something less than a fifth of a second. A quasi-steady treatment of the response is thus well justified.

In contrast, although the instantaneous yawing moments also depend on the instantaneous shear, these moments are small and the inertia of the nacelle high, so that an excursion in shear lasting one or two seconds, however intense, has little effect on the position of the nacelle. Yawing activity is determined by changes in wind direction and (less obviously) in windspeed with a timescale of five seconds or more. Over this time the average vertical component of wind shear is well-defined and the average horizontal component small.

To summarise: we assume that those variations in windspeed and direction that drive cone angle flapping are too rapid to change the nacelle position and that the variations that drive yawing activity correspond to spatial derivatives of velocity too slight to excite flapping. Such a separation greatly simplifies the problem and allows considerable progress without the use of numerical methods.

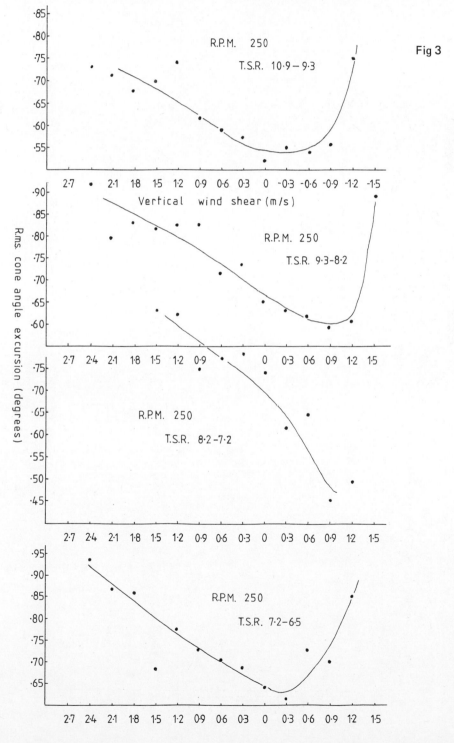

Fig 3

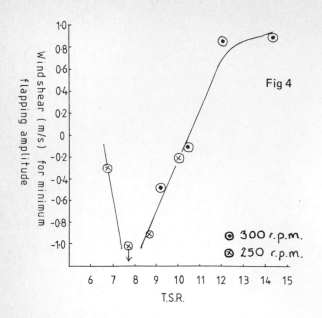

Fig 4

⊙ 300 r.p.m.
⊗ 250 r.p.m.

velocity leads directly to a flapping angle

$$\theta_{rms} = \frac{1}{\sqrt{2}} \left[\left(\sqrt{2}\,\theta_{min} \right)^2 + \left(\frac{V_{shear} - V_{min}}{2\omega R} \times 57.2^2 \right)^2 \right]^{1/2}$$

This formula is compared with measurements in Fig.5. The close agreement can be taken to justify the assumptions (a) that the magnitude of vertical and horizontal shear components are statistically independent and (b) that only aerodynamic damping is important.

The weight of the blades produces a small variation in angle of attack but the consequent yawing moments are small for all but the lowest windspeeds.

The first effect of the flapping response to a sheared wind on the yawing moment is that the imbalance in tangential aerodynamic forces (those that allow power production) between blades when they are vertical is removed.

3.3 Yawing Moment on a Rotor in a Sheared Flow

We consider first the blade dynamics of an unyawed rotor in a flow with a small linear vertical shear. The steady solution is a harmonic oscillation in cone angle with the rotational frequency, such that the velocity of the blade relative to the distant stream is unchanged during rotation: there are no changes in aerodynamic forces with blade azimuth. If the shear is small and linear, this is clearly possible for all radial stations simultaneously. Changes in centrifugal force provide the conservative restoring moments when the blades are horizontal. Deviations from this solution are rapidly damped.

If the average horizontal shear is also constant, the requirement of constant relative

Three new possible contributions to the yawing moment arise, however:
(i) The component of the centrifugal forces normal to the chord tending to restore the equilibrium cone angle, which are anti-symmetrically distributed across the rotor when the blades are horizontal. Because the hinge is offset 15cm from the rotation axis the reaction at the root to these normal forces may constitute a couple. For freely hinged blades this turns out to be zero (Appendix 1). (ii) The large component of centrifugal force along the blade also

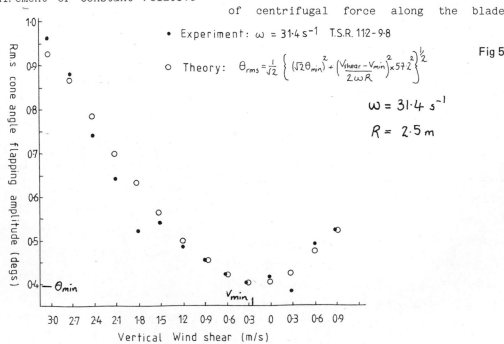

• Experiment: ω = 31·4 s⁻¹ T.S.R. 112–9·8

○ Theory: $\theta_{rms} = \frac{1}{\sqrt{2}} \left\{ \left(\sqrt{2}\,\theta_{min} \right)^2 + \left(\frac{V_{shear} - V_{min}}{2\omega R} \times 57.2^2 \right)^{2} \right\}^{\frac{1}{2}}$

Fig 5

$$\omega = 31.4\ s^{-1}$$
$$R = 2.5\ m$$

constitutes a yawing moment when the coning angle of the two blades is unequal (See Fig.6).

(iii) Coriolis forces in the tangential direction when the blades are vertical. Because the average coning angle is non-zero, the blade flapping changes the angular momentum of the blade about the rotation axis. If, for example, the windshear is positive, the blade flaps so as to flee the wind at the top. For a downwind machine such a change in cone angle decreases the angular momentum of the blade and implies the action of a Coriolis force as shown in Fig.6.

Since the blades cannot flap in the lead-lag direction and are assumed stiff, any nett tangential force is transferred directly to the hub and when the blades are vertical constitutes a yawing moment.

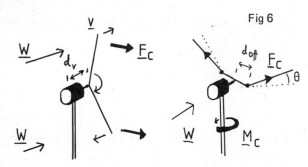

Fig 6

W Wind velocity
v Blade velocity
Fc Coriolis force

W Wind velocity
Fc Cent'g'l force
Mc   "   moment

For the given values of d_v and d_{off}, the average yawing moment of centrifugal forces is partly offset by that of Coriolis forces at all tipspeed ratios. The difference is typically of the same order of magnitude and in the opposite sense to that associated with tower shadow ; it is the balance of these two effects that determines to what extent and in what direction the turbine yaws. The balance shifts with tipspeed ratio with the combination of centrifugal and Coriolis (inertial) forces dominating at high tipspeed ratio. This is because:

(a) The inertial forces vary with ω and aerodynamic forces with $\overline{V}$ (average windspeed)

(b) The Coriolis part of the inertial moment which decreases its magnitude is proportional to $\sin \beta$ which is small at high tipspeed ratio.

3.4 Tower Shadow

The tangential forces on a blade are approximately constant along the span, whereas the out-of-plane flapping moment is determined by the flow near the blade tip. Because the depth of the shadow and proportion of the rotational period over which it is effective increase rapidly towards the root, the influence of tower shadow on the total sideways force on the rotor, and thus on the yawing moment, is much greater than that on the flapping response. Changes in flapping amplitude caused by tower shadow are ignored below.

The shadow caused by the tower fairing has been measured in a wind tunnel (Ref.3) and the deficit immediately behind it was found to be about 40%. Unfortunately the most important section for yawing moments is near the hub, where the fairing had to be truncated to allow passage for the blades. The deficit is presumably a little greater here. The reduced frequency of these changes in windspeed is high and the Kutta condition only a convenient approximation. However, steady aerofoil characteristics may be used to find the order of magnitude of impulsive yawing moments on the rotor and, with more certainty, to determine how these moments vary with tipspeed ratio.

3.5 Blade Dynamics in Yaw

Given the shadowing effect of the tower and the flapping response to an average wind shear, there is in general a yawing moment on the nacelle. The second half of the problem is the calculation of the restoring couple on a yawed rotor.

One effect of yaw is to induce a flapping with a phase similar to that caused by wind shear. This response is analysed in Appendix 2. Such flapping immediately leads to a yawing moment through the action of Coriolis forces, as explained above. Allowing for the sense of the flapping, this turns out to be a restoring torque. However, this response is not so simple, from the aerodynamic point of view, as flapping in response to a shear

191

where variations in angle of attack can be simultaneously removed at all radial stations. In a yawed rotor both angle of attack and relative velocity are changed to first order in the yaw angle and the effect on normal and tangential forces is different. Whereas the cyclic thrusts on a yawed rotor may be balanced by aerodynamic damping of the flapping motion, this flapping does not in general also remove variations in the tangential force. If the blade is unstalled, variations in both angle of attack and relative velocity due to yaw leave the tangential lift force unchanged (Appendix 2). The drag increases somewhat with the relative velocity. This leaves the effect on the tangential lift forces of the induced flapping, which produces an additional restoring moment. The problem is more complicated if a sizeable fraction of the blade is stalled or the drag coefficient is large.

Finally the yaw and consequent cross-stream thrust lead to an asymmetry in the windspeed across the rotor. Again for freely hinged or teetered rotors these variations in thrust are removed by a suitably oriented flapping response. If the variation in velocity across the rotor is linear, this response is completely effective in nullifying the yawing moment.

This horizontal gradient of velocity can be understood as an effect of the constraint of incompressibility on the deflected airstream. Fig. 7 shows a plan view of the flow at hub height through a yawed turbine.

Fig 7

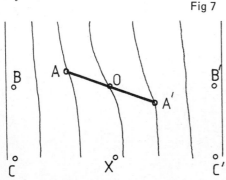

Consider areas S and S' bounded by OXCB and OXC'B' with OB and OB' large. In the steady state the nett flux through the boundaries of both S and S' is zero. Fluxes through CB and C'B' are zero and those through XC into S and XC' into S' equal. However, fluid flows from S' to S along OX so that

it follows that the sum of velocities normal to OB' along that line exceeds that along OB. The calculation of the distribution of this velocity deficit along OB will be difficult but it is clear that the velocity at A will be less than that at A'. Reference to Fig.6 shows that for a turbine with cantilevered blades, or for a solid disc, this velocity difference would lead to a restoring moment. This effect might be large in magnitude and explain, for example, the strong stability in yaw of the Koldby Turbine (Ref.4).

3.6 Predictions of Yawing Activity

We are now in a position to explain qualitatively the form of the variation of yaw error with tipspeed ratio (Fig.4) and to test some qualitative predictions of the preceding theory.

Consider first a rotor at high tipspeed ratio initially aligned with the wind. Flapping has eliminated variations in the aerodynamic forces in both normal and tangential directions. In response to the average yawing moment the turbine yaws and the flapping amplitude decreases. The effect of this on the yawing moment is twofold.
(i) The combined moment of the centrifugal and Coriolis forces decreases.
(ii) The wind shear is only partly offset by flapping and the aerodynamic forces that appear constitute a restoring moment.

The yawing continues until the flapping provides just sufficient moment through combined centrifugal and Coriolis forces to balance that part of the wind shear and tower shadow not eliminated by flapping. As mentioned previously, the yaw itself produces little variation in tangential force, as almost all sections are unstalled. Because the tipspeed ratio is high, inertial forces are large relative to those of aerodynamic origin and this flapping amplitude is small. We therefore expect that when the instantaneous wind shear takes its average value, the flapping amplitude will be small. We do in fact observe that at high tipspeed ratio the shear for minimum cone angle excitation tends to a limit close to the average shear.

Secondly, we expect that at lower tipspeed ratio the yawing moments on an initially aligned turbine due to tower shadow,centrifugal and Coriolis forces may become equal so that the average yaw error remains zero.

3.7 Balance of Yawing Moments

We calculate the yawing moments due to forces on one blade averaged over a single revolution.

(a) Tower shadow : we assume an average 40% velocity deficit behind the fairing.

$$\delta F_{tang} = \frac{1}{2} \cdot \frac{1}{2} \rho \pi R^3 \bar{V}^2 \frac{1}{x} \frac{dD}{dx}(x) \frac{\Delta V(x)}{\bar{V}}$$

where $\frac{dD}{dx} = 2 \frac{dC_Q}{dx} - \lambda_0 \frac{d}{d\lambda} \frac{dC_Q(x)}{dx}$, $\Delta V(x)$ tower shadow

According to Anderson (Ref.5) the variation of $\frac{dD}{dx}$ is a roughly linear increase with x (the fractional blade radius) to $\frac{dD}{dx}(x) = 0.15$ at x = 1. This crude approximation gives for the average yawing moment:

$$M_{tower} = d_v \cdot \frac{1}{4} \rho \pi R^2 \bar{V} \, 0.15 \int_{hub, x=0.1}^{1} \Delta V(x) \, f(x) \, dx$$

$f(x)$ fraction of period shadow effective , d_v distance from hub to veer axis.

(b) Coriolis forces

mean cone angle

$$M_{coriolis} = d_v \cdot \frac{\bar{V}_{shear}}{2} . M . x_{cM} \, \omega \sin \beta$$

(c) Centrifugal forces   fractional radius of C.O.M

$$M_{centrifugal} = - d_{off} \cdot \frac{\bar{V}_{shear}}{4} . M . x_{cM} \, \omega$$

With $\bar{V} = 7.5 m/s$ and $\omega = 31.4$ (300 RPM) and appropriate values for $\sin\beta$, x_{cM} etc, these formulae give

$$M_{Tower} = +0.35 \text{ Nm}$$

$$M_{Coriolis} = +0.77 \text{ Nm}$$

$$M_{Centrifugal} = -1.01 \text{ Nm}$$

Given the uncertainties in the calculation of the effect of tower shadow it is plausible that these moments are in balance at tipspeed ratio 10.5, as is observed. The important things about the expressions are; that the moment due to tower shadow varies with $\bar{V}$ whilst that of centrifugal and Coriolis forces on the blades varies with ω ; and the dependence of the Coriolis force on $\sin\beta$ .We assume that both the magnitude of the tower shadow and the average wind shear are proportional to the average windspeed.

Fig.4 shows a further decrease in the yaw angle at tipspeed ratios below 10.5 corresponding to an increase in the impulsive sideways forces caused by passage through the tower shadow. The minimum yaw angle is reached at a tipspeed ratio between 8 and 8.5. This is to be expected. Clearly the tipspeed ratio at which the increase in tangential force for a given increment in velocity is a maximum depends on the radial position. However, contributions to the total sideways force are weighted very heavily towards the root where the tower shadow is most effective. A calculation for these blades using Anderson's performance prediction program (Ref.5) shows that if the blade is divided into 20 equal sections, the three nearest the rotation axis but clear of the hub first enter stall at tipspeed ratios 8.9, 7.9 and 7.1 in order of increasing distance from the hub. As one or more of these segments enter stall, the decrease in tangential force caused by tower shadow will level off and then start to decline. At tipspeed ratios so low that a sizeable proportion of the blade is stalled, the decrease in velocity behind the tower may lead to an increase in the tangential force i.e. the yawing moment of the tower is reversed.

A quantitative understanding of the variation in yawing moments in this range of tipspeed ratios is clearly too complex for analytic methods and requires particularly accurate knowledge of the tower and nacelle shadow near the hub, characteristics in stall, and unsteady aerodynamic effects.

3.8 Forces on the Tower

The total force on the rotor is approximately oriented along the rotation axis and if the yaw error is non-zero this does not coincide with the wind direction.

Fig.9 shows the effect on yaw error of a relatively prolonged increase in windspeed. This begins at t=0 and lasts some 30 seconds (many times the characteristic period of the variations in cone angle flapping amplitude, as shown in Fig.1). During the first 20 seconds the wind

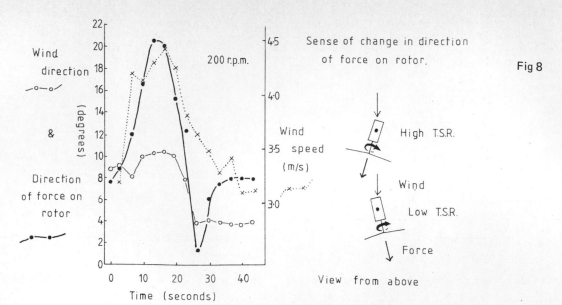

Fig 8

Sense of change in direction of force on rotor.

High T.S.R.

Wind

Low T.S.R.

Force

View from above

direction remains fairly constant, but because of the change in yaw error with tipspeed ratio the direction of the force on the tower changes. At about 20s the wind veers by some 5 and thereafter remains constant. The yaw error appropriate to the tipspeed ratio is regained in about 5 seconds.

Unfortunately no proper zero relation between the direction of the force on the tower and wind direction was obtained. However the results do indicate that a change in tipspeed ratio from 11 to 14.5 corresponds to a change in yaw error of some 13° in the sense predicted by theory. To compare the magnitude with theory we follow Fig.4 and the calculation of section 7 in assuming that tipspeed ratio 11 corresponds to very small yaw. At high tipspeed ratio (Appendix 2)

$$\sin \gamma_{eff} \sim \frac{V_{tip} \cdot \frac{V_{shear}}{2}}{U_{wind}^2} \qquad V_{tip} = 52 \, m/s$$

$$\gamma_{eff} \sim 27^{\circ} \qquad V_{shear} = 0.4 \, m/s$$
$$U_{wind} = 3.3 \, m/s$$

The apparent discrepancy between this γ_{eff} and the measured yaw error may reflect the fact that γ_{eff} arises partly from a misalignment of the rotor axis with the direction of the free stream and partly from a deflection of the airstream (Appendix 2).

4. CONCLUSIONS

The relations between blade dynamics and yawing activity for a free yawing, independently hinged rotor in a sheared wind have been examined. Such rotors are particularly simple because the flapping response eliminates variations in

aerodynamic forces in a wind with a linear shear.

Predictions of the variation of yaw error with tip speed ratio have been compared with measurements on the Cambridge MkIV turbine.

(1) At high tip speed ratio the main part of the yawing moment on a turbine initially aligned with the wind comes from centrifugal forces on the flapping blades partially offset by Coriolis forces.

(2) At lower tip speed ratio the effect of impulsive yawing moments as the blades pass through the tower becomes more important.

5. APPENDIX 1: REACTION AT THE ROOT OF A FREELY HINGED BLADE

c.o.M   Forces on blade

R y $F_{(y)}$

$\leftarrow x \rightarrow$

If the inertia of the hub is large,

$$R_{tot} = \int_{blade} F_{(y)} \left[\frac{xyM}{I_{cm}} - 1 - \frac{x^2 M}{I_{cm}} \right] \cdot \left(1 + \frac{x^2 M}{I_{cm}} \right)^{-1}$$

where the set of normal forces $F(y)$ are the centrifugal forces tending to restore the equilibrium cone angle, which are anti-symmetrically distributed across the rotor when blades are horizontal and may therefore lead to a yawing torque.

$$F_{(y)} = m_{(y)} \, \omega^2 \, y \, (\theta - \theta_{eqm}) \, dy$$

which gives for a reaction at the hinge:

$$R = \left[\int_0^R \frac{m(y)\,\omega^2 y^2 x M}{I_{cm}} - \int_0^R m(y)\,\omega^2 y \left(1 + \frac{x^2 M}{I_{cm}}\right) \right] \cdot \frac{\theta - \theta_{eqm}}{1 + \frac{x^2 M}{I_{6M}}}$$

$$= M x \omega^2 \left[\frac{I_{hinge}}{I_{cm}} - \left(1 + \frac{M x^2}{I_{cm}}\right) \right]$$

$$\sim 0 \quad \text{since } I_{hinge} = I_{cm} + M x^2$$

Therefore this contribution to the yawing moment is small.

6. APPENDIX 2: CYCLIC BLADE LOADS IN YAW

In a yawed rotor with untwisted blades, the components of relative wind velocity normal and parallel to the chord are modified as follows:

$$U_N \rightarrow U_N \cos \gamma \qquad \gamma \text{ yaw angle}$$
$$U_p \rightarrow U_p + U_N \sin \gamma \sin \psi \qquad \psi \text{ azimuthal blade position}$$

If the blade is <u>unstalled</u> the lift forces normal and parallel to the chord are changed by approximately

$$\Delta F_N = \tfrac{1}{2} \rho a \, c(r)\,dr \left[\frac{U_N}{U_p + \sin \psi \sin \gamma} \cdot \left((U_p + \sin \gamma \sin \psi)^2 + U_N^2 \right) - \frac{U_N}{U_p} U_p^2 \right]$$

$$= \tfrac{1}{2} \rho a \, c(r)\,dr \sin \psi \sin \gamma \, U_N^2 + O(\gamma^2)$$

$$\Delta F_p = \tfrac{1}{2} \rho a \, c(r)\,dr \left[\left(\frac{U_N}{U_p + \sin \psi \sin \gamma} \right)^2 \left((U_p + \sin \gamma \sin \psi)^2 + U_N^2 \right) - \left(\frac{U_N}{U_p} \right)^2 U_p^2 \right]$$

$$= O(\gamma^2)$$

If we wish to calculate the value of γ such that the flapping response to a shear is exactly stilled, we note that

$$\Delta F_N(\text{shear}) = \tfrac{1}{2} \rho a \, c(r)\,dr \, V_{shear}(r) \, U_p(r) \sin \psi$$

and require that

$$\sin \gamma = \tfrac{1}{2} \frac{V_{smax} \cdot U_{pmax}}{U_N^2} \cdot \frac{\int_0^{R_0} c(r)\,r^3\,dr}{\int_0^{R_0} c(r)\,r\,dr} \cdot \frac{1}{R_0^2} = \tfrac{1}{2} \frac{V_{shear}^{max} \cdot U_p^{max}}{U_N^2} \cdot \frac{0.019}{0.041}$$

γ is not in general the yaw error because the deflection of the airstream adds to the difference in the relative velocity parallel to the chord when the blades are vertical (Fig.6). In fact

$$\gamma = \gamma_{yawerror} + \gamma_{deflection}$$

$\gamma_{deflection}$, which may vary across the rotor, is the direction of the velocity at a given point on the rotor relative to the free stream. To obtain a crude estimate we assume that there is no wake expansion so that

$$U_{rotor} = U_0 \left(1 - \frac{C_T}{2} \right)$$

Similarly we assume that only air parcels that pass through the turbine are accelerated sideways, so that

$$V_{rotor} = \frac{U_0 C_T}{2} \cdot \gamma_{yaw\,error} \quad \text{(an overestimate)}$$

If $C_T \sim 1$ then $\gamma_{def} \sim \gamma_{yaw\,error}$ ie $\gamma \sim 2 \gamma_{yaw\,error}$

8. ACKNOWLEDGEMENT

The authors wish to thank Mr. R.P.Flaxman for his work on the Cambridge turbines.

7. REFERENCES

1. D.M.A.Wilson,E.J.Fordham and S.J.R.Powles : A 5m Diameter Horizontal Axis Wind Turbine Using a Commercially Available Alternator and Gearbox. Proc. 6th BWEA Wind Energy Conference,Reading,1984,CUP

2. F.Rasmussen and T.Pedersen : Measurements and Calculations of the forces on a Stall-regulated HAWT. 4th Int.Symp. Wind Energy Systems. 1982 BHRA

3. S.M.B.Wilmshurst,S.J.R.Powles and D.M.A.Wilson : The Problem of Tower Shadow. 7th BWEA Wind Energy Conference Oxford 1985. MEP

4. H.Petersen : Calculations on the DWT Wind Turbine 29.3/3/D, the Koldby-Windmill, in yaw. 7th BWEA Wind Energy Conference, Oxford 1985, MEP

5. M.B.Anderson : A Theoretical and Experimental Study of Horizontal Axis Wind Turbines. Ph.D Thesis, University of Cambridge.

Turbulence induced response of the Orkney 20 m diameter WTG

D C QUARTON, J G WARREN, R S HAINES and D LINDLEY
Taylor Woodrow Construction Ltd

Synopsis   The impact of wind turbulence upon the measured blade strain and fatigue damage of the WEG 20m wind turbine rotor is considered. Processing of the strain data to provide extracted deterministic and stochastic components for comparison with prediction is demonstrated.   Brief descriptions of the theoretical methods adopted by WEG for computation of deterministic and stochastic bending moments are given, and comparisons of prediction and measurement presented.   The comparisons are discussed with reference to their discrepancies.

1.   INTRODUCTION

It has become increasingly clear over recent years that a major problem in the design of wind turbines is the inability to predict accurately the fatigue life of the principal structural components. One reason for this is the lack of definitive data describing the fatigue properties of the commonly used materials and structural elements. However, a further, and somewhat more fundamental cause of the problem, has been the inadequacy of methods used in design for the calculation of component loads and stresses. Such inadequacy is associated in particular with the calculation of wind turbulence induced loads, the methods used for prediction of deterministic response being relatively well researched, and validated for most operational conditions.

The Wind Energy Group (WEG) is concerned to develop models for the prediction of structural response and fatigue lifetime, which may be used economically, and with confidence of their accuracy, during design. The procedure adopted has been to consider the problem firstly from basics, in terms of the steady aerodynamic loads on the wind turbine structure, building up the realism and complexity of the theoretical models in step with their validation by means of measured data. Such measured data, obtained from the Orkney 20m turbine, MS1, by means of the Burgar Hill monitoring system (1), has therefore been invaluable for this process of model development.

Previous reports of the analysis of MS1 monitored data, (2, 3, 4), have been concerned chiefly with the deterministic response of the machine, although the validation of predicted turbulent wind

pressure and teeter spectra is presented in (4). The main subject of this paper is the impact of wind turbulence upon blade bending moments, and following a brief description of the principal features of the theoretical model, predicted and measured responses are compared in terms of spectra. The analysis of measured data discussed in the paper is typical of that carried out at WEG as part of the MS1 Monitoring Project.

2.   STRAIN DATA COLLECTION AND PROCESSING

The strain data considered in this paper has been measured during steady, rigid hub, operation of the MS1, by means of strain gauges mounted at 33% radius on blade B. The data was collected at a sampling frequency of 125 Hz for a campaign duration of 240 secs. Further data describing the meteorological conditions during the campaign was also recorded, and this is summarised in Table 1.   It is clear that the hub height windspeed, turbulence intensity and yaw offset were approximately constant throughout the campaign.

2.1 Fatigue Damage Assessment

The strain data of particular interest when studying the effects of wind turbulence, is that measured in the flapwise direction, the edgewise strain being dominated by the gravity induced cyclic response at 1P. A 20 sec. sample of the flapwise strain time history recorded at the 33% radius location is shown in Figure 1. The time history shows clearly the stochastic nature of the strain response. By means of the fatigue damage assessment program, FATIMAS, the cycle distribution for the whole 240 sec. strain time history may be computed using rainflow techniques. The

results of such a calculation are
presented as a histogram in Figure 2,
giving the percentage of the total number
of cycles within a specified stress mean
and range bin. The calculation may then
be taken a stage further to obtain the
equivalent fatigue damage distribution
presented in Figure 3. Here the
histogram gives the percentage of total
damage accumulated in the 240 sec. time
history, associated with each stress mean
and range.

The cycle and damage distributions
display an important characteristic in
that a high proportion of the total
damage is associated with a relatively
low number of large range stress cycles.
In fact, it is only the upper 5.5% of
cycles which cause 61% of the total
damage during the campaign. This clearly
indicates the extremely damaging effect
of wind turbulence.

The analysis may be extended. The
underlying deterministic strain response
during the campaign, excited by cyclic
pertubations of the incident wind
velocity due to such as wind shear, yaw
and tower shadow, may be extracted from
the total measured signal by means of
azimuth binning. This procedure has been
described in some detail in (4) and
consists essentially of dividing the
rotor disk into a number of azimuthal
bins, and binning the measured data such
that the strain at a particular blade
azimuth is stored in the bin which spans
that azimuth. The mean strain in each
bin then gives the underlying
deterministic response. This has been
carried out for the data discussed above,
and the deterministic signal is presented
in Figure 4. The fatigue damage
associated with this deterministic strain
during the 240 sec. campaign has been
calculated at 0.025×10^{-6} and is to be
compared with the damage accumulated by
the total strain at 0.18×10^{-6}. Assuming
continuous operation under these
meteorological conditions, failure of the
blade according to the deterministic
strain would occur after approximately
300 years, whereas according to the total
measured strain, failure would occur
after 43 years. Although this comparison
may not be wholly meaningful in practice,
it nevertheless gives a clear indication
of the critical impact of turbulence upon
blade fatigue lifetime.

2.2 Spectral Analysis

As described in Section 3.2 of this
paper, the method adopted by WEG for
prediction of turbulence induced response
of a wind turbine, is based upon
frequency domain analysis. The method
enables the computation of the spectral
density of the turbulent fluctuations of
blade bending moment. In order to
validate the method, equivalent measured
data is required. This is not however
straightforward. Direct spectral
analysis of measured strain or bending

moment time histories such as that
discussed above, will result in the
spectrum containing not only turbulent
response but also that due to
deterministic excitation. The solution
is therefore to firstly azimuth bin the
measured time history to obtain the
deterministic response, subtract this
from the total signal leaving the purely
turbulent response, and perform the
spectral analysis on this data for
comparison with prediction. Although
this process seems rather involved, it
nevertheless offers the best method for
step by step validation of the theory.

The campaign data considered in section
2.1, has been analysed in this way. The
measured strain time history was,
however, firstly transformed into
out-of-plane bending moment, the
deterministic response then extracted,
and the turbulent spectrum computed as
shown in Figure 5. The spectrum exhibits
turbulent bending moment response at the
rotational frequency and its harmonics up
to 8P, and also at the rigid body drive
train, tower fore-aft and first two
blade flapwise modal frequencies. It
should be noted that the spectral density
is plotted on a logarithmic scale, and
hence the response at frequencies above 8
Hz is approximately two orders of
magnitude less than the dominant
responses at frequencies close to 1P and
4P.

3. BENDING MOMENT PREDICTION

The prediction of dynamic blade bending
moments, has been approached in terms of
separate computation of deterministic and
stochastic response. Validation of the
method used for predicting deterministic
bending moments has been reported in (3,
4), and only a brief outline of the
theoretical model is given here. The
method used for calculating turbulence
induced response is based upon the
frequency domain approach described in
(5, 6).

3.1 Determinstic Response

The prediction of bending moment response
to deterministic excitation is based upon
the method derived by Garrad (7).
Standard strip theory is used together
with 2-dimensional lift and drag
characteristics to model the
aerodynamics, and this is capable of
analysing a comprehensive range of
ambient flow perturbations including wind
shear, yaw and tower shadow. The
dynamics model is based upon a modal
representation of the MS1 rotor and has
the degrees of freedom indicated in
Figure 6. For the analysis reported in
this paper, based upon the measured
campaign data described in Section 2, the
teeter degree of freedom is suppressed
due to the rigid hub operation of the
machine. The mode shapes and frequencies
used as input to the predictive analysis
are those determined by an experimental

modal survey of the rotor.

The deterministic content of measured blade bending moment time histories has been compared with prediction for a wide range of operational load cases at all gauged rotor radii. In general these comparisons have indicated good agreement between measurement and theory providing substantial validation of the predictive method. The out-of-plane bending moment response at the 33% rotor radius, measured during the experimental campaign described in Section 2, is compared with prediction in Figure 7 in terms of time histories and Fourier harmonic magnitudes.

These results are typical of the overall analysis to date. The dominant components of the deterministic response are clearly at frequencies of 1P and 4P. The magnitude of the 1P harmonic may be explained in terms of the rigid hub operation of the machine, the response being much reduced for teetered operation. The significant response at 4P is due to the proximity of the harmonic frequency to that of the first blade flapwise mode and is excited principally by tower shadow. The model of tower shadow assumed by the predictive analysis is based upon a revision of the commonly used potential flow idealisation. Such a revision was carried out in view of blade surface pressure data collected by means of a transducer mounted at 7.0 m radius at the 17% chord position. This data, azimuth binned to yield the deterministic response, indicated a real tower shadow much less impulsive in form than that predicted by the potential flow solution (Refer (4)).

3.2 Stochastic Response

The frequency domain approach to the prediction of turbulence induced wind turbine dynamic response has been described in detail by Madsen et al (5), and Garrad and Hassan (6).

Longitudinal turbulence is assumed to be described by the von Karman correlation function which, by means of Taylor's Frozen Turbulence hypothesis, can be used to relate windspeed fluctuations at two points separated in time and space. Adjustment of the model to a frame of reference rotating with the blades, has the effect of transferring turbulent energy from low frequencies to higher ones associated with the rotational frequency and its harmonics.

The dynamic behaviour of the rotor is assumed to be described in terms of its natural modes, and the turbulent forcing of the modes is calculated by means of double integrals involving modeshapes, local aerodynamic influence functions and the rotating frame cross correlation function referred to above. These forcing terms are then developed in the frequency domain by means of 'fast Fourier transforms' to give the matrix of modal force spectral densities, $[S_{pp}]$. The response to turbulent forcing is subsequently computed by application of $[S_{pp}]$ to a transfer function representation of the rotor aerodynamics and dynamics, $[H]$, where

$$[H] = \left\{ ([K] - w^2 [M]) + iw([C_s] + [C_a]) \right\}^{-1}$$

and the modal response matrix is given by,

$$[S_{QQ}] = [H]^* [S_{pp}] [H]^T$$

Here, $[M]$, $[C_s]$, $[C_a]$ and $[K]$ are the modal mass, structural damping, aerodynamic damping and modal stiffness matrices, and w is the frequency of response. It should be noted that the aerodynamic damping matrix is generally full due to coupling between rotor modes. Having evaluated $[S_{QQ}]$, the calculation of spectral density of real blade deflections and inertial bending moments is relatively straightforward.

For prediction of the out-of-plane bending moment spectrum to compare with the measured data in Figure 5, only the first two blades modes have been included in the dynamic model. (The data was collected during rigid hub operation of the MS1 and hence the teeter degree of freedom has clearly not been included). The predicted spectrum is shown in Figure 8.

It is apparent that although the predicted bending moment response at the first blade modal frequency is in close agreement with that measured, the response at the rotational frequency of the rotor and its first harmonic is significantly less than measurement. The discrepancy is due to the calculation of the bending moment spectrum being based only upon the inertial component of blade load. This is generally considered a reasonable approximation for problems of this type, since stochastic excitation tends to cause response of a system near its resonant frequencies where the inertial loads will greatly exceed the applied loads. However, in the case of a wind turbine rotor 'slicing' through the turbulent structure of the wind, there is significant applied aerodynamic load contribution to bending moments at frequencies away from resonance associated in particular with the rotational speed and its harmonics.

A solution to the problem is to compute directly the bending moment spectrum due to turbulent fluctuations of applied aerodynamic load. This may then be summed with the inertial response to give the total bending moment spectrum.

This procedure has been carried out for the case in hand, and Figure 9 and 10 show the aerodynamic and total bending moment spectra respectively. It should be noted that for modelling of a teetered rotor, calculation of the aerodynamic bending moment spectrum is in general not necessary. This is because the rigid body teeter mode has a resonant frequency of 1P, the frequency of the dominant component of the applied load, and hence the inertial bending moment response is far greater than that due to aerodynamic loading of the blade.

The predicted total bending moment spectrum in Figure 10 may be compared with that measured in Figure 5 in terms of the logarithmic spectral density. A comparison of predicted and measured linear spectral density is given in Figure 11. It is apparent that prediction and measurement are in close agreement although discrepancies do exist in terms of an underprediction of the responses at 1P and 2P, and the non-prediction, due to the neglect from the theoretical model, of the rigid body drive train and tower fore-aft responses. Such discrepancies should not, however, be over-emphasised, the degree of predictive accuracy exhibited by the results of Figure 11 is very encouraging given the complex nature of the problem.

A comparison of the standard deviations of the predicted and measured responses, both in terms of deterministic and stochastic components, confirms the good agreement as shown in Table 2. These standard deviations also indicate that for this set of meteorological conditions at least, the deterministic and stochastic blade bending moments are of comparable magnitude.

4. CONCLUDING REMARKS

It is now well known that wind turbulence has a critical impact upon the fatigue lifetime of wind turbine structural components. This has been demonstrated by means of a simple damage analysis of typical strain data recorded during an experimental campaign.

Further use of the data has been made to provide the extracted deterministic and stochastic bending moment components for comparison with prediction. Methods for prediction of deterministic response are now well developed, and validated for a wide range of operational load cases.

The procedure for prediction of turbulence induced bending moments has been described, and comparison of prediction with typical measured data demonstrated. The agreement obtained to date is particularly pleasing, and encourages further effort towards a comprehensive validation of the predictive method.

A further step to be taken in the near future is the development of methods for combining predicted deterministic and stochastic components of blade load, This step is particularly important since it is the combination of the two components which is required as a realistic input to fatigue damage calculations during design.

ACKNOWLEDGEMENTS

The authors acknowledge the support of the UK Department of Energy and the directors of WEG for their financial support and their permission to publish this paper. (WEG is the Wind Energy Group, a consortium of Taylor Woodrow Construction Ltd., British Aerospace plc., and GEC Energy Systems Ltd.).

REFERENCES

(1) HASSAN, U., HENSON, R.C. & PARRY, E.T. Design of a micromputer based data acquisition and processing system for the Burgar Hill, Orkney, wind turbines. 4th International Symposium on Wind Energy Systems, Stockholm, Sweden, September, 1982, 2, pp 137-152.

(2) GARRAD, A.D., HASSAN, U. & LINDLEY, D. Monitoring results and design validation of the Orkney 20m wind turbine generator. 6th British Wind Energy Association Conference, Reading, UK, March 1984, pp 76-84.

(3) GARRAD, A.D., QUARTON, D.C. & LINDLEY, D. Performance and operational data from the Orkney 20m diameter WTG. European Wind Energy Conference, Hamburg, W. Germany, October 1984, pp 170-176.

(4) WARREN, J.G., QUARTON, D.C. & LINDLEY, D. An evaluation of the measured dynamic response of the Orkney 20m Diameter WTG. 7th British Wind Energy Association Conference, Oxford, UK, March 1985, pp 259-268.

(5) MADSEN, P.H., FRANDSEN, S., HOLLEY, W.E. & HANSEN, J. Dynamic analysis of wind turbine rotors for lifetime prediction. Report, RISO Nat. Lab., 102-43-51, 1983.

(6) GARRAD, A.D. & HASSAN, U. Turbulence induced loads on a wind turbine rotor. 5th British Wind Energy Association Conference, Reading, UK, March 1983.

(7) GARRAD, A.D. An approximate method for the dynamic analysis of a 2-bladed horizontal axis wind turbine system. 4th International Symposium on Wind Energy Systems, Stockholm, Sweden, 1982.

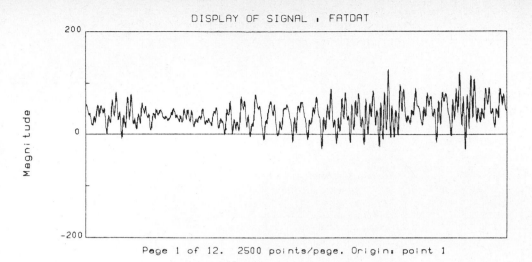

Fig 1   Flapwise strain at 3.28 m on Blade B (20 s)

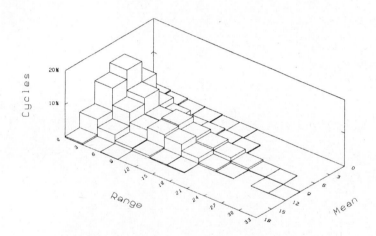

Fig 2   Cycle distribution for flapwise strain at 3.28 m

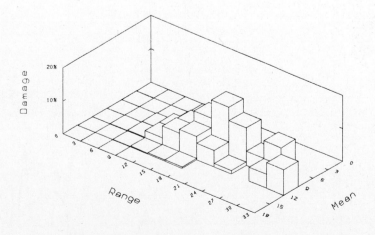

Fig 3   Damage distribution for flapwise strain at 3.28 m

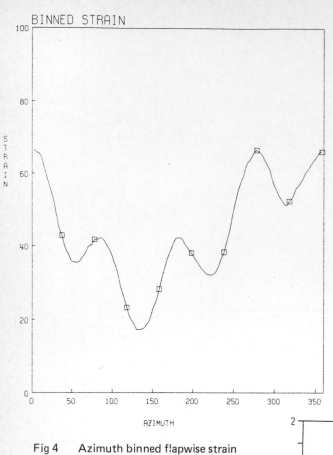

Fig 4   Azimuth binned flapwise strain

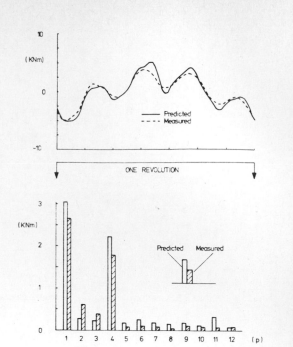

Fig 7   Out-of-plane bending moment at 3.28 m rigid hub.
13 m/s windspeed

AEROELASTIC MODEL
(6 DEGREES OF FREEDOM)

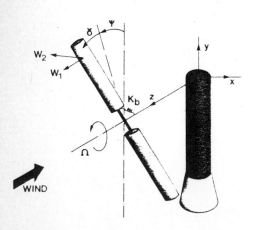

o   Teeter δ
o   Azimuthul perturbation Ψ
o   Flapwise bending (1st and 2nd modes) W_1
o   Edgewise bending (1st and 2nd modes) W_2

Fig 6

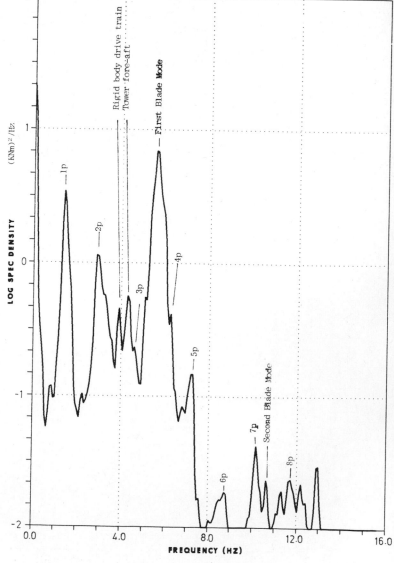

Fig 5   Turbulent bending moment — measured at 3.28 m

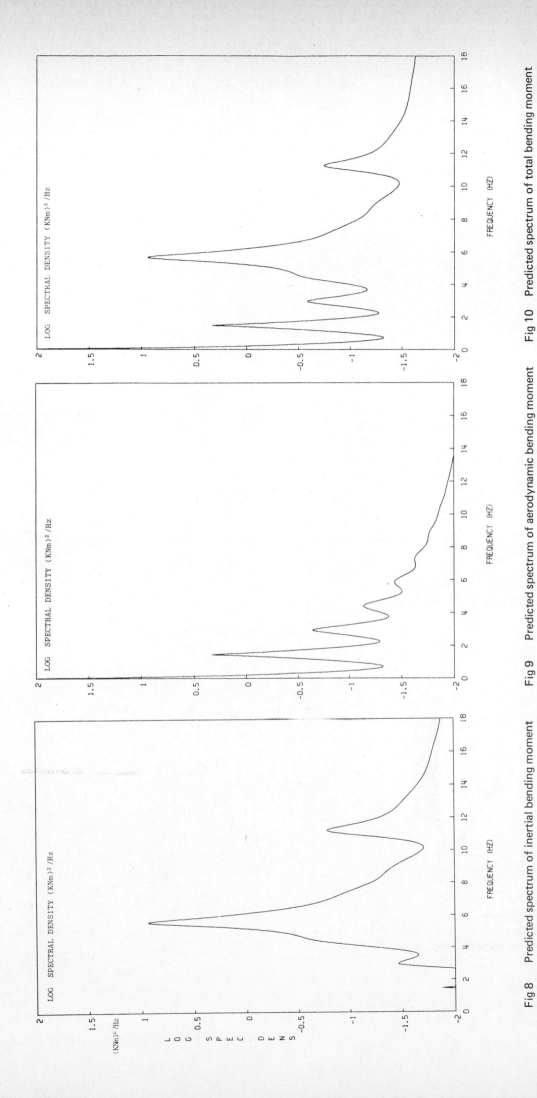

Fig 8   Predicted spectrum of inertial bending moment   Fig 9   Predicted spectrum of aerodynamic bending moment   Fig 10   Predicted spectrum of total bending moment

203

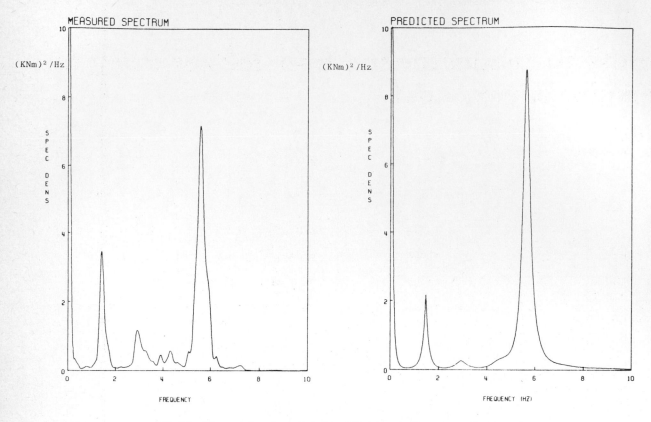

Fig 11   Measured and predicted bending moment spectra -- linear scales

60 sec. period	Mean Windspeed (m/s)	Windspeed Standard Deviation (m/s)	Turbulence Intensity (%)	Mean Yaw (deg.)
1	13.39	0.98	7.3	-4.3
2	12.79	0.95	7.4	-3.5
3	12.89	0.91	7.1	-4.0
4	13.01	0.94	7.2	-3.6

Table 1   Meteorological Conditions During Campaign   - Initiated at 21:07:59 on 12.7.85

Standard Deviation of Bending Moment (KNm)	Measured	Predicted
Deterministic	2.32	2.70
Stochastic	2.65	2.62

Table 2   Comparison of Measured and Predicted Bending Moment Standard Deviations

Windfarm conductor and transformer sizing to minimize energy cost

L D STAUDT
Enertech Corporation

SYNOPSIS: A methodology for selecting wind farm power cables and transformers is presented. The method accounts for wind speed distribution, turbine power curve and electrical characteristics, and cable and transformer electrical and economic data. Example cases are included.

1 INTRODUCTION

With the advent of reliable wind energy conversion hardware at reasonable installed costs, increasing attention is being focussed on those factors that produce a less dramatic effect on the cost of energy from a wind power station. Items such as conductor and transformer sizing, maintenance operations, and turbine control strategies are being given greater attention as the economic implications of better methods become apparent. This phenomenon has occurred in other industries where, after initial technical hurdles were cleared, it was possible to improve efficiencies by percentage points and then fractions of a percent. The wind industry is still at the stage where advances improve efficiencies by percentage points. To put this in perspective, a typical ten megawatt wind farm in California would increase its revenue by $20,000 annually for every percentage point of improvement in efficiency.

Miles of cable are installed in a typical wind farm, and although larger cable adds significantly to project cost, it is shown that cable significantly larger then that of minimum ampacity is economically justified. In a similar way it is shown that relatively expensive but more efficient transformers within the wind farm are a good investment.

2 WIND FARM ELECTRICAL LAYOUT

Wind turbines usually generate electricity at a low voltage (typically 480V in California wind farms) and power is brought to a distribution transformer via low voltage cable. The low voltage cable is usually buried in conduit and quite often more than one turbine feeds a distribution transformer. The transformer then raises voltage to distribution levels (tens of kilovolts), and depending on the size of the farm the distribution transformers may feed a dedicated substation where the voltage is raised to transmission levels (hundreds of kilovolts).

The example case that will be used in this paper will be a 10Mw windfarm consisting of 168 60Kw turbines. These turbines are connected to fourteen 750Kva transformers in groups of twelve (see Figure 1). A typical wind farm requires some 1150 feet of power cable per turbine, so for this example we require 36.6 miles of low voltage cable in 12.2 miles of conduit.

3 CABLE SELECTION METHODOLOGY

The methodology is summarised in Figure 2. First wind turbine electrical data and the power curve are used to develop an amps-squared vs. wind speed curve. This curve is multiplied together with the site wind speed distribution curve and then integrated to get total amps-squared-hours at the site. This is multiplied by the resistance of candidate cables to get the expected annual energy loss due to I^2R losses for each cable.

Having calculated energy loss for each cable we now must determine the present value of that energy loss over the life of the project, as well as the present value of construction funds spent on the cable. This is done with each cable starting from the least expensive (thinnest) to the most expensive (thickest). We know that it does not make sense to invest in the next largest cable when the incremental increase in cable present value is no longer exceeded by the incremental decrease in loss present value. To say it another way, keep choosing a larger cable until the energy savings no longer pay for the additional cable cost.

Note that this analysis assumes 100% availability. (Losses will decrease in approximate proportion to availability.) The analysis also assumes faithful adherence to the power curve, and that the wind distribution used is the actual distribution seen by the turbine (i.e. in spite of array wind speed reduction effects). The analysis does not account for the fact that at a given wind speed the power output varies about (but is centered on) the power curve data point. This variation will give a slight upward bias to the amps-squared vs. wind speed curve.

4 EXAMPLE CASE - CABLE SELECTION

Wire data for candidate cables is given in Table 1. The minimum size is that with the minimum ampacity to safely do the job. This is the size that would be used by the installation contractor if the job was left to him. The conductors under review are all aluminum, since they offer significant installed cost savings compared to copper.

5 TRANSFORMER SELECTION METHODOLOGY

Transformer losses are typically divided into no-load (iron) losses and I^2R (copper) losses. The no-load losses turn out to be quite significant since the transformer is seldom fully loaded - indeed in out example case the turbines connected to it are not producing power at all twenty-two percent of the time.

Table 1 - Candidate wire data (per 1000 feet)

THHN Al wire size	Wire cost	Conduit size/cost	Installation cost	Total cost	D.C. resistance
2	$548	1¼"/$545	$700	$1793	.2666
1	$638	1½"/$658	$800	$2096	.2114
0	$704	1½"/$658	$900	$2262	.1676
00	$794	2"/$870	$1000	$2664	.1329
000	$938	2"/$870	$1100	$2908	.1054
0000	$1097	2"/$870	$1200	$3167	.0836
250MCM	$1208	2½"/$1394	$1400	$4002	.0708

We will assume a 7 meter/second Rayleigh-distributed sight (Figure 3) corrected by the 1/7 power rule from 33 feet to a turbine hub height of 80 feet. Turbine power and power factor data are shown in Figures 4 and 5. We will use 17.5 Kvar of power factor correction capacitance located at the turbine tower.

The analysis is done on a per-1000-foot basis. The results are shown in Table 2. The best cable choice would appear to be the 4/0 cable, since at that point the additional cost of the next largest cable is not recovered by the value of the additional energy savings resulting from its use. It is interesting to note that the chosen cable has one-third the resistance (3X the area) of the minimum cable.

We assume that the transformers are always connected to the grid, such that the no-load losses are continuous throughout the life of the project. No-load losses are assumed to be constant and can therefore be calculated by multiplying the no-load power loss by 8760, the number of hours in a year. Transformer copper losses are specified at full load and can then be calculated at other loads since they are a square function of current. From this plus the current characteristics of single turbines, the number of turbines per transformer, and the wind speed distribution, we can calculate the annual I^2R losses.

Once again candidate transformers are evaluated from least expensive to most expensive. It doesn't make sense to invest in a more

Table 2 - Cable selection economic analysis

Assumptions:
1. Wire cost is the total installed cost (including conduit and labor)
2. All data is per thousand feet
3. Construction loan interest - 14%/10 years
4. Project life - 20 years
5. Buyback rate - 6 cents per kwh
6. Buyback rate increase matches inflation (3%)

Wire size	Wire cost	Annual energy loss	Wire present value	Loss present value	Change in wire PV	Change in loss PV
2	$1793	13558 kwh	$2932	$16305	-	-
1	$2096	10774 kwh	$3427	$12928	$495	$3376
0	$2262	8542 kwh	$3699	$10250	$271	$2678
00	$2664	6773 kwh	$4356	$8127	$657	$2122
000	$2908	5372 kwh	$4755	$6446	$399	$1681
0000	$3167	4260 kwh	$5179	$5112	$423	$1334
250MCM	$4002	3608 kwh	$6544	$4329	$1365	$782

Table 3 - Transformer selection economic analysis

Assumptions: Same as in Table 2

Transformer	Cost	No-load losses	I^2R losses	Annual energy loss	Transformer present value	Loss present value
No. 1	$8500	12483 kwh	19444 kwh	31927 kwh	$13900	$38312
No. 2	$9100	13122 kwh	14899 kwh	28021 kwh	$14881	$33625
No. 3	$10300	7805 kwh	10853 kwh	18658 kwh	$16844	$20172

efficient transformer when the increase in cost doesn't cover the additional value of energy saved.

6 EXAMPLE CASE - TRANSFORMER SELECTION

Table 3 gives data for three Westinghouse 750Kva transformers. Once again if one was only concerned with doing the job safely at minimum construction cost, No.1 would be the obvious choice. Once again the least expensive one would be the wrong choice.

Twelve turbines of the type used in the previous example are connectied to each transformer, each with 17.5 Kva of power factor correction. A 7 meter/second Rayleigh wind speed distribution is once again assumed.

The results of the analysis in Table 3 show that the clear choice is transformer No.3, since for $3000 in additional investment we reduce loss value by $18,000. Figure 7 shows transformer energy loss distribution.

7 COMMENTS AND CONCLUSION

We need to know wind data, turbine data, and electrical component data to perform this analysis. The key is to develop an amps-squared vs.wind speed relationship for the component in question, and then with wind speed distribution and resistance data, energy loss can be calculated.

Had we used minimum wire size and the least expensive transformer on our 10Mw wind farm, we would have saved $186,000 in construction costs. However by using the more efficient wire and transformer we save $975,000 over the life of the project.

Electrical losses amount to 5.2% of the farm output with the least efficient components and 2.1% of the farm output using the optimum wire and transformer. This assumes a turbine output of 150Mwh per turbine annually.

We can draw the general conclusion that is makes sense to use better than minimum standard wire and transformers on wind farm projects.

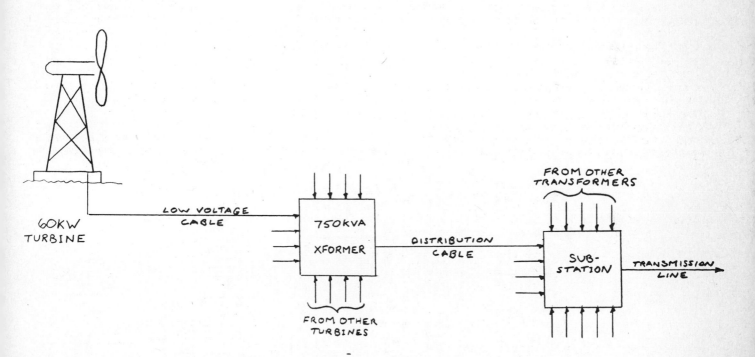

Fig 1   Windfarm electrical layout

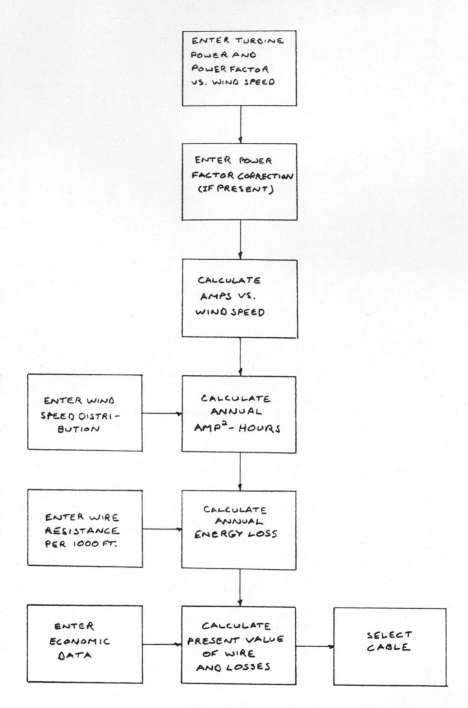

Fig 2   Cable selection methodology

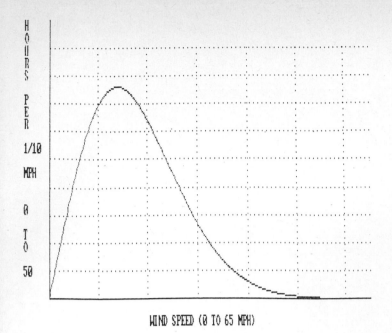

WIND SPEED (0 TO 65 MPH)

Fig 3   7 m/s wind distribution

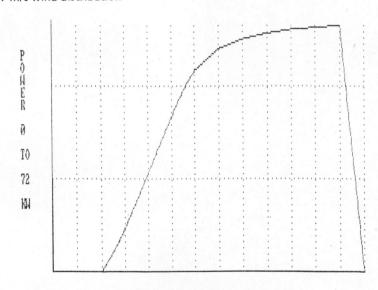

HUB HEIGHT WINDSPEED (0 TO 65 MPH)

Fig 4   60 kW turbine power curve

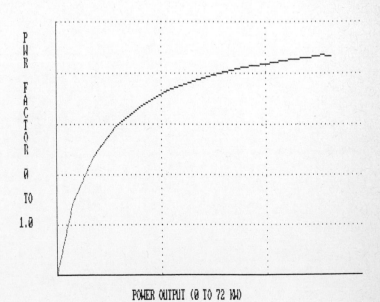

POWER OUTPUT (0 TO 72 KW)

Fig 5   Turbine power factor

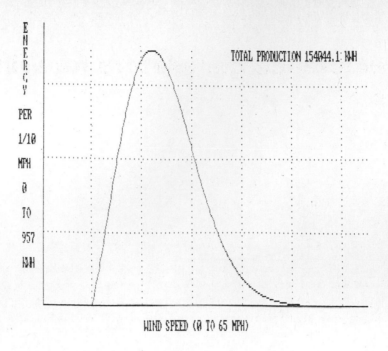

Fig 6   Turbine production distribution

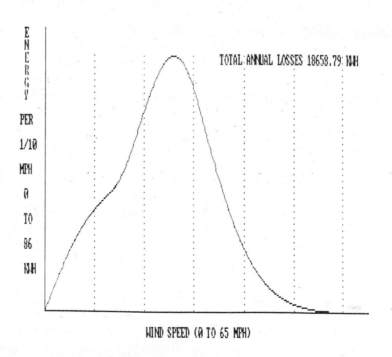

Fig 7   Transformer loss distribution

Switched reluctance generators for wind energy applications

N N FULTON and **S P RANDALL**
S R Drives Ltd, Leeds

SYNOPSIS

Recent developments in both electrical machines and electronics have enabled the capabilities of switched reluctance machines to be realised in a variety of applications. Generating schemes based on such machines are discussed and it is concluded that they are ideally suited to frequency-wild operation, having robust construction, wide range of control, high efficiency and low initial and running costs.

1 INTRODUCTION

The object of a wind energy scheme is usually to convert wind energy to electrical energy. This conversion process is normally accomplished via an intermediate stage of mechanical energy being transmitted from a rotating turbine (which captures the wind energy) along a shaft to a generator (which produces the electrical output). A gearbox is often used to increase the speed of the generator and hence reduce its size and weight. In the majority of cases, the electrical energy is required for loads which may be supplied either from the wind energy scheme or from the public utility; or indeed it may be wished to export the energy to a public utility, as in the case of a wind farm or a smaller installation with spare capacity. These constraints require the electrical output to be in the form of alternating currents and voltages with a frequency equal to (or, in some cases, nearly equal to) that of the utility's supply. For these reasons, and because of their widespread role in more conventional electrical generation, synchronous generators are generally employed, though induction generators have found favour in small sizes in the USA.

Other electrical systems have been employed in some installations and this paper reviews the characteristics of the different types of generators available, considers their suitability for wind energy schemes and discusses the advantages of a new machine – the switched reluctance generator. This is a form of electrical machine which has recently been shown to be capable of previously unrealised power densities and has benefitted particularly from the rapid, recent developments in electronics technology.

2 CONVENTIONAL ELECTRICAL GENERATORS

The three types of electrical machines normally used as generators are the direct current (dc) machine, the induction machine and the synchronous machine. Each of these has its own constructional and performance characteristics : these are reviewed in this section of the paper.

2.1 The dc generator

This is probably the type of electrical machine most associated with the term "generator". It is familiar in its roles as an old-fashioned car dynamo and as the engine-driven fairground generator. Its construction is also well-known, partly because of the similarity to domestic power drills: a stator carrying a few poles with simple coils (called the "field") around them; a rotor (normally referred to as the "armature") with many coils embedded in slots around the periphery and connected to the segments of a commutator at one end. The commutator acts as a mechanical switch – as the armature revolves it continuously connects and re-connects the coils the correct way round to allow current in the external load to flow in the same direction all the time, hence the name "direct current" machine. The connections to this commutator are, of necessity, sliding connections and take the form of carbon brushes.

If a current is passed through the field coils, a magnetic flux is set up in the machine. With the machine running at a constant speed, a voltage is developed across the armature terminals. If a load is connected, this unidirectional voltage causes unidirectional current to flow through the load. In the normal working range of the machine, the armature voltage is approximately proportional to the field current supplied, so the output voltage (and hence current, if the load is linear) is easily controlled by a simple feedback system. The armature voltage is also proportional to the speed of rotation, so fluctuations in speed must be compensated by corresponding changes in field current if the voltage is to be kept constant.

The dc generator is well suited for those situations where the load requires only direct current and where the prime mover does not run (or cannot easily be constrained to run) at fixed speed since simple feedback circuits can compensate for changes in load or speed (both of which cause voltage variations). Where the load demands an alternating voltage, an inverter must be introduced between the generator and the load. The inverter is an arrangement of (electronic) switches which will change the generated voltage (and hence the current) from unidirectional to bidirectional. This is the technique which is used, for example, at the receiving end of a high-voltage dc link (eg between France and England) and is well-established technology.

The problems with dc machines lie mainly in their relatively complex construction and hence high cost. The rotating armature winding, which carries the full load current, experiences large thermal stresses and hence is vulnerable to insulation failure. The carbon brushes wear, deposit carbon particles around the machine and brushgear (which is not conducive to good insulation levels) and require periodic inspection and replacement. The commutator is always a difficult area, especially a high speeds, and needs very careful machining to ensure that it runs true. The machine does not cope well with overspeeds or electrical overloads, both of these being liable to affect the commutator. The cost of repairs to such machines is generally high, normally requiring the removal of the armature to a repair shop.

In summary, though the dc machine has characteristics which fit well with a wind-generation system, it is costly and does not have the ruggedness and serviceability which is also desirable. These serious shortcomings have been noted by other authors, eg Freris (1), Gipe (2).

2.2 The induction generator

The induction machine is in the family of alternating current (ac) machines. By contrast to the dc machine, the induction machine has no brushes. The winding which carries the load current is now stationary and is distributed in many slots around the stator. The rotor carries a closed winding (which has no external connections) and is generally very rugged in construction. Induction machines account for over 90% of all electrical machines.

The principle of operation is quite different to that of a dc generator. If the machine is connected to a supply (generally a 3-phase supply), a rotating magnetic field is set up inside the stator bore. The speed of rotation of this field, the synchronous speed, is fixed by a combination of the winding geometry and the supply frequency. The rotor, if free to rotate, will run up to a no-load speed slightly less than the synchronous speed. Energy is converted if the rotor is forced (mechanically) to run at any other speed, so if an accelerating torque is now applied to the shaft and its speed increased above synchronous speed, the machine will act as a generator and power will be passed to the electrical supply.

At first sight this machine is very attractive for a wind energy scheme, on account of its ruggedness and simplicity, but the difficulties are that :

(1) it will only generate over a narrow speed range, the bottom of which is at the synchronous speed of the machine,

(2) it will not normally deliver power into a passive load, ie it requires connection to a supply network before power transfer can take place,

(3) for a given design, the magnitude of the torque curve is fixed by the voltage of the supply into which the machine generates, so matching to the torque characteristic of a turbine is difficult and does not allow efficient energy capture,

(4) the efficiency, while good at full load, falls off rapidly above and below full load and

(5) although the cage rotor is normally thought of as being very rugged, in applications where vibration levels and/or thermal cycling are high, cage fractures are a common failure mode.

2.3 The synchronous generator

The synchronous generator is the other ac machine commonly used in power generation and, as its name suggests, runs at a speed which corresponds precisely to the frequency of the supply. Constructionally, the synchronous machine is rather like the induction machine. It has a similar stator, with windings disposed in slots and carrying the load current. The rotor is different, having a field winding comprising a few pairs of coils, often on salient poles - rather like an outside-in version of a dc machine stator. This winding is supplied either by a pair of carbon brushes running on slip rings or by an equivalent, brushless, device which transmits the power for the field across an airgap in an auxiliary machine coupled to the main generator.

It is this class of machine which is employed in power stations to generate electricity for the national grid. The term "synchronous" is used because the frequency of the alternating voltages and currents in the machine correspond exactly to the speed of the rotor. If, therefore, the generator is connected to a supply network, its rotational speed has to be exactly constant. If the generator is used in a stand-alone system, it can be driven at any convenient speed, with a corresponding change to the output frequency.

The basic operating characteristic of the synchronous generator is relatively simple: if the machine is considered to be connected to a supply and to have the rotor running at the synchronous speed which corresponds to the frequency of the supply, then the rotor will supply power to any load which tries to slow it down and absorb power from any prime mover which tries to speed it up. (This characteristic only applies, of course, within a working range – too great a torque will cause the rotor to pull out of synchronism with the supply frequency and stable running will then be lost.)

It will be realised that, although synchronous generators have been widely used in wind-energy schemes (primarily because of their good efficiency at full load, but also because of a lack of any real opposition from dc and induction machines) their operating characteristics are not particularly suitable. The machine must run at a fixed speed if it is connected to an ac supply and this speed has to be held within a very small tolerance while the machine is being connected. The wind turbine is therefore constrained to run at a fixed speed.

2.4 Variable frequency inverters

In the above descriptions of the two ac machines, the importance of the supply frequency is evident. This frequency largely dictates the operating speeds of the generators. Means have been sought to use electronic devices and circuits to allow these machines to run at variable speeds (as convenient to a wind turbine) and yet provide a constant frequency of output. In both cases, this can be done by taking the natural, alternating-voltage output of the machine, rectifying it to a direct voltage and then, as was described in Section 2.1 for the dc machine, inverting this voltage to an alternating voltage of the desired frequency. In the case of induction generators, however, this process introduces other technical problems. The added complexity and cost of these systems do not appear to give any advantage in efficiency of energy conversion, so there is little prospect of their widespread use.

3 THE SWITCHED RELUCTANCE GENERATOR

The switched reluctance (SR) machine is not a recent invention – its use has been recorded as early as 1842

– but only recently have its full capabilities been realised. The late arrival of drives based on the SR machine is due to relatively recent advances in other fields. The first of these is the availability of high-performance, digital computers to aid the electromagnetic design work. The second, and more important, is the availability of solid-state power switches (transistors and thyristors) used to control the currents in the machine.

The SR generator has several attributes which make it ideally suited to wind energy schemes. These include extreme robustness, easy matching to frequency-wild turbines, high efficiency of energy conversion and ability to work over very large speed ranges.

3.1 Construction

Fig 1 shows the cross-section of one form of an SR machine : both the rotor and stator have salient poles; and coils on diametrically opposite stator poles are connected in series to form "phase" windings. The machine shown has four phases, though many other configurations are possible: the choice of configuration and phase number depends largely on the application. Fig 2 shows the stator of a 50kW machine and illustrates the simplicity of the windings. The rotor of this machine is shown in Fig 3, together with examples of smaller SR rotors. Note the complete absence of any rotor windings or magnets and the consequent ability to tolerate mechanical abuse.

The principle of operation is also simple in concept: it can be seen that if the winding on poles D-D′ is energised, the rotor, if free to move, will be pulled clockwise into line with the axis of the magnetic field. If now the winding on poles A-A′ is energised, the rotor will continue to move in the same direction. Continuous motion is produced by monitoring the position of the rotor with an angular position sensor and switching the current on and off at the correct times. The machine is acting as a motor converting electrical energy into rotor movement. Generator action is obtained simply by changing the angular position at which current is applied to the phase winding. If the rotor is moving, say, clockwise from the position shown and poles B-B′ are energised, generating torque will be produced. All of this is achieved by correctly timing the connection of the windings to the supply. Further details of the design and operation of SR machines can be found in references 3 and 4.

It can be seen, therefore, that the switched reluctance machine requires a power converter (which switches the winding currents on and off) and a control system to synchronise the switches to the rotor position and produce the required performance. This is illustrated in Fig 4.

3.2 Power electronics

The power electronics consists essentially of two parts. The first part, the converter, switches the currents in the machine and is therefore an integral part of the SR drive. Solid-state switches are invariably used, eg thyristors, GTO's or transistors. Many different circuits are possible, depending on the application, and an important advantage of the SR drive is the simplicity of the power converter. This arises from the fact that (unlike ac machines) the windings need carry current in one direction only. Thus only one switch is required per phase. Furthermore, the switches are all in series with motor windings, resulting in a fault-tolerant system which is easily made self-protecting.

The study of the power converter is a very full subject in its own right and has only been outlined here: a fuller treatment is to be found in References 5-7.

The second part of the power electronics is a standard 3-phase inverter which converts the dc output of the switched reluctance generator into an ac output to the power system. If a dc output is required, this stage may, of course, be omitted.

3.3 Control system

The control system controls the power switches in synchronism with the rotation of the rotor. This is done in such a way that the machine always runs at its best efficiency - whatever the speed and load. Correctly designed SR systems exhibit a remarkable ease and range of control in both motoring and generating modes. It is this flexibility of control (accompanied by the simplicity of the power converter and the rugged nature of the generator) which makes the switched reluctance machine ideal for wind energy systems.

The SR drive has no fixed power/speed characteristics - because the control system is able to turn on and turn off the switches which allow current into the phase windings, any characteristic (within the thermal rating of the machine) can be produced. In this respect, the SR drive can be considered to be the equivalent of a dc machine. The types of characteristic which can be achieved are discussed in the following section: while these are mainly associated with motoring, it must be emphasised that the SR machine, as the physical symmetry of its geometry implies, is as capable of acting in the generating mode as in the motoring mode.

4 OPERATING CHARACTERISTICS OF SR DRIVES

SR systems have been designed and built for a wide variety of applications - power outputs from 10W to 200kW and speed ranges up to 10000 rev/min. Design studies have included much larger sizes, up to several megawatts. This Section gives examples of some areas where the technology has already been applied and discusses the suitability of SR generators for wind energy schemes.

214

4.1 Examples of applications of SR systems

SR drives are commercially available for use as general-purpose, variable-speed drives in a range of sizes from 4-22kW and these have performance characteristics typically as shown in Fig 5. This shows that the torque can be controlled up to 110% of its rated value between 0 and 1500 rev/min and that, in addition, 150% torque is available at standstill for a few seconds, to cope with high stiction loads. Note that the efficiency of the drive remains high over a large working range of both torque and speed. These drives are applied to a very wide range of products - chemical processing machinery, wire drawing equipment, textile machinery, fan and ventilation systems, etc.

The flexibility inherent in the SR concept is demonstrated by the way in which a set of the general-purpose drives described above has been modified to provide a demonstration project for rail traction. GEC Traction, in conjunction with Blackpool Transport, have fitted to a modern tram 4 drives supplied by SR Drives Ltd. These drives are modified 22kW sets and their revised characteristic is shown in Fig 6. It should be emphasised that this characteristic has been obtained not by changing the motor in any way, nor by changing the basic power converter, but by changing the control scheme of the drive.

The work on all the drives described above grew out of a development project on SR traction drives for battery-electric vehicles. This was sponsored by Chloride Technical Ltd at the Universities of Nottingham and Leeds. Part of this work included the commissioning of a demonstrator vehicle which was used most successfully around Nottingham, mainly as a test-bed for control system development. The control scheme, part of which is shown in Fig 7, was 4-quadrant, ie both motoring and generating were available in either direction. The shape of this characteristic was produced purely by control of the pulses of current in the windings of the machine. The regenerated energy was fed back to the lead-acid battery. Although only an early prototype, the vehicle was very manoeuvrable and again demonstrates the flexibility of control of SR systems. A 50kW drive, designed for a 7.5 tonne vehicle, had the power/speed characteristic shown in Fig 8. Not only is the peak efficiency remarkably high, but stays within 3% of this peak over 50% of the power/speed plane. At the power level shown, these efficiencies are the equal of a dc machine (normally regarded as unmatchable!) and are better than any other ac drive.

4.2 Suitability for wind energy schemes

The problem of extracting the maximum amount of energy from a wind turbine has been studied by many authors and there is general agreement on the following points:

(a) The turbine should at all times be operated at, or near, the optimum tip-speed ratio, Cp_{max}. This immediately implies variable-speed (frequency-wild) operation.

(b) When this criterion is followed, the power developed by the turbine is proportional to the cube of the wind velocity and leads to maximum energy extraction. This is elegantly stated by Buering and Freris (8).

(c) Because of the temporal distribution of wind speeds, the turbine will spend much of its time in the lower part of its speed range, where, because of the V^3 relationship, it will only develop a small fraction of its rated power. It is therefore very important to have a generator which has high efficiency at these low speeds.

The particular features of SR systems which make them appropriate to wind energy schemes are summarised under the following points.

(1) Being variable-speed systems, they are capable of being controlled to load the turbine so that its speed is held at any desired operating point. This speed can be selected by reference to a wind-speed monitoring device, so that operation at the optimum value of Cp is always achieved. This principle has been outlined, for induction generators, in Ref 9. This enables turbines to run at Cp_{max} regardless of wind speed (leading to increased efficiency). Furthermore it relaxes the constraint on turbine designers to produce turbines with flat-topped Cp characteristics and allows them to concentrate on raising the absolute value of Cp_{max}. The increase in energy yield will vary between installations and will depend on several parameters, not least the maximum working speed of the turbine. Simple calculations which compare the outputs of fixed-speed and variable-speed systems for a typical wind pattern indicate that up to 30% more energy would be extracted by working the turbine over a wide speed range. Even if the top speed of the variable-speed system were limited to that of the fixed-speed system, some 15% improvement would be expected.

(2) They maintain very high values of efficiency over very wide working ranges and hence can give high rates of energy extraction over a wide range of turbine speeds.

(3) They have a rapid response to a change in control signals. Because the control system is continuously monitoring both the rotor position (and hence speed) and the currents in the windings, corrective action (eg because of a change in wind conditions) can be taken within a fraction of a revolution.

(4) They are exceptionally rugged, having no winding at all on a very simple rotor.

(5) They are brushless and require little or no routine maintenance.

(6) They have no significant speed limitations, so that, if required, any gearing up from the turbine could be high and would give a corresponding reduction in the machine size. This would reduce space and strength requirements in the nacelle.

(7) They also offer the possibility of gearbox elimination, though this would result in an increase in generator size.

(8) They have a low capital cost (due to their inherent simplicity both in machine construction and electronic circuitry), a low maintenance cost (due to 4 and 5 above) and a high return (due to their exceptionally high operating efficiency).

5 CONCLUSIONS

The characteristics and qualities of an SR generator make it well-suited to wind energy schemes. The ease of control and range of operating speeds yield a generation scheme which matches the requirements of a frequency-wild turbine having a non-linear power output. They are of rugged construction; they exhibit high efficiencies over a wide working range; they have fast rates of response to external disturbances; and they are economic to install and operate. SR systems have already been applied in many industrial applications and it is confidently expected that they will be as successful in wind energy schemes.

6 ACKNOWLEDGEMENTS

The contribution of Mr J C Riddell in discussion and guidance is gratefully acknowledged. The authors are part of a team, now SR Drives Ltd, which has been involved in the development of SR systems for some years and they are grateful for the help and contributions given by various members.

7 REFERENCES

1 Freris, L L : A plain man's guide to electrical aspects of WECS, Proc BWEA Conf, Oxford, 1985, pp 299-304

2 Gipe, P : Wind energy - how to use it, Stackpole Books, 1983

3 Lawrenson et al : Variable-speed switched reluctance motors, Proc IEE, Vol 127, Pt B, 1980, pp 253-265

4 Stephenson, J M & Čorda, J : Computation of torque and current in doubly salient reluctance motors, Proc IEE, Vol 126, No 5, 1979, pp 393-396

5   Ray, W F & Davis, R M : Inverter drive for doubly salient reluctance motor, IEE J EPA, 1979, pp 185–193

6   Davis et al : Inverter drive for switched reluctance motor, Proc IEE, Vol 128, Pt B, 1981, pp 126–136

7   British Patents   22891, 13415, 13416, 22892, 22893

8   Buering, I K & Freris, L L : Control policies for wind–energy conversion schemes, Proc IEE, Vol 128, Pt C, 1981, pp 253–261

9   Wertheim, M M & Herbermann, R J : Wind turbine maximum power tracking device, US Patent 4,525,633, June 1985

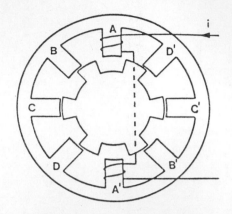

Fig 1   SR machine showing a phase winding

Fig 2   Typical stator construction of an SR machine

Fig 3   Selection of rotors from SR machines ranging from 50kW @ 750 r/min to 10 W @ 10 000 r/min

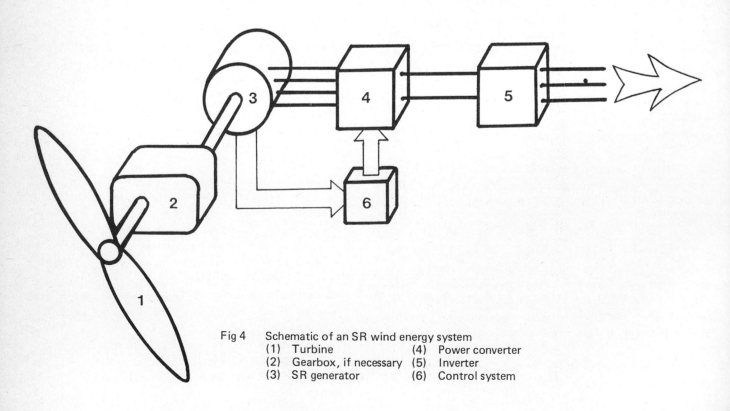

Fig 4   Schematic of an SR wind energy system
          (1)   Turbine   (4)   Power converter
          (2)   Gearbox, if necessary (5)   Inverter
          (3)   SR generator   (6)   Control system

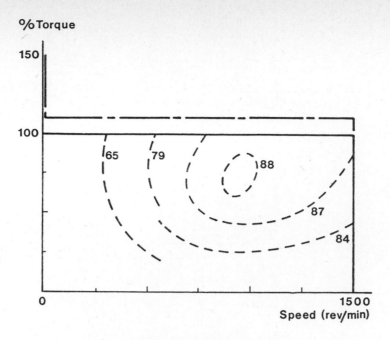

Fig 5   Performance characteristic of an 'Oulton' SR drive,
         manufactured by Tasc Ltd. 22kW output at 1500
         r/min, 3-phase, 415 V, 50 Hz supply
         ———————— Rated torque
         ——— · ——— Overload rating
         — — — — Constant efficiency contours

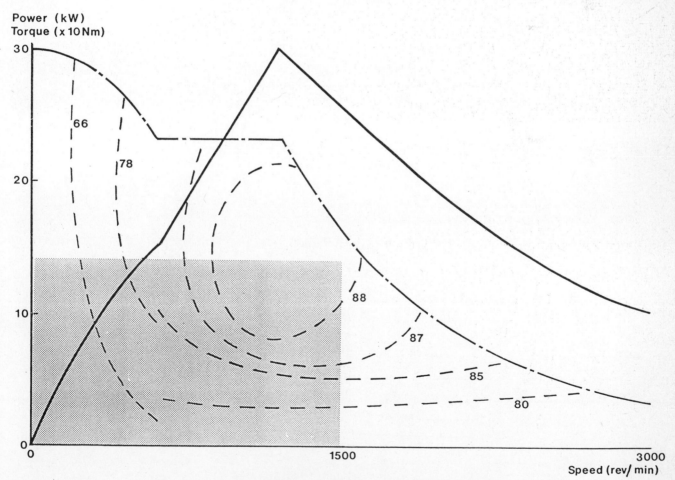

Fig 6   Enhanced output from 22 kW drive. 575 V dc supply
         ——— · ——— Torque
         ———————— Power
         — — — — Constant efficiency contours
         The shaded area represents the original torque/speed
         envelope

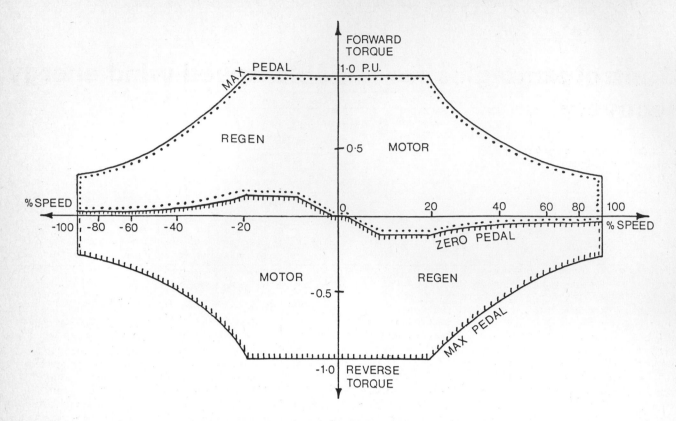

Fig 7   Four quadrant torque/speed control envelope for
SR-powered vehicle

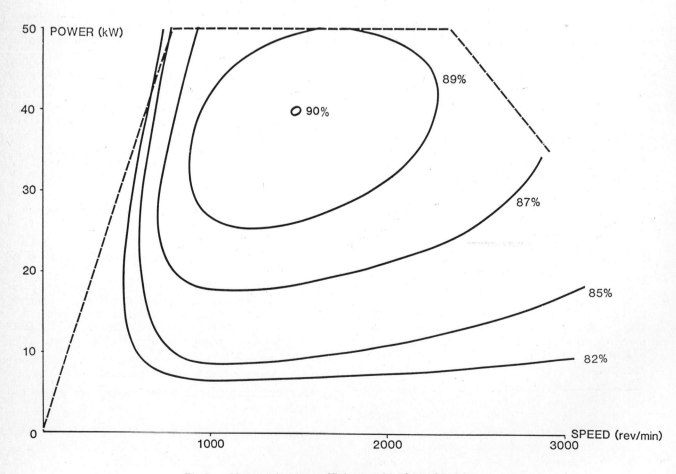

Fig 8   Measured system efficiency of a 50 kW SR drive

Control strategies for variable-speed wind energy recovery

D GOODFELLOW and G A SMITH
University of Leicester
G GARDNER
CEGB

1.0 INTRODUCTION

This paper describes control strategies for variable-speed wind energy generators considered as part of an on-going research activity. The research has concentrated on the application of doubly fed slip-ring induction generators in which the control of slip-energy recovery allows variable-speed generation into the national grid network. Three types of power electronic conversion equipment to control the slip-energy have been designed, constructed and tested in the laboratory. These methods are based on two techniques,

(a) the d.c. link inverter
  (i) Kramer system: generation over the range synchronous speed to twice synchronous speed.
  (ii) Scherbius system: generation from virtually standstill to twice synchronous speed.

(b) the cycloconverter
  (i) Scherbius system: generation over the range $\pm$ 20% about synchronous speed.

In these systems the power conditioning equipment need only be rated for the slip-energy being recovered which can be considerably less than the total power being delivered to the grid. The generator torque is electronically controlled to provide the desired operating torque:speed characteristic.

Before a test programme could be finalised an investigation was necessary into the relative merits of various alternative operating strategies. When comparing these strategies it is necessary to consider the rating limits of the turbine. The maximum rotational speed of a wind turbine is limited by constraints of tower vibration, blade stressing and gearbox rating. (The maximum torque in the system will depend again on mechanical design limitations.) There may be a possibility of limited short-term overspeeding and torque overloads.

The electrical generator and the slip energy recovery equipment must be carefully chosen and rated to meet the operating conditions present in the strategy being considered.

Some benefits of variable-speed operation are generally assumed to be:-

(i) increased energy capture,
(ii) reduced stressing of components,
(iii) greater operational flexibility.

In this paper the benefits are assessed for five operating strategies using a cycloconverter slip recovery system which will allow $\pm$ 20% variation about synchronous speed.

2.0 OUTLINE OF THE CONTROL STRATEGIES

When turbines operate with variable-speed the effect of system inertia must be taken into account when considering the dynamic behaviour. Further, the variable-speed slip-energy systems control torque as a function of speed. Therefore it is necessary to consider the operation of the system with reference to the torque-speed characteristics of the turbine. These characteristics can be derived from the C_p curve for the turbine being investigated.

For a particular wind speed the power and torque characteristics are shown in Fig.1. A family of such curves can be drawn for a range of wind speeds as shown in Fig.2. The points of maximum C_p occur along a square-law torque characteristic.

The control strategies can be considered in two groups:

(i) Power limitation by pitch angle control. Three methods are considered and shown in Fig.3.
  (a) Fixed speed variable pitch.
  (b) Variable speed variable pitch.
  (c) Variable speed at max C_p up to power rating.

  Note that mode (c) exceeds the speed of the other systems.

(ii) Power limitation by stalling. Three methods are considered and shown in Fig.4.
  (a) Fixed speed fixed pitch.
  (b) Variable speed fixed pitch, speed reduction at constant torque.
  (c) Variable speed fixed pitch, speed reduction at constant power.

  Note that (c) exceeds the continuous torque limit of the other systems.

More detailed torque speed curves are shown in Figs. 5 and 6.

3.0 THEORETICAL BASIS FOR PERFORMANCE COMPARISON

3.1 Energy capture calculations

The energy captures have been calculated under steady state conditions, assuming that the turbine instantly achieves its operating point for any particular windspeed. The results are obtained by calculating the number of hours per year a particular windspeed occurs, and multiplying this value by the power in Watts obtained by the turbine at that windspeed. Summing the results over the range of operating windspeeds gives the number of Whrs/annum obtained by the turbine. The wind model is a Rayleigh distribution:

$$p(>\nu) = \exp\left[-\frac{\pi}{4}\left(\frac{\nu}{\nu_m}\right)^2\right] \qquad (1)$$

where $p(>\nu)$ is the probability that the windspeed exceeds ν and $\nu_m = \nu_{mean}$.

The ν_{mean} used in this report is 9.0 m/s ($\simeq 20$ mph). The mechanical power output is given by the torque x rotational speed for each windspeed in the mode of operation being considered. These comparisons have been made for a 40 m radius turbine with an electrical rating of 2.5 MW. The electrical output of the turbine is calculated with the following equation with running efficiency and standing losses typically associated with a 2.5 MW machine.

$$Pe(\nu_w) = 0.97\ Pm(\nu_w) - 55 \times 10^3$$

where, $Pe(\nu_w)$ = electrical power at windspeed ν_w.

$Pm(\nu_w)$ = mechanical power at windspeed ν_w.

The total energy captured per annum is given by the following

$$E = \int_{\nu_{wi}}^{\nu_{wo}} Pe(\nu_w)\ p(\nu_w)\ 8766\ d(\nu_w)\ \text{Whrs/annum}$$

ν_{wi} = cut-in windspeed (5 m/s)

ν_{wo} = cut-out windspeed (30 m/s)

8766 = number of hours/year

The energy capture calculations were based on C_p curves derived from the Nibe system. Curves for $-1°$ and $-4°$ pitch angle were considered but the energy capture figures are always greater for $-1°$ pitch and so these figures were used in the comparison of operating schemes.

Most wind turbines are fixed-speed variable pitch machines and so this mode of operation was used as the datum for comparison with other modes. The fixed speed for this case was chosen to optimise the energy capture for this wind regime (eqn.1), C_p curve and power rating. The calculations showed this speed (ω_1 in Figs. 3 and 4) to be 1.8 rad/s.

3.2 Torque rating calculations

In addition to consideration of the energy recovery it is important to determine the torques developed at the generator for different operating modes.

The torque rating under normal operating conditions is calculated, and also a figure for the maximum intermittent torque seen due to gusting or rapid ramping wind conditions occurring at rated power. For the variable speed systems a figure for the inertia of the blade system is required, in this analysis an inertia of 22.5 Kg m^2 is used (c.f. Boeing MOD2, r=45 m, J=28.2 Kg m^2).

The gust parameters have been obtained from ref.(1), recommended as an estimation of discrete gust patterns. The gusts are modelled as (1-cosine) excursions in wind speed, and the amplitudes are calculated from the equation given in the reference.

4.0 THE CONTROL STRATEGIES

The energy captures for each of the following control strategies are given in Table 1.

4.1 Fixed Speed Variable Pitch (F.S.V.P.)

This type of wind turbine operates with a synchronous or low slip induction generator. The power limitation is achieved by altering the pitch of either the whole blade or part of the blade. Compared with the fixed pitch system this can operate at a higher rotational speed (see Figs.3 and 4), which will bring the C_p max operating point up to a higher wind speed. The extra available power at even higher wind speeds is limited to 2.5 MW by the pitch control mechanism. Power limiting is shown at point r in Fig.5, curve a. This technique increases the energy capture for this method over the fixed pitch system.

The control strategy for the pitch operation can either have power, torque, or incident windspeed as its input. The pitch mechanism is usually a hydraulic system requiring connections along the blade itself and through the shaft hub. Partial blade angle control also implies a mechanical weakness at the pivoted connection. There will be a maximum operating rate at which the pitch mechanism can deploy. During operation at rated power there is a possibility of both torque and power exceeding their continuous rating due to gusting. The pitch mechanism may be fast enough to limit the excess torque fluctuations produced by low ramp rate gusting, but not to limit the effects of rapid gusting. Blade angle control to limit gusting will greatly increase the wear on the control mechanism.

The load torque rating is obtained from the effect of a 2 second time period gust occurring at rated power. It is assumed that for longer period gusts the variable pitch mechanism will be able to smooth the torque pulsations seen by the blades. Operation with a synchronous generator, in which the speed is not allowed to rise, implies that the torque developed by the generator measured at the low speed shaft must equal the torque developed by the wind. Operation with high slip and low slip induction generators, will allow limited amounts of speed increase which will limit the torque excursion seen by the generator to a greater or lesser degree. These generators have been modelled simply as loads which operate upon steep torque-speed curves. The gradients used are 70 and 50 MN m/rads-1 for the low and high slip generators respectively, each having a rated running torque of 2.5 MNm.

4.2 Variable Speed Variable Pitch (V.S.V.P.)

In this mode of operation the maximum rotational speed is set at the speed of the fixed speed system above. The torque-speed characteristic is shown in Fig.5, curve b, and shows that this system achieves maximum C_p over a range of windspeeds up to the windspeed at which the fixed speed system also attains max C_p (point S). Thereafter the control scheme will produce an almost vertical torque gradient which will mimic the action of a fixed speed machine. This gradient allows speed increase to occur and so provides for some torque damping at the generator. However, inevitably gusting will produce torque and power fluctuations for this steep torque operating region just below rated windspeed.

The control system can maintain the steep torque curve above the rated operating point and so some of the torque produced by gusting wind conditions during operation at rated power can be absorbed as an increase in rotational stored energy as the speed increases. During this time the pitch mechanism which has a limited speed of response is also operating. The gradient of this torque line can be set electronically to give either a large torque increase with small increase in speed or vice versa.

In the case presented here, the inertia of the turbine is so large that the torque can simply be held at the rated value and for a 2 second gust the speed does not increase excessively, as can be seen in Table 1. If the variable pitch were not to be used for suppression of long period gusts then this electronically controlled torque characteristic would have to be of some definite gradient in order to prevent excessive speed increase.

4.3 Variable Speed at max C_p to Rated Power (V.S.Max C_p)

Figure 5, curve b, shows the variable speed variable pitch scheme outlined previously, which was limited, for the purposes of this report to 1.8 rad/s maximum rotational speed. As can be seen, the torque speed characteristic upon which the generator would have to operate has abrupt changes, due to the requirement of not exceeding the maximum speed limit. A much simpler control scheme would be to allow the turbine to continue increasing speed on the maximum C_p line until it reached maximum power shown in Fig.5, curve c. This would not require the pseudo fixed speed range at 1.8 rad/s and the torque speed characteristic is greatly simplified. Gust limitation could be achieved by merely extending this curve beyond the new speed limit.

This system would therefore need to have its maximum speed set at the speed where the blades are operating at maximum C_p and the power reaches its limit, point p on the diagram. For the energy capture calculations the system is assumed to produce power at max C_p until rated power is reached. High wind speeds again require aerodynamic action to restrict a further increase in speed. As the generator control system continues to provide operation along the $C_{p\,max}$ curve then the increase in speed can be used to initiate and control the aerodynamic braking method. As long as the speed is limited, the electronic control will ensure that the torque produced maintains power at 2.5 MW.

This method of power limitation could be a simpler form of control than the previously mentioned systems. Possibilities are tip spoiling and centrifugally operated flaps (which may already be used to prevent overspeeding in emergencies). It may be possible to devise a system whose operation is a compromise between this system and a fully variable pitch system, in which the variable speed range would allow the use of a less complex variable pitch mechanism.

If we assume that there is no method of limiting the gusting torque fluctuations aerodynamically then the load torque rating must be calculated for a 10 s gust. However, the speed limiting system may be able to suppress these gusts and so a 2 s gusting figure is also given for comparison. The torque-speed characteristic for

gusting conditions at rated power is simply an extension of the square-law $C_{p\,max}$ curve.

4.4 Fixed Speed Fixed Pitch (F.S.F.P.)

In this mode of operation the turbine is either connected to a synchronous generator, or a low slip induction generator. These essentially hold the speed constant, operating at $C_{p\,max}$ at only one windspeed. The power rating is reached as the blade stalls and so rated power occurs at only one windspeed. The torque-speed characteristic for this system is shown in Fig. 6, curve a. Due to the broad C_p characteristic used in these systems there is a large range of windspeeds between the $C_{p\,max}$ point and the stalling point (the distance r-s on the figure). This means that for a particular rated windspeed, the $C_{p\,max}$ is achieved at a low windspeed. This limits the energy capture of this system, but is widely used on small turbines due to its simplicity, requiring no control scheme.

For this study the rotational speed was set at 1.45 rad/s, to provide stall regulation at 2.5 MW. The load torque limit is then the torque at rated power (point r). Extra torque may be produced during rapid gusting at rated power due to the time taken for the blade to stall. The evaluation of this phenomena is, however, beyond the scope of this report.

4.5 Variable Speed with Speed Reduction

It is possible that by a suitable electrical operating scheme a variable speed system could avoid the need for blade angle control, with consequent saving in the cost of the blades. This must be achieved by reducing the rotational speed of the system for high windspeeds.

Below rated windspeed the control system for the method is the same as for the variable speed pitch system outlined in 4.2. When maximum speed is reached and the power limit is detected, either by measurement of power or torque, the control scheme must try to obtain an operating point at reduced rotor speed. Two systems are investigated here; speed reduction with constant torque, and speed reduction with constant power.

4.5.1. Speed Reduction at Constant Torque (V.S.C.T.)

From Fig.6, curve b, it can be seen that the power limit is reached at a windspeed of 12.5 m/s (point x). For an increase in windspeed to 14 m/s. a reduced speed operating point could be obtained at rated torque, corresponding to point y. In order to make the turbine slow down, the generator must provide a greater (loading) torque than the blade (accelerating) torque. When the windspeed is at 14 m/s and the turbine is still operating at its maximum rotational speed, the torque produced on the shaft by the wind is shown at t_g. The generator must provide a torque exceeding this value in order to reduce the turbine speed. The amount of extra torque required would depend upon the rate at which the system is required to slow down and reach its new operating point.

The turbine would operate on the constant torque line with its speed decreasing with increasing windspeed for windspeeds from 12.5 m/s to 15 m/s, and then after reaching the stall point at 15 m/s (point z), it would require to speed up for an increase in windspeed, eventually reaching its speed limit again at 26 m/s (point x). For further increase in windspeed the generator could be made to mimic a low slip induction generator, corresponding to a vertical line from x down to V. For all windspeeds during the speed reduction phase the generator is producing less than rated power, i.e. it is producing rated torque at less than rated speed.

4.5.2. Speed Reduction at Constant Power (V.S.C.P.)

This scheme would operate in a similar way to the system above, except it would reduce speed along a torque line as shown in Fig.6, curve c, which corresponds to operation at rated power.

This scheme would in steady state energy capture terms, produce the same energy capture as a variable speed variable pitch system. However it would require a higher continuous torque rating at z'.

4.5.3. Torque Rating for Speed Reduction Systems

For the non-speed reduction systems the torque produced by wind gusts at rated power at the generator can be suppressed by allowing an increase in speed. However in the speed reduction systems the speed increase must not be too great as the "gust" may resolve into a ramp increase, in which case the turbine will have to slow down again. The largest loading torques will occur during a rapid ramp increase in windspeed, when the generator must load the blades with a greater torque than the blades provide in order to slow the turbine down. The worst case has been chosen as a 2 m/s increase in windspeed in 2 seconds, as used in a MOD 2 test analysis, ref.(2), though it must be noted that in the event of increases >2 m/s the torque rating must be increased accordingly. The braking torque has been chosen so as to be able to reduce the speed to its new operating point in 10 seconds from the time of its application. For the constant power system this extra braking torque above that produced by the wind is 750 kNm, and for the constant torque system is 1200 kNm. This duty will be reflected in the cost of the speed control equipment though it may not incur any additional penalties for the generator or drive train.

5.0 TORQUE RATING/ENERGY CAPTURE RESULTS

These results have been obtained for a C_p curve derived from the Nibe system, with a radius of 40 m, and rated at 2.5 MW.

The energy captures per annum and the peak torque rating of each system are shown in Table 1.

The amount of energy capture per annum per unit windspeed is plotted in Fig.7 for the systems being considered and shows how the different control strategies are more or less effective at different windspeeds. Note in particular the very considerable energy loss of the VSCT system due to the speed reduction needed to keep the torque constant. The figures are calculated for 1 m/s increments in windspeed, but are plotted as a continuous line for clarity.

Only the VSCP achieves the same energy as the VSVP without incurring higher speed operation and this is only achieved with a possible 20% increase in torque rating, with additional control difficulties.

VS Max C_p achieves higher energy but at the expense of 17% higher operating speed which may result in a significant increase in the cost of the complete WTG installation.

6.0 EFFECT OF INERTIA

6.1 Energy Capture

The theoretical energy capture results for variable-speed wind turbines have been based upon the assumption that the turbine has negligible inertia and therefore instantly achieves its steady state operating point for each windspeed. In practice the system inertia will prevent this and so the dynamic results will differ from the theoretically derived figures.

The dynamic results have been obtained by running a real-time computer simulation of a wind turbine. Two second wind data was input to the simulation and the W-hours of energy generated calculated. The computer programme allowed the simultaneous comparison of performance for the following three operating situations.

Table 1

SECTION NO.	SCHEME	Steady State Conditions				Gusting Wind Conditions		
		ω (rads^{-1})	E (mWhrs/ann)	T (kNm)		worst case	T_m (kNm)	ω_m (rads^{-1})
4.1	F.S.V.P. { S.M. / 2% IM / 5% IM }	1.8	8932 (100%)	1463		{ 2 s Gust }	2632 / 2520 / 2276	1.8 / 1.815 / 1.827
4.2	V.S.V.P.	1.8	9226 (103%)	1463		2 s Gust	1463	1.865
4.3	V.S.Max C_p	2.15	9231 (104%)	1330		{ 10 s Gust / 2 s Gust }	1717 / 1456	2.387 / 2.198
4.4	F.S.F.P.	1.45	8119 (91%)	1811		–	1811	1.45
4.5.1	V.S.C.T.	1.8	8095 (91%)	1463		2 s Ramp	3243	1.81
4.5.2	V.S.C.P.	1.8	9226 (103%)	1811		2 s Ramp	2793	1.81

In the table for the fixed speed variable pitch system
S.M. refers to Synchronous Machine and IM refers to Induction Machines

(i) Fixed speed variable pitch

This is used as the datum for the comparison. The rotational speed is held constant at 1.8 rad s^{-1}. If the generated power is calculated to be greater than 2.5 MW then it is recorded as 2.5 MW on the assumption that it would have been so limited by the action of the pitch mechanism.

(ii) Variable speed variable pitch - zero inertia.

For all windspeeds for which the system could run at a rotational speed giving $C_{p\,max}$ then the C_p values was set instantly at $C_{p\,max}$. The steep torque curve preventing a speed increase above 1.8 rad s^{-1} was assumed to be a vertical line, i.e. giving the same generated power as (i) above. It should be noted that this corresponds to the way in which the theoretical energy figures in Table 1 were calculated.

(iii) Variable speed variable pitch - including inertia.

This time the turbine takes time to accelerate or decelerate towards its new steady state operating point. Once again the speed is limited by the steep torque: speed curve. The pitch controls limit power by limiting the speed (see Section 4.3). During acceleration below the speed limit, the blade torque may exceed the torques corresponding to 2.5 MW generation at rated speed.

Comparative results have been obtained for three different sets of wind data each containing 16 minutes of data at 2 second intervals with linear interpolation every 0.4 s. Results for three different wind regimes are shown in Table 2.

Table 2

Wind Sample	Fixed Speed	Control Strategy	Inertia (kg m^2 x 10^6)		
			0	10	50
1	155.01	VSVP	179.21	177.47	177.72
		Max C_p	179.21	177.46	177.47
2	343.11	VSVP	352.35	349.66	348.49
		Max C_p	353.89	352.88	350.08
3	573.4	VSVP	573.92	572.81	582.4
		Max C_p	580.5	590.13	610.81

all figures in kWhrs.

It should be noted that for the wind samples used the velocities are:

(1) 5 to 10 m/s - operation is mainly along the $C_{p\,max}$ curve.
(2) 6 to 15 m/s - operation is partly along the $C_{p\,max}$ curve with operation along the steep torque limit curve and power limiting region.
(3) 8 to 15 m/s - operation mostly in the power limiting region.

To ensure that the net transfer of energy to the rotating inertia is zero each computer run was arranged to begin and end at the same windspeed and steady state operating point.

Comparison of the actual energy obtained and the theoretical energy available shows that for windspeeds predominantly below the rating, a system obtains less than its theoretical energy capture. This is due to the time taken for the turbine speed to reach its new operating point, during which it is not operating at maximum C_p. However, when the windspeeds are fluctuating

about the rated speed, the effect of an inertia is to hold the system at the rated point, and not allow the turbine to slow down during short-term drops in windspeed. This seems to enable the turbine to obtain more than the theoretically available energy.

6.2 TORQUE PULSATIONS

Although the inertia does not significantly affect the energy captures, its effect upon the torque fluctuations seen at the generator is marked.

Figure 8 shows the blade and variable-speed generator torques for a $\approx$ 10 minute wind sample. If the generators used were restricted to a fixed speed operation, then it would have to accept the complete fluctuations in blade torque, with corresponding power fluctuations back into the grid. From the figure it can be seen that allowing a speed change reduces the torque fluctuations seen by the generator quite considerably and so also reduces the power fluctuations into the grid. Note that in the V.S.V.P. case shaft torques greater than $\approx$ 0.75 MNm corresponds to operation on the steep torque region, and so have a greater fluctuation. This steep torque characteristic is not present in the 'max C_p' case.

7.0 DISCUSSION

Various studies (refs. 3, 4) have concluded that the advantages of variable speed wind energy generation are

(i) Increased energy capture.
(ii) Reduction in intermittent torque levels on the generator, and hence reduction of power fluctuations into the grid.
(iii) Reduction in stressing of the structure.
(iv) Flexibility of operation for differing site conditions.
(v) Motoring from startup/assisted braking during shutdown.
(vi) Operation at demanded C_p, e.g. operation with constant power output.

This work has concentrated on points (i) and (ii) for assessing different operating strategies applied to an existing wind turbine. However, some implications for the other points have been obtained.

(i) The extra energy captured by the variable speed facility calculated for a Rayleigh distribution wind regime is 3 to 4% per annum (see Table 1). Over a range of windspeeds at which $C_{p\,max}$ operation is possible, the extra energy captured can be approximately 15% (Table 2, wind sample 1).
(ii) Reduction of intermittent torque levels depends on the ability to absorb blade input torque into the inertia as an increase in speed (see Fig. 8). This will depend on the proposed operating strategy and its associated speed limit. Continuous and peak intermittent torque levels are shown in Table 1.
(iii) Reduction of stressing on the turbine structure will depend on the torque and rotational speeds the system will have to accept. The assessment of these effects, however, is beyond the scope of this report.

(iv)   Whereas fixed speed machines obtain $C_{p\,max}$
         at only one windspeed, and are thus opti-
         mised for one particular site, variable
         speed schemes obtain $C_{p\,max}$ over a range
         of windspeeds and so can be used to advan-
         tage over a range of site conditions.

(v)   The slip recovery scheme outlined in this
         report will allow motoring from start-up.
         Electrical braking may be used to supple-
         ment other braking techniques. This would
         require the generator torque, as controlled
         by the slip energy recovery equipment, to
         be greater than the accelerating torque
         from the blades. This may be possible
         during low windspeed conditions. However,
         shutdown in this way during high windspeeds
         will require the generator to provide very
         large torques and the slip recovery dev-
         ices will have to be rated to accept the
         resulting high currents.

(vi)   Operation at any desired λ (or C_p) can be
         achieved by means of the appropriate
         torque-speed characteristic. Control of
         speed to produce constant power output is
         equivalent to the speed reduction with
         constant power scheme in the report
         (Section 4.5) which may be difficult to
         implement.

8.0 CONCLUSIONS

Despite the relatively low improvement in energy
capture (3-4%), variable speed operation has
sufficient advantages to merit consideration for
the next generation of wind turbines. This study
has shown the importance of including all operating features in a realistic assessment. Designs which allow increases in speed or torque or
power to increase energy capture are liable to
inclur added costs which may not be justified by
the extra energy gained. The reduction in severe
torque fluctuations may prove to be the main
advantage of variable speed operation. Of the
five systems considered the technique giving
operation at max C_p to the power rating (Section
4.3) has the best overall potential. The turbine
must be capable of operation of about 20% overspeed but this will give the best energy capture,
greatest smoothing of torque fluctuations and can
be implemented by a relatively unsophisticated
control system.

9.0 REFERENCES

1)   Synchronization of wind turbine generators
       against an infinite bus under gusting wind
       conditions. Hwang, H.H. & Gilbert, L.T.
       IBEE PES Summer Meeting, Mexico City,
       July 17-22, 1977.

2)   Stability simulation of wind turbine systems.
       Anderson, P.M. & Rose, A. IEEE PES 1983.
       Summer Meeting, Los Angeles, California,
       July 17-22, 1983.

3)   Variable rotor speed for wind turbines:
       objectives and issues. Hinrichson, E.N.
       Power Technologies, Inc., New York.

4)   Application of broad range variable speed
       generators to large horizontal axis wind
       turbines. Doman, G.S. United Technolo-
       gies Corp.

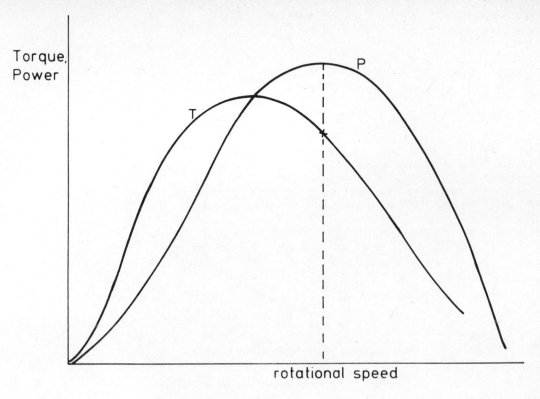

Fig 1

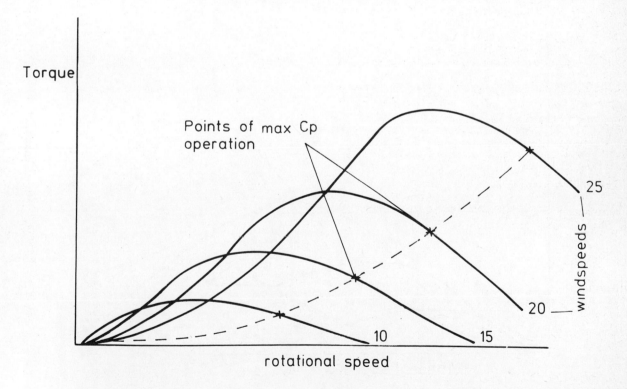

Fig 2

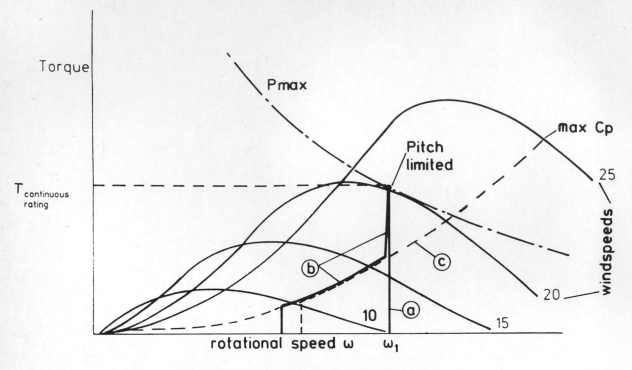

Fig 3   (a)   Fixed speed variable pitch (41)
            (b)   Variable speed variable pitch (42)
            (c)   Max. Cp to power rating (43)

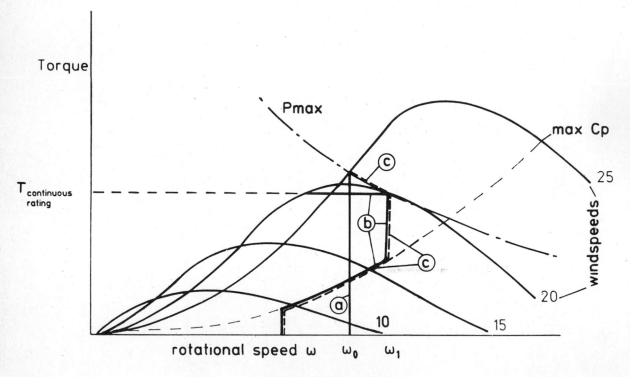

Fig 4   (a)   Fixed speed fixed pitch (44)
            (b)   Constant torque speed reduction (451)
            (c)   Constant power speed reduction (452)

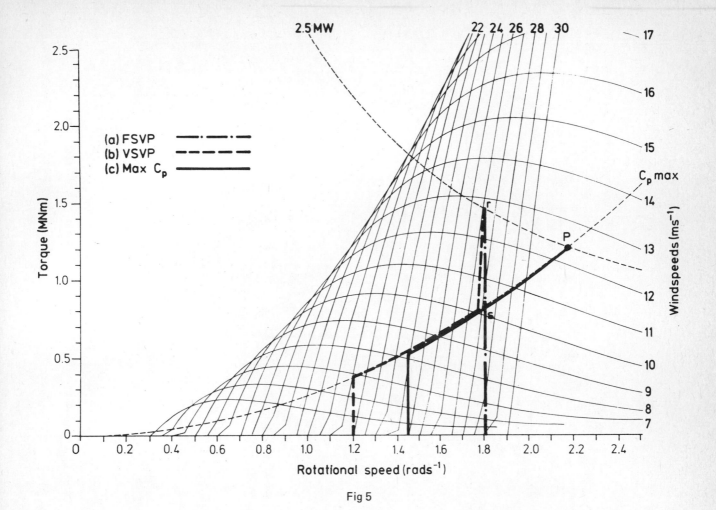

Fig 5

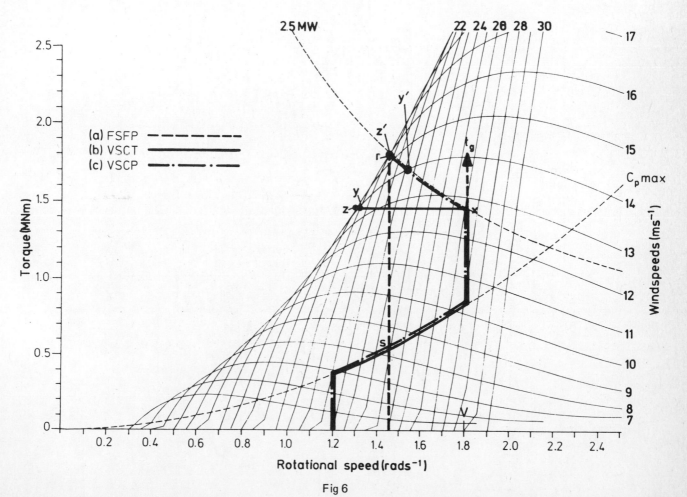

Fig 6

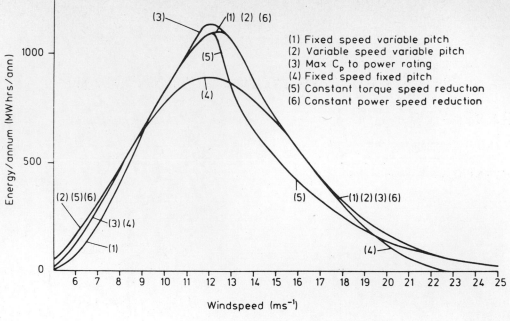

(1) Fixed speed variable pitch
(2) Variable speed variable pitch
(3) Max C_p to power rating
(4) Fixed speed fixed pitch
(5) Constant torque speed reduction
(6) Constant power speed reduction

Fig 7

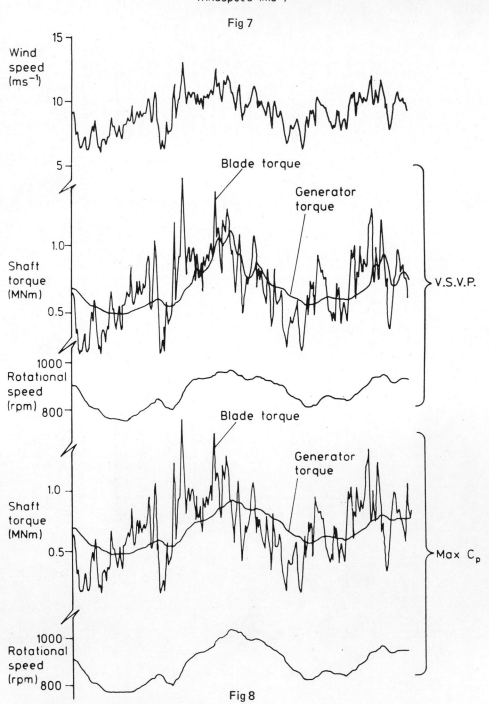

Fig 8

Layman's guide to the aerodynamics of wind turbines

D J SHARPE
Department of Aeronautical Engineering, Queen Mary College, University of London

1 INTRODUCTION

A wind turbine is a device for extracting kinetic energy from the wind but a sudden step change in velocity is neither possible nor desirable because of the enormous accelerations and forces this would require. Pressure energy can be extracted in a step-like manner, however, and all wind turbines, whatever their design, operate in this way. The turbine first causes the approaching air gradually to slow down which results in a rise in the static pressure. Across the turbine swept surface there is a drop in static pressure such that, on leaving, the air is below the atmospheric pressure level. As the air proceeds downstream the pressure climbs back to the atmospheric value causing a further slowing down of the wind. Thus, between the far upstream and far wake conditions, no change in static pressure exists but there is a reduction in ordered kinetic energy (Fig. 1.1)

1.1 The Actuator Disc

The mechanism described above accounts for the extraction of kinetic energy but in no way explains what happens to that energy: it may well be put to useful work but some may be spilled back into the wind as turbulence and eventually be dissipated as heat. Nevertheless, we can begin an analysis of the aerodynamic behaviour of wind turbines just by considering the energy extraction process. The general device which carries out this task is called an ACTUATOR DISC.

The streamtube which just contains the disc has a cross-sectional area smaller than that of the disc upstream and an area larger than the disc downstream. This is because the mass flow rate must be the same everywhere within the streamtube.

$$\rho_\infty A_\infty V_\infty = \rho_D A_D V_D = \rho_W A_W V_W \qquad 1.1$$

It is usual to consider that the actuator disc induces a velocity variation which must be superimposed on the free stream velocity. The streamwise component of the induced flow at the disc is given by aV, where a is called the axial flow induction factor, or the inflow factor. At the disc, therefore, the streamwise velocity is

$$V_D = (1-a)V_\infty \qquad 1.2$$

1.2 Momentum Theory

The air which passes through the disc undergoes an overall change in velocity, $V_\infty - V_W$ and a rate of change of momentum equal to the overall change of velocity times the mass flow rate.

$$\text{Momentum rate} = (V_\infty - V_W)\rho_D A_D (1-a) V_\infty \qquad 1.3$$

The force causing this change of momentum comes entirely from the pressure difference across the actuator disc because the streamtube is otherwise surrounded by air at atmospheric pressure, which gives zero nett force.

Therefore $\quad (p_D^+ - p_D^-)A_D = (V_\infty - V_W)\rho_D A_D (1-a)V_\infty \qquad 1.4$

To obtain the pressure difference $(p_D^+ - p_D^-)$ Bernoulli's equation is applied separately to the upstream and downstream sections of the streamtube: separate equations are necessary because the total energy is different upstream and downstream. Bernoulli's equation states that, under steady conditions, the total energy in the flow, comprising kinetic energy, static pressure energy and gravitational potential energy, remains constant provided no work is done on or by the fluid. Thus, for a unit volume of air,

$$\tfrac{1}{2}\rho V^2 + p + \rho g h = \text{const} \qquad 1.5$$

Upstream, therefore, we have

$$\tfrac{1}{2}\rho_\infty V_\infty^2 + p_\infty + \rho_\infty g h_\infty = \tfrac{1}{2}\rho_D V_D^2 + p_D^+ + \rho_D g h_D \qquad 1.6$$

Assuming the flow to be incompressible $(\rho_\infty = \rho_D)$ and horizontal $(h_\infty = h_D)$ then,

$$\tfrac{1}{2}\rho V_\infty^2 + p_\infty = \tfrac{1}{2}\rho V_D^2 + p_D^+ \qquad 1.6a$$

Similarly, downstream

$$\tfrac{1}{2}\rho V_W^2 + p_\infty = \tfrac{1}{2}\rho V_D^2 + p_D^- \qquad 1.6b$$

Substracting these equations we obtain

$$p_D^+ - p_D^- = \tfrac{1}{2}\rho(V_\infty^2 - V_W^2)$$

Equation 1.4 then gives

$$\tfrac{1}{2}\left(V_a^2 - V_w^2\right) = \left(V_\infty - V_w\right)\left(1 - a\right)V_\infty \qquad 1.7$$

and so

$$V_w = \left(1 - 2a\right)V_\infty \qquad 1.8$$

That is, half the axial speed loss in the streamtube takes place upstream of the actuator disc and half downstream.

The force on the air becomes, from 1.4

$$F = \left(P_D^+ - P_D^-\right)A_D = 2\rho A_D V_\infty a(1-a) \qquad 1.9$$

As this force is concentrated at the actuator disc the rate of work done by the force is F.V and hence the power extraction from the air is given by

$$\text{Power} = 2\rho A_D V_\infty a(1-a) \qquad 1.10$$

A power coefficient is then defined as

$$C_p = \frac{\text{Power}}{\tfrac{1}{2}\rho V_\infty^3 A_D} \qquad 1.11$$

where the denominator represents the power available in the air, in the absence of the actuator disc.

Therefore $\qquad C_p = 4a(1-a)^2 \qquad 1.12$

The maximum value of Cp occurs when dCp = 4(1-a)(1-3a) = 0, which gives a value of dCp/da a = 1/3

Hence $\qquad C_{P_{max}} = \dfrac{16}{27} = 0\cdot593 \qquad 1.13$

This is known as the Betz limit after Albert Betz the German aerodynamicist.

Cp is, perhaps, more fairly defined as

$$C_p = \frac{\text{Power}}{\text{Actual Power Available}}$$

$$= \frac{\text{Power}}{16/27 \times \tfrac{1}{2}\rho V_\infty^3 A_D} \qquad 1.14$$

The power extracted, then, varies with a in a cubic manner.

The variation of power coefficient and axial force coefficient with a is shown in figure 1.2. The force coefficient is defined as

$$C_T = \frac{F}{\tfrac{1}{2}\rho V_\infty^3 A_D} = 4a(1-a) \qquad 1.15$$

which has a maximum value of 1 at a = 1/2.

A problem arises for values of a> = 1/2 because the wake velocity, given by $V_\infty(1-2a)$, becomes zero, or even negative. In these conditions the momentum theory, as described, no longer applies and an empirical modification

has to be made.

1.3 Turbine Design

The manner in which the extracted energy is converted into usable energy depends upon the particular turbine design. Some designs are more efficient than others but no turbine so far built has exceeded the Betz limit, although values of Cp greater than 0.5 have been reported. Most turbines employ blades with aerofoil cross sections because this is the most efficient way of developing the necessary pressure difference across the disc, formed by the rotating blades, and converting it into useful torque. For the case of vertical axis machines the disc becomes a surface of revolution. Other designs use aerodynamic drag, which is far less efficient than aerodynamic lift, but provides a much larger starting torque.

To investigate wind turbine aerodynamics further it will be necessary to understand the basic principles of lift and drag.

2 AERODYNAMIC DRAG

The forces acting on a body moving relative to a fluid, neglecting bouyancy, can be resolved into streamwise (drag) and normal (lift) components. Both of these forces are caused by the viscosity of the fluid but the link is not necessarily direct.

In a very slowly moving fluid the drag on a body may be directly attributable to the viscous, frictional shear stresses set up in the fluid due to the fact that, at the body wall, there is no relative motion. In this case the effects of viscosity extend a long way from the body. This type of flow is known as Stokes' flow after Sir George Stokes who, along with Navier, was the first to introduce the effects of viscosity into the general equations of fluid motion. Stokes used his equations to determine the drag force on a sphere in creeping flow.

$$\text{Drag} = 6\pi\mu r V \qquad 2.1$$

where μ is the fluid viscosity. Curiously, the Stokes' flow pattern is very similar to the theoretical, inviscid (potential) flow pattern for which the drag is zero but which is never realised in practice. (Figure 2.1)

In a real fluid, when the viscosity is low and the velocity is relatively high, the drag force that exists is due primarily to an asymmetric pressure distribution, fore and aft. This is caused by the fact that the fluid does not follow the boundary of the body but separates from it leaving low pressure, stagnant fluid in the wake. On the upstream side, where the flow remains attached, the pressure is high because the fluid has been slowed down and pushed aside by the body obstructing its path. At the nose of the body the flow is brought exactly to rest and this is called the stagnation point. (Fig. 2.2)

2.1 The Reynolds Number

The nature of the flow around a body at any given velocity depends upon the relative magnitudes of the inertia and the viscous forces. The ratio of these forces is known as the Reynolds number after Osborne Reynolds who first introduced its use.

In a steady flow the inertia force required to accelerate a unit volume of fluid, from a point where the velocity is u to a point where the velocity is u + du, is

$$F_I = \rho u \frac{\partial u}{\partial x} \qquad 2.2$$

Whereas the nett viscous force is

$$F_v = \frac{\partial \tau}{\partial y} = \mu \frac{\partial^2 u}{\partial y^2} \qquad 2.3$$

Equation 2.3 derives from Newton's law of viscous flow which states that the frictional shear stress is proportional to the transverse velocity gradient.

$$\tau = \mu \frac{\partial u}{\partial y} \qquad 2.4$$

For a given flow pattern past a body of diameter d, both $\partial u/\partial x$ and $\partial u/\partial y$ are proportional to V/d.

Therefore $Re = \dfrac{F_I}{F_v} = \dfrac{\rho u \, \partial u/\partial x}{\mu \, \partial^2 u/\partial y^2} = \dfrac{\rho V^2/d}{\mu V/d^2} = \dfrac{\rho V d}{\mu} \quad 2.5$

which is the definition of the Reynolds number Re. If the size of the body is changed then the flow speed must also change for the flow patterns to remain geometrically similar.

2.2 The Drag Coefficient

If Stokes' drag equation for the sphere is re-arranged, giving

$$D = 6\pi \mu r V = 24 \frac{\mu}{\rho V d} \times \tfrac{1}{2}\rho V^2 \times \pi r^2 \qquad 2.6$$

it is then in the standard form of drag co-efficient x dynamic pressure x frontal area. The drag coefficient is then defined as

$$C_D = \frac{\text{Drag force}}{\tfrac{1}{2}\rho V^2 A} \qquad 2.7$$

Note that $C_D = 24/Re$, a function only of the Reynolds number. This turns out to be valid for all bodies in incompressible flow but the functional relationship is not usually as simple as in this case.

2.3 The Boundary Layer

The reason for the separated flow at the higher Reynolds numbers is the existence of a thin boundary layer of slow moving fluid, close to the body surface, within which viscous forces predominate. Outside of this layer the flow behaves almost inviscidly. The drag on the body due to viscosity is quite small but the effect on the flow pattern is profound.

The potential flow pattern around a cylinder requires that the pressure distribution be as shown in figure 2.3.

Fore and aft the pressure is high; above and below the pressure is low. The fluid on the downstream side is slowing down against an adverse pressure gradient and, right at the wall, it slows down exactly to a standstill at the rear stagnation point. In the real fluid the boundary layer, which has already been slowed down by viscosity, comes to a halt well before the stagnation point is reached and the flow begins to reverse under the action of the adverse pressure. At this point, where the pressure is still low, the boundary layer separates from the body surface forming a wake of stagnant, low pressure fluid. (Fig. 2.4)

The high pressure acting at and around the forward stagnation point is no longer balanced by the high pressure at the rear and so a dragwise pressure force is exerted.(Fig. 2.5)

A boundary layer grows in thickness from the forward stagnation point, or leading edge. Initially, the flow in the layer is ordered and smooth but, at a critical distance l from the stagnation point, characterised by Re = $\rho V l/\mu$, the flow begins to become turbulent. This turbulence causes mixing of the boundary layer with the faster moving fluid outside resulting in re-energisation and delaying the point of separation. The result is to reduce the pressure drag (because the low pressure stagnant rear area is reduced), increase the viscous drag and increase the boundary layer thickness.

The coefficient of drag, therefore, varies with Re in a complex fashion. For small bodies, at low speeds, the critical Re is never reached and so separation takes place early. For large bodies, or high speeds, turbulence develops quickly and separation is delayed. (Fig. 2.7)

Turbulence can be artificially triggered by roughening the body surface or by simply using a 'trip wire'. General flow turbulence tends to produce turbulent boundary layers at Reynolds numbers ostensibly below the critical value and this certainly seems to happen in the case of wind turbine blades.

A sharp edge on a body will always cause separation. For a flat plate broad side on to the flow the boundary layer separates at the sharp edges and the Cd is almost Re independent, but is dependent upon the plate's aspect ratio.

So-called streamlined bodies taper gently in the aft region so that the adverse pressure gradient is small and separation is delayed until very close to the trailing edge. This produces a very much narrower wake and a very low drag.

3 AERODYNAMIC LIFT

Lift (normal force) is also a phenomenon which only occurs because of the existence of the thin boundary layer. However, the connection is more subtle than in the drag case and it will be useful first to consider the lift on a rotating cylinder. Because of the boundary layer, a rotating cylinder in a fluid at rest entrains the surrounding fluid to rotate as well. The nature of the fluid rotation is such that the tangential velocity is inversely proportional to the distance from the centre of the cylinder (Fig. 3.1).

This type of rotational motion is known as a vortex and is quite unlike rigid body rotation in which the tangential velocity is directly proportional to the radial distance.

If the fluid also has a uniform velocity V past the cylinder, then the resulting flow field is as shown in Fig. 3.2. Adding velocities vectorially, the velocity above the cylinder is increased, and so the static pressure there is reduced. Conversely, the velocity beneath is slowed down, giving an increase in static pressure. There is clearly a normal force upwards on the cylinder, a lift force.

This is known as the Magnus effect after its original discoverer and explains, for example, why spinning tennis balls veer in flight.

The lift force is given by

$$L = \rho V \Gamma \qquad 3.1$$

where Γ is the circulation or vortex strength around the cylinder, defined as the integral

$$\Gamma = \oint v \, ds \qquad 3.2$$

around any path enclosing the cylinder and v is the velocity tangential to the path s. Choosing a circular path of radius r around the centre of the vortex, and assuming the general flow velocity V is zero, then v = k/r, where k is a constant. At the cylinder wall r = a, so $v = \Omega a$. Therefore, $k = \Omega a^2$. The circulation Γ, which is the same for all circles and every other path enclosing the cylinder, is given by

$$\Gamma = \oint v \, ds = \oint \frac{\Omega a^2}{r} ds = 2\pi \Omega a^2 \qquad 3.2a$$

3.1 The Aerofoil

An aerofoil works in a similar manner and does so because of its sharp trailing edge. Consider an aerofoil at a small angle of incidence to the oncoming flow. The potential flow pattern around the aerofoil is as shown in fig. 3.3a.

The inviscid flow is such that no force on the aerofoil exists. In real flow separation occurs at the sharp trailing edge, causing the flow to leave it smoothly, with no flow around the edge: figure 3.3b. The flow is therefore largely attached above and below the aerofoil but the flow pattern is now such that there is a nett circulation (equation 3.2), which increases the velocity above and reduces it below the aerofoil, thus resulting in a lift force.

The same pattern can be formed if we replace the aerofoil by a series of vortices with axes lying in the plane of the chordline of the aerofoil forming a vortex sheet (Fig. 3.4).

In a manner similar to the Magnus effect, a pressure difference occurs across the vortices and the overall circulation can be shown to be $\pi V c \sin \alpha$, where c is the chord of the aerofoil. The velocities above and below the aerofoil at the trailing edge must be the same because the pressures there must be the same (Bernoulli), but the particles which meet there are not the same ones that parted company at the leading edge; the particle which travelled above the aerofoil reaches the trailing edge first.

A lift coefficient is defined as

$$C_L = \frac{Lift}{\frac{1}{2}\rho V^2 A} \qquad 3.3$$

In this case, A is the plan area of the wing. For an infinitely long wing (two-dimensional flow)

$$C_L = \frac{Lift \ per \ unit \ span}{\frac{1}{2}\rho V^2 c} \qquad 3.4$$

$$= \frac{\pi V c \sin \alpha}{\frac{1}{2}\rho V^2 c} = 2\pi \sin \alpha$$

In practice, $C_L = a_0 \sin \alpha \qquad 3.5$

where a called the lift-curve slope ($dC_L/d\alpha$) is about 5.73 (0.1/deg.), rather than 2π.

It has become conventional to use the plan area A, or the chord c, to define the drag coefficient as well.

3.2 The Stalled Aerofoil

If the incidence angle exceeds a certain critical value (10 to 16 deg, depending on the Re. No.), separation of the boundary layer on the upper surface takes place. This causes a wake to form from above the aerofoil, reduces the circulation, reduces the lift and increases the drag. The flow past the aerofoil has then stalled.

A flat plate will also develop circulation and lift but will stall at a very low incidence angle because of the sharp leading edge. Arcing the plate will improve the stalling behaviour but a much greater improvement can be obtained by giving the aerofoil thickness and a well rounded leading edge.

At low Re. No's ($<10^5$) laminar separation of the boundary layer occurs close to the leading edge but soon re-attaches again, as a turbulent boundary layer, forming a 'bubble' of almost stagnant, re-circulating fluid. As incidence increases the bubble spreads towards the trailing edge until the aerofoil is completely stalled. Loss of lift is gradual but there is a sharp rise in drag as soon as the separation bubble forms.

At higher Re. No's the bubble spreads very suddenly and a sharp drop in lift takes place. This is the region in which most medium size wind turbines operate.

At Re No's greater than about 2×10^6 no laminar separation occurs before the boundary layer becomes turbulent. Separation of the turbulent boundary layer then takes place at the trailing edge. The abruptness of this trailing edge stall depends upon the shape of the rear section of the aerofoil. This type of stall will also occur with thick aerofoils at intermediate Re. No's.

Figure 3.6 shows the form of the C_L-α curve for the three cases described above. Notice that the simple relationship of equation 3.5 is only valid for the pre-stall region.

In the post-stall region (angles of incidence greater than 10-16 degrees) there is still some remaining circulation, and hence some remaining lift, but the drag is very high. Beyond about 35 degrees, the flow becomes almost Re No. independent.

Figures 3.7 and 3.8 show the variation of C_L and C_D with α, for several values of Re No., for the symmetrical NACA0015 aerofoil ('00' means zero camber and '15' indicates a maximum thickness/chord ratio of 15%). Notice that beyond about 30 degrees of incidence the coefficients are virtually independent of Re No. because the leading and trailing edges fix the boundary layer separation points.

It is common to present the C_L and C_D data in one graph called the Lift-Drag Polar' because it represents the non-dimensional resultant force on the aerofoil as the incidence varies.

The chordwise pressure distribution is such that, below the stall, most of the lift comes from the region of the leading edge and this is largely due to the low pressure on the upper surface. The variation of pressure coefficient, defined as,

$$C_p = \frac{p - p_\infty}{\frac{1}{2}\rho V_\infty^2} \qquad 3.6$$

for α = 10 deg. is shown in figure 3.10a. The centre of pressure is at the 1/4 chord position for all values of below the stall. Above the stall the pressure on the upper (stalled) surface is much more uniform and the centre of pressure moves towards the mid-chord position. Pressure distributions for 30 deg. and 50 deg. of incidence are shown in figures 3.10b and 3.10c, respectively.

Cambered aerofoils, such as the NACA4415, shown in figure 3.11a, have curved chord lines and this allows them to produce lift at zero incidence without the same increase in drag that is usual with symmetrical aerofoils. The behaviour of the NACA4415 is shown in figure 3.11b for angles of incidence below and just above the stall.

The centre of pressure is behind the 1/4 chord position on cambered aerofoils and moves with incidence. However, if a fixed chordwise position is chosen then the resultant force through that point is accompanied by a pitching moment (nose-up positive, by convention).

If a pitching moment coefficient is defined as

$$C_m = \frac{\text{Pitching moment per unit span}}{\frac{1}{2}\rho V^2 c^2} \qquad 3.7$$

then there will be a position, called the Aerodynamic Centre, for which $dC_m/d\alpha = 0$. Theoretically, the aerodynamic centre lies at the 1/4 chord position and is close to this point for most practical aerofoils. The value of C_m depends upon the degree of camber but for the NACA4415 the value is -0.1. Above the stall there is no aerodynamic centre, as defined, and so the pre-stall position continues to be used to determine the pitching moment coefficient, which then becomes dependent upon α.

4 THE HORIZONTAL AXIS WIND TURBINE

The blades of a modern horizontal axis wind turbine have aerofoil cross-sections and act in just the same way as an aircraft wing. Lift is generated by virtue of the pressure difference across the aerofoil (figure 3.10) and this is exactly the step change in pressure required of the actuator disc, to which the rotating blades approximate in a time averaged manner (Fig. 4.1).

4.1 Torque and the Tangential Flow Induction Factor

The force on the blades not only has a component in the flow direction but, as a shaft torque has to be developed, there must be a component in the tangential direction also. The reaction to this torque must be to impart an equal but opposite rate of change of angular momentum to the wind. The streamlines in the wake, therefore, follow a helical path resulting from the superposition of the streamwise and rotational velocities (Fig. 4.2).

The transfer of rotational motion takes place entirely across the thickness of the disc. Because of the bound vorticity associated with the blades, which is spread out over the disc in a time averaged manner, the disc becomes a surface of tangential velocity discontinuity, or rather, a disc of tangentially shearing flow of finite thickness (see Fig. 4.3).

The change in tangential velocity is expressed in terms of a tangential flow induction factor a. Upstream of the disc the tangential velocity is zero. At the aerofoil chord line the tangential velocity is $\Omega r a'$. Downstream of the disc the tangential velocity is $2\Omega r a'$.

Thus, torque = rate of change of angular momentum
= mass flow rate x change of tangential velocity x radius
= $A\rho(1-a)V_\infty \times 2a'\Omega r \times r$

If A is taken as being the area of an annular ring of radius r and radial thickness dr, then $A = 2\pi r dr$ and so the total torque is given by

$$Q = \int_0^R 2\pi r \rho (1-a) V_\infty \times 2a'\Omega r \times r dr \qquad 4.1$$

giving $\quad Q = \pi R^3 \rho (1-a) a' V_\infty \Omega R \qquad 4.2$

4.2 Blade Element Theory.

Having information about how the aerofoil characteristic coefficients C_L and C_D vary with the angle of incidence it is now possible to determine the forces on the blades for given values of a and a'. If the axial force on the blades is set equal to the rate of change of axial momentum of the wind and the torque on the blades is put equal to the rate of change of angular momentum then it is possible to determine the values of the flow induction factors in terms of the blade geometry and the ratio of blade tip speed to wind speed.

Consider a turbine with N blades of radius R, each with chord c, which may vary along the blade, and a set pitch angle α measured between the aerofoil zero lift line and the plane of the disc, which also may vary.

Let the blades be rotating at Ω and let the wind speed be V_∞ Figure 4.4a shows the velocities relative to the blade chord line at radius r.

The resultant relative velocity at the blade is

$$W = \left(V_\infty^2(1-a)^2 + (\Omega r)^2(1+a')^2\right)^{\frac{1}{2}} \qquad 4.3$$

which acts at an angle φ to the plane of rotation, such that

$$\tan\varphi = \frac{V_\infty(1-a)}{\Omega r(1+a')} \; ; \; \sin\varphi = \frac{V_\infty(1-a)}{W} \; ; \; \cos\varphi = \frac{\Omega r(1+a')}{W} \quad 4.4$$

The angle of incidence α is then given by

$$\alpha = \varphi - \beta \qquad 4.5$$

The lift force per unit length, normal to the direction of W, is therefore

$$L = \tfrac{1}{2}\rho W^2 \tilde{c} \, C_L(\alpha) \qquad 4.6$$

and the drag force parallel to W is

$$D = \tfrac{1}{2}\rho W^2 \tilde{c} \, C_D(\alpha) \qquad 4.7$$

where $\tilde{c}$ = the number of blades (N) x actual b blade chord c

The force per unit length of blade, acting in the wind direction, is

$$f = \tfrac{1}{2}\rho W^2 Nc (C_L \cos\varphi + C_D \sin\varphi) \qquad 4.8$$

$$\text{Force on the rotor} = F = \int_{R_R} f \, dr \qquad 4.9$$

where R_R is the root radius.

The result of equation 4.9 can be equated to the axial rate of change of momentum given by equation 1.9 with $A_D = \pi(R^2 - R_R^2)$

Hence
$$a(1-a) = \int_{R_R}^{R} \frac{NcR}{4\pi(R^2 - R_R^2)} \frac{W^2}{V_\infty^2} C_L \cos\varphi \frac{dr}{R} \qquad 4.10$$

The drag term has been left out of this equation on the grounds that drag is largely due to the shearing stresses in the boundary layer and the loss of momentum is confined to the thin vortex sheets shed from the trailing edges of the blades. Such loss of momentum is therefore not a general feature of the flow.

The tangential force per unit length acting on the blades is given by

$$t = \tfrac{1}{2}\rho W^2 Nc (C_L \sin\varphi - C_D \cos\varphi) \qquad 4.11$$

and the torque on the complete rotor is

$$Q = \int_{R_R}^{R} t \, r \, dr \qquad 4.12$$

Equating 4.12 with the rate of change of angular momentum in 4.4 gives

$$a'(1-a)\frac{\Omega R}{V_\infty} = \int_{R_R}^{R} \frac{NcR}{2\pi(R^4 - R_R^4)} \frac{W^2}{V_\infty^2} C_L \sin\varphi \frac{r}{R} \frac{dr}{R} \qquad 4.13$$

Again, the drag term has been left out because its effect is confined to the thin sheets of shed vorticity.

4.3 Tip Speed Ratio

An extremely important parameter which so far has not been mentioned is the ratio of the tangential speed of the blade tips to the undisturbed wind speed. This is known as the TIP SPEED RATIO

$$\lambda = \frac{\Omega R}{V_\infty} \qquad 4.14$$

The tip speed ratio dictates the operating condition of a turbine and has a direct effect on the values of the flow induction factors.

4.4 Turbine Solidity

The primary non-dimensional factor which describes the geometry of the turbine is the turbine solidity. Solidity is defined as the ratio of the blade area to the area of the disc but it is conventional to define it simply as

$$\sigma = \frac{N\bar{c}}{2\pi R} \qquad 4.15$$

where $\bar{c}$ is the mean chord of the blades.

4.5 Iterative Solution Procedure

The equations to be solved, in order to determine the power output of a turbine operating at a given tip speed ratio, are not readily amenable to analytical techniques and are best treated by a numerical approach. If measured aerofoil data are used then a numerical solution is the only option.

The solution involves an iteration procedure, based upon equation 4.10, to calculate the value of a, the axial flow induction factor. Commencing with zero initial values for both a and a′, the value of the resultant relative velocity W is found at each of the chosen radial stations.

$$\frac{W}{V_\infty} = \left((1-a)^2 + \lambda^2 \frac{r^2}{R^2}(1+a')^2 \right)^{1/2} \qquad 4.3a$$

The angle this velocity vector makes with the plane of rotation is then found from

$$\sin\varphi = \frac{V_\infty}{W}(1-a) \; ; \; \cos\varphi = \lambda \frac{V_\infty}{W}\frac{r}{R}(1+a') \qquad 4.4a$$

Knowing the set pitch angle at each station the angle of incidence may be found.

$$\alpha = \varphi - \beta \qquad 4.5$$

From the tabulated aerofoil data the values of C_L and C_D are looked up for the current value of using linear interpolation where necessary.

A new value of a can then be calculated from

$$a_i = \frac{\bar{F}}{(1-a_{i-1})} \qquad 4.16$$

where $\bar{F}$ is the right hand side of equation 4.10 and i signifies the current iteration stage. Finally, the tangential flow induction factor is obtained.

$$a'_i = \frac{\bar{Q}}{(1-a_i)\lambda} \qquad 4.17$$

$\bar{Q}$ is the right hand side of equation 4.13.

Shaft torque and power can now be calculated using equations 4.11 and 4.12, which include the effects of drag, and the relevant non-dimensional coefficients are obtained.

$$\text{Torque Coeff. } C_Q = \frac{\text{Torque }(Q)}{\frac{1}{2}\rho V_\infty^2 A_D R} \qquad 4.18$$

$$\text{Power Coeff. } C_P = \frac{\text{Power }(Q \times \Omega)}{\frac{1}{2}\rho V_\infty^3 A_D} \qquad 4.19$$

Similarly, the actual force on the turbine disc can be determined from equations 4.8 and 4.9 and the force, or thrust, coefficient is then given by

$$\text{Thrust Coeff. } C_T = \frac{\text{Axial Thrust}}{\frac{1}{2}\rho V_\infty^2 A_D} \qquad 4.20$$

4.9 Performance of a Typical Turbine

The performance of a three bladed turbine of 4 metre diameter will be analysed. The blades are tapered, but untwisted, and have a NACA4415 aerofoil section. The rotor operates upwind of the tower and so is unconed. The blades are set at a constant pitch angle of 4 degrees to the zero lift line.

The principal indicator of turbine performance is the plot of power coefficient v. tip speed ratio. The C_P-λ curve for the turbine is shown in figure 4.5 .

The full line curve in figure 4.5 represents the power actually transferred to the shaft and should be compared with the power lost by the airstream which is shown in the dotted line and was calculated by integrating the momentum power loss (equation 1.12) over the whole disc. The differences between the two curves shows the losses which occur due to the effects of stall and drag. The tip losses are shown by differences between the dashed curve, obtained by integrating equation 1.12 directly, and the dotted curve.

The maximum value of C_P reached by the turbine is 0.41 at a tip speed ratio of 5 which is well short of the ideal Betz limit of 0.593 but is in fact as much as can be expected from a turbine of this size operating at relatively low Reynolds numbers.

The aerofoil data used in this analysis applies to a Re of about 500,000 which is of the right order of magnitude but, ideally, the aerofoil data should be sufficiently detailed to allow the Re to be matched to local blade element conditions.

The variation of the torque and thrust coefficients with tip speed ratio are shown in figure 4.6 .

The maximum value of C_T reached is 0.82 compared to 1 which is given by the simple momentum theory (equation 1.14). This shows clearly that the value of a is far from uniform along the blade.

4.10 Optimum Blade Design

The best performance can be obtained by designing the variation of chord and pitch to maximise the energy extraction everywhere along the blade. This, unfortunately, leads to complex blade designs which are difficult to manufacture. However, compromise solutions can be obtained and the turbine of the previous section was designed for maximum output with the constraints of zero twist and uniform taper.

5. THE VERTICAL AXIS WIND TURBINE

The vertical axis turbine is probably even older than its horizontal counterpart but, in terms of modern developments, it is still regarded as an innovation. The primary aerodynamic advantage of the vertical axis is that it can receive the wind from any direction without the need to yaw. This simplifies the turbine design significantly but, even in steady winds, it means that the torque output varies considerably during every revolution. There are other aerodynamic differences between the two designs but the VAWT still extracts energy from the wind on the actuator 'disc' principle.

5.1 The Actuator Cylinder

Vertical axis blade configurations are many but, for the purposes of this exercise, we shall assume that the blades are straight and vertical. Such a machine sweeps out a cylinder, instead of a disc, and so intersects any given streamtube twice. Early theoretical analyses ignored this fact and, from the momentum theory aspect, were identical to the horizontal axis theory. More recent analyses have taken account of the double intersection and have been able to provide much greater detail about the aerodynamic behaviour.

5.2 Momentum Theory

The momentum theory of section 1, can be applied directly to each intersection of the airstream with the actuator cylinder. However, conditions vary greatly around the cylinder and so it is common to consider a multiplicity of streamtubes which pack together to fill the cylinder volume. The momentum theory is applied, separately, to each intersection in each streamtube. Figure 5.1 shows a plan view of the cylinder and a typical streamtube.

The two intersections are treated as two actuator discs in tandem. The disc areas are different because of the expansion of the streamtube and, although the discs are not normal to the flow direction, these areas are taken to be of the normal cross-sections.

It is assumed that at a point somewhere between the discs the pressure rises through the atmospheric level p_∞ and at this point the streamtube velocity is V_a. By the momentum theory, therefore, at the upstream disc,

$$V_u = V_\infty (1 - a_u) \qquad 5.1$$

and

$$V_a = V_\infty (1 - 2a_u) \qquad 5.2$$

The rate of change of momentum for the upstream part of the streamtube is then

$$F_u = 2 A_u \rho V_\infty^2 (1 - a_u) a_u \qquad 5.3$$

The speed V_a now becomes the upstream velocity (instead of V_∞) for the downstream disc and hence

$$V_d = V_a (1 - a_d) \qquad 5.4$$

and

$$V_w = V_a (1 - 2a_d) \qquad 5.5$$

so the rate of change of momentum is

$$F_d = 2 A_d \rho V_a^2 (1 - a_d) a_d \qquad 5.6$$

A_u and A_d are the respective normal cross-sectional areas.

5.2 Blade Element Theory

Each blade has an aerofoil cross-section and produces lift which has a component in the tangential direction, thus providing a torque. The torque is not constant but varies with blade position and, when the blades are few in number, this means that the shaft torque fluctuates. Figue 5.2 shows the blade element forces and velocities at points in each quadrant.

As can be seen, the lift always has a component in the forward direction but the blade surface facing the wind changes between the upstream and downstream passes. This means that the angle of incidence changes sign and so it would seem that the aerofoil should be symmetrical. A cambered aerofoil would give an increased torque on one pass but a decreased torque on the other and experiment has shown that the latter predominates which ever way the blade is cambered. Pitching the blades nose in or nose out should, in principle, give similar results but a small advantage can be obtained with a little nose out pitch, especially at very low tip speed ratios. This is useful because vertical axis machines, generally, have low starting torques, although there are exceptions. For simplicity, however, we shall assume zero set pitch.

Figure 5.3 shows the forces on a blade element in the first quadrant, measuring the azimuth angle clockwise from the downstream direction.

The angle θ is not the blade azimuth but the angle between the radius vector and the local streamline. This streamline is assumed to be straight as it crosses the turbine and so the angle θ is the same at both actuator discs/blade elements.

The forces, resolved into the local streamwise sense, give

$$F = (L \cos\alpha + D \sin\alpha)\cos\theta - (L \sin\alpha - D \cos\alpha)\sin\theta \qquad 5.7$$

The terms in brackets are the normal (N) and chordwise (T) components of the resultant force on the aerofoil and it is usual to use these rather than L and D (Fig. 5.4).

$$\left. \begin{array}{l} N = L \cos\alpha + D \sin\alpha \\ T = L \sin\alpha - D \cos\alpha \end{array} \right\} \qquad 5.8$$

As C_L and C_D are known functions of α then C_t and C_n can be calculated and used instead.

$$C_t = \frac{T}{\frac{1}{2}\rho w^2 c} \quad ; \quad C_n = \frac{N}{\frac{1}{2}\rho w^2 c} \qquad 5.9$$

Note that N and T are, like L and D, forces per unit length of blade. (from this point on only the coefficient C_n will be referred to so the symbol N can revert to the number of blades).

Figures 5.5 to 5.8 show the variation of C_t and C_n with α for a range of Reynolds numbers for the NACA0012 aerofoil. It will be seen that C_t is positive over most of the range 0 - 180 degrees if incidence.

Equation 5.7 can therefore be written as

$$F = \tfrac{1}{2}\rho W^2 \tilde{c}\left(C_n \cos\theta - C_t \sin\theta\right) \qquad 5.10$$

If θ it is considered to be positive in the 1st and 2nd quadrants and negative in the 3rd and 4th quadrants the equation 5.10 applies to all blade positions.

As written, F is an instantaneous force and what is required is a time averaged force. To obtain this the chord length $\tilde{c}$ is replaced by

$$\tilde{c} = \frac{Nc\,\Omega\delta t}{2\pi} \qquad 5.11$$

where $\frac{\Omega\delta t}{2\pi}$ is the probability of a given blade being in any given position during a time interval δt and $\delta t\Omega = \delta\beta$, the angle of arc of the actuator disc.

5.4 Torque and Power

The torque can be determined by taking moments of the forces acting on a unit length of blade about the axis of rotation.

$$Q = \tfrac{1}{2}\rho W^2 \tilde{c}\; C_t R \qquad 5.12$$

The performance coefficients C_P C_Q and C_T are defined as in equations 4.18 - 4.20 using as A the frontal projected area 2RL.

The results for a turbine of solidity $(Nc/R) = .2$, with blades of aspect ratio 14 and using the NACA0012 aerofoil data given in figures 5.6 - 5.6 at a Reynolds number = $(\rho\Omega Rc/\mu) = 3 \times 10^6$ are shown in figures 5.7 and 5.8.

The general form of the performance curve is much the same for the HAWT. One noticeable difference is the lower tip speed ratios which is a direct result of the double intersection of the streamtube. The maximum power coefficent attained is about 0.4, which is rather low but is attributable to the use of NACA0012 data; a thicker symmetrical aerofoil would have shown a better performance.

6. BIBLIOGRAPHY

(1) Durand,W.F. Aerodynamic Theory, Section 4
     19   Dover Publications.

(2) Glavurt, M. The elements of aerofoil and
     airscrew theory, Second Ed., 1959.
     Cambridge University Press.

(3) Prandtl, L. and Tietjens, O.G. Applied
     Hydro-and aeromechanics. 1957. Dover
     Publications.

(4) Abbott, I.H. and von Doenhoff, A.E.
     Theory of Wing sections 1959. Dover
     Publications.

(5) Wilson, E.W., Lissaman, P.B.S. and
     Walker, S.N. Aerodynamic performance
     of wind turbines. June, 1976. Oregon
     State University, Carvallis, Oregon.

(6) Wind Energy for the Eighties. 1982.
     British Wind Energy Association

(7) Van Dyke, M. An album of fluid motion.
     1982. The Parabolic Press, Stamford,
     California.

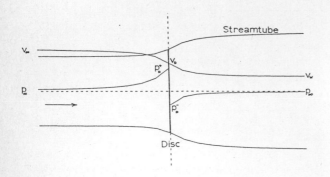

Fig 1.1   An energy extracting actuator disc and streamtube

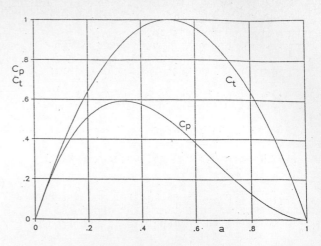

Fig 1.2   Variation of C_p and C_T without induction factor

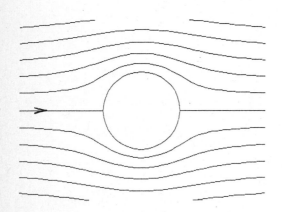

Fig 2.1   Potential flow past a cylinder

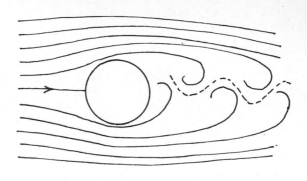

Fig 2.2   Separated flow past a cylinder

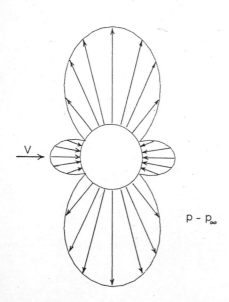

Fig 2.3   Potential flow pressure distribution around a
cylinder

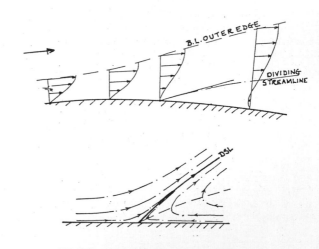

Fig 2.4   Separation of a boundary layer

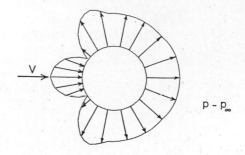

Fig 2.5   Separated flow pressure distribution around a
cylinder

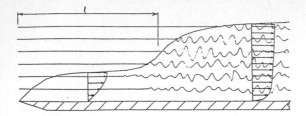

Fig 2.6　Laminar and turbulent boundary layer

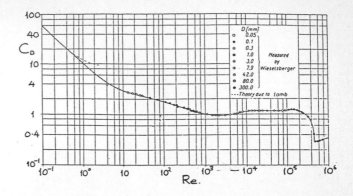

Fig 2.7　Variation of C_D with Re No. for a long cylinder

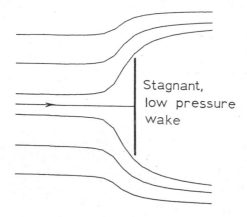

Fig 2.8　Separated flow past a flat plate

Fig 2.9　Flow past a streamlined body

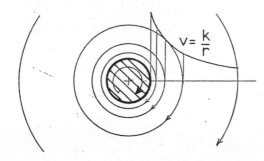

Fig 3.1　Vortex flow round a rotating cylinder

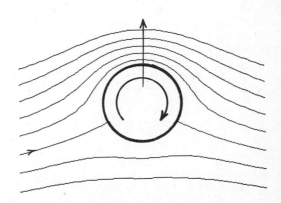

Fig 3.2　Flow past a rotating cylinder

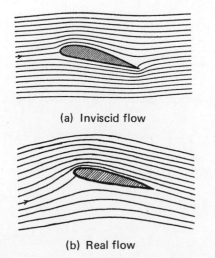

(a) Inviscid flow

(b) Real flow

Fig 3.3　Flow past an aerofoil at a small incidence

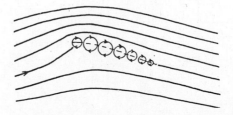

Fig 3.4　Inclined aerofoil modelled as a vortex sheet

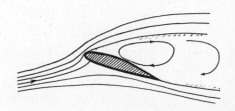

Fig 3.5　Stalled flow around an aerofoil

239

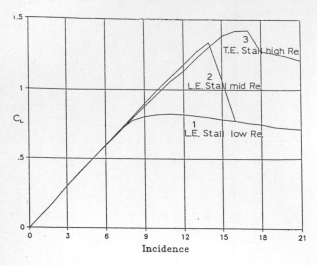

Fig 3.6 Typical $C_L-\alpha$ curves for an aerofoil

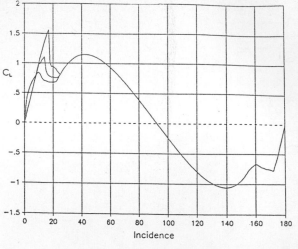

Fig 3.7 Variation of C_L with incidence for a NACA0015 aerofoil

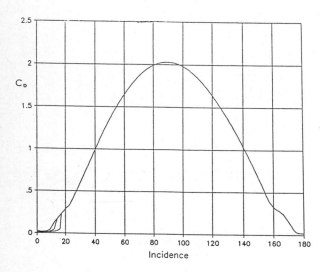

Fig 3.8 Variation of C_D with incidence for a NACA0015 aerofoil

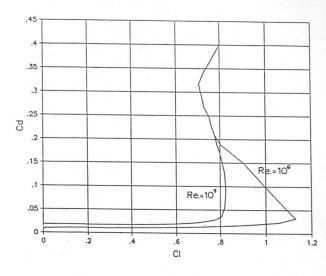

Fig 3.9 Lift-drag polar for NACA0015 at Re No = 10^5 and 10^6

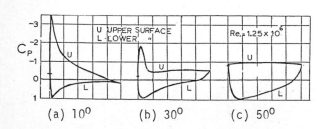

Fig 3.10 Pressure distributions around the NACA0015 aerofoil

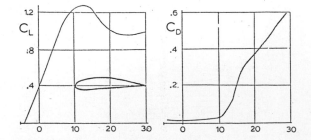

Fig 3.11 The characteristics of the NACA4415 aerofoil

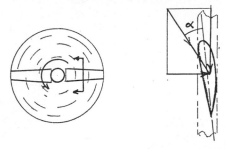

Fig 4.1 Rotating blades of a HAWT form an actuator disc

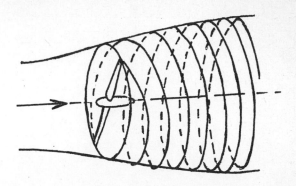

Fig 4.2 Helical wake of a HAWT

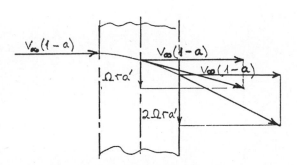

Fig 4.3 Tangential velocity grow across the disc thickness of a HAWT

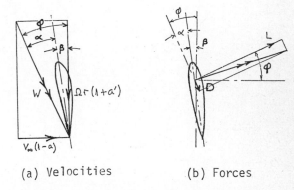

(a) Velocities (b) Forces

Fig 4.4 Blade element velocities and forces

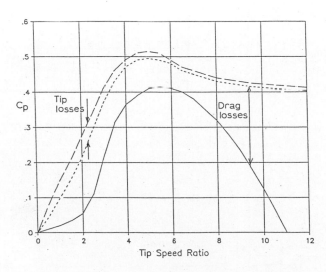

Fig 4.5 $C_p - \lambda$ Performance curve for a HAWT

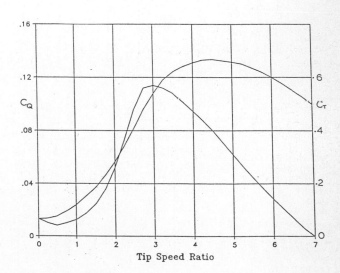

Fig 4.6 $C_Q - \lambda$ $C_T - \lambda$ curves for a HAWT

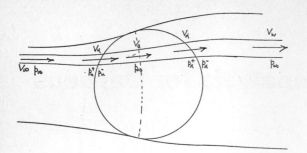

Fig 5.1   Plan view of an actuator cylinder

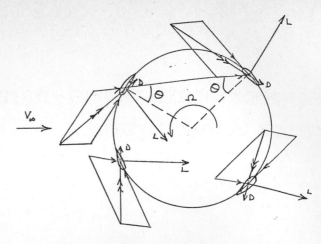

Fig 5.2   Blade forces and relative velocities

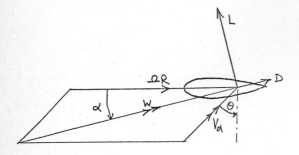

Fig 5.3   Forces on a blade element

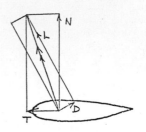

Fig 5.4   Normal and chordwise components of force on an aerofoil

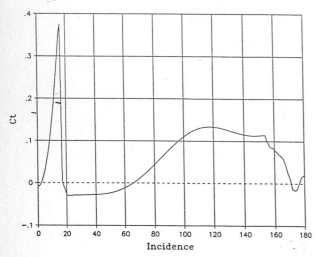

Fig 5.5   Variation of C_t with incidence for a NACA0012 aerofoil

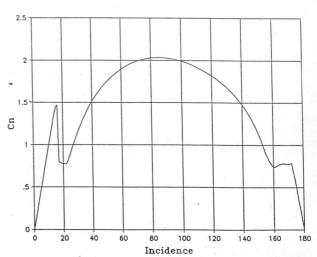

Fig 5.6   Variation of C_n with incidence for a NACA0012 aerofoil

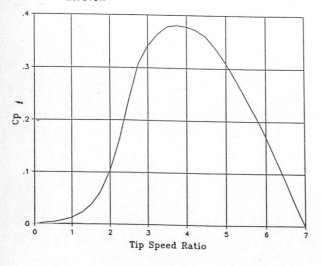

Fig 5.7 $C_p - \lambda$ performance curve for a VAWT

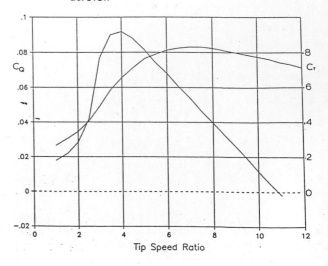

Fig 5.8 $C_Q - \lambda$ and $C_T - \lambda$ curves for a VAWT

Approximate aerodynamic analysis for Darrieus rotor wind turbines

E W BEANS

Department of Mechanical Engineering, The University of Toledo, Toledo, Ohio, USA

SNYOPSIS An Approximate analysis, which is based on the assumption of constant aerodynamic coefficients and the trigonometry of small angles, is presented. The analysis produces a performance
model which consists of a series of closed-form equations. A result of the analysis is an interference factor which varies with the azimuthal angle.

    The model predicts a power coefficient which agrees well with wind tunnel data. The region of
validity for the model is from the cut-in speed to rotor stall. The model predicts a power coefficient which is relatively constant with planform shape. The model also predicts a minimum in performance for a planform which is basically a troposkien.

NOTATION

A	rotor area	R	constant
a	local interference factor	r	rotor reference radius
b	blade length	r'	local radius
C_D	airfoil drag coefficient	S	constant
C_f	local force coefficient	S_L	lift coefficient slope
C	cosine integral	V	wind velocity
C_L	lift coefficient	V_n	normal velocity
C_N	normal force coefficient	V_t	tangential velocity
C_n	local normal force coefficient	x	tip speed ratio
C_p	power coefficient	x_o	tip speed ratio at cut-in
C_t	local tangential force coefficient	α	angle of attack
$C\alpha$	blade coefficient	β	blade parameter
c	chord	γ	radius ratio
F	force	δ	drag parameter
G	constant	θ	azimuthal angle
H	constant	ρ	density
h	rotor semi height	ϕ	blade angle
I	integral	ω	angular velocity
J	constant		
K	constant	**Subscripts**	
m	number of blades	c	curved section
N	blade normal force	D	downwind
n	normal force parameter	i	index
P	power	m	maximum or maximum condition
p	power parameter	s	straight section
		U	upwind
		θ	any azimuthal angle

1 INTRODUCTION

The aerodynamic analysis of a Darrieus rotor
wind turbine is more complex than that for a
horizontal axis wind turbine because (a) the
flow field varies with azimuthal angle (b) the
downwind blade passes thru the wake of the upwind blade, and (c) the blades are curved. The
approach used to analyze the Darrieus rotor is
either the multiple streamtube model by Strickland (1) and Sharpe and Read (2) or the vortex
model by Strickland, Webster and Nguyn (3).

    This paper presents an approximate aerody-
namic analysis which follows the approach presented by Beans (4). The analysis results in a
closed form solution, which is readily solvable
with a programmable calculator. Hence, the importance of the various design parameters can be
assessed, which makes the model well suited for
preliminary design studies.

2 ASSUMPTIONS

The approximate analysis is based on the following assumptions:

1. The airfoil can be characterized by a con-
    stand slope for the lift curve and a constant
    value for the profile drag.
2. The blade has a constant chord and can be
    approximated by a circular arc and a straight
    line.

3. The turbine angle of attack can always be treated as a small angle.
4. The airfoil never stalls.

Assumption 1 is reasonable for airfoils at small angles of attack. Assumption 2 is made to simplify the analysis. The blade shape for a true Darrieus rotor is a troposkien which Sharpe (5) used. The circular arc-straight line shape is used by Sandia (3). Parashivoiv and Delclaux (6) showed that the difference in performance between the troposkien and the circular arc-straight line shape is small.

Assumptions 3 and 4 are the key to the approximate analysis as they were in Ref. 4. The assumptions are valid up to stall. The analysis breaks down in the post stall region as it did in Ref. 4.

3   DEVELOPMENT OF EQUATIONS

The blade geometry for a Darrieus rotor is illustrated in Fig. 1. The relationships of the local radius of the blade are:

$$r'/r = 1 - \gamma(1-\cos\phi) \qquad 0 \leqslant \phi \leqslant \phi_m \qquad (1)$$

$$r'/r = r_m/r - b\sin\phi_m \qquad \phi_m = \text{const} \qquad (2)$$

where $\gamma = r_c/r$. Other relationships which describe the blade geometry are:

$$r_m/r = 1 - \gamma(1-\cos\phi_m) \qquad (3)$$

$$(h/r)\sin\phi_m = \gamma + (1-\gamma)\cos\phi_m \qquad (4)$$

$$\frac{A}{r^2} = 2 \{ \gamma^2\phi_m + \gamma\sin\phi_m (1 - \gamma) +$$

$$\frac{h}{r} [1 - \gamma (1-\cos\phi_m)] \} \qquad (5)$$

Figure 2 shows the upwind and downwind blade elements at any azimuthal angle. From Fig. 2 the normal and tangential velocity components to the blade element are:

$$V_n = V(1-a) \cos\phi \cos\theta \qquad (6)$$

$$V_t = \omega r' \mp V(1-a)\sin\theta \qquad (7)$$

The minus-plus sign in Eq. 7 refers to the upwind and downwind velocities, respectively. The turbine angle of attack is:

$$\tan \alpha = \frac{V(1-a) \cos\phi \cos\theta}{\omega r' \mp V(1-a)\sin\theta} \qquad (8)$$

The force coefficient on the upwind element in the direction of the wind is:

$$C_{fU} = C_n \cos\theta \cos\phi + C_t \sin\theta$$

Applying assumptions 1 and 3, the force coefficient in terms of the lift and drag coefficients is

$$C_{fU} = (S_L + C_D)\alpha \cos\phi \cos\theta - CD \sin\theta \qquad (9)$$

In a similar manner, for the downwind element

$$C_{fD} = (S_L + C_D)\alpha \cos\phi \cos\theta + C_D \sin\theta \qquad (10)$$

The terms S_L and C_D are the aerodynamic characteristics which are a function of the blade airfoil. Once selected, they are assumed to be constant along the blade regardless of orientation.

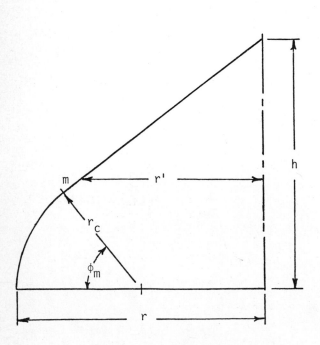

Fig 1   Blade geometry

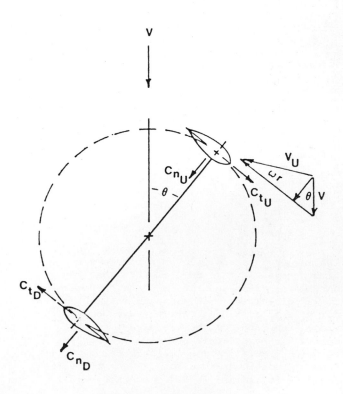

Fig 2   Blade element pair

The force on the pair of blade elements in Fig. 2 in the direction of the wind is:

$$dF_\theta = dF_{\theta U} + dF_{\theta D}$$

$$= \frac{c\rho}{2} \frac{(V_{tU}^2 \, C_{fU} + V_{tD}^2 \, C_{fD})}{\cos^2 \alpha} \, db \qquad (11)$$

Substituting Eqs. 7 thru 10 into Eq. 11 and assuming that the angle α is small, one obtains

$$dF_\theta = 2(C_\alpha \cos^2 \phi \, \cos^2 \theta + 2_{CD} \sin^2 \theta)$$

$$\frac{c\rho \omega r'}{2} V(1-a)db \qquad , \qquad (12)$$

where $C_\alpha = S_L + C_D$.

From impulse theory for a pair of elements at any azimuthal angle

$$dF_\theta = 4a (1-a) \frac{\rho V^2}{2} (2 \, r' \cos \phi)db \qquad (13)$$

Equating the two expressions, one obtains the relationship for the interference factor

$$a = \frac{m(C_\alpha \cos^2 \phi \, \cos^2 \theta + 2 \, C_D \sin^2 \theta)}{8 \cos \phi} \times (c/r) \quad (14)$$

The term m is the number of blades and is limited to two in this paper.

Equation 14 is an important conclusion of the analysis. Equation 14 shows that the interference factor varies with the azimuthal angle and blade angle. The equation also shows that for a constant blade angle, the interference is constant along the blade at any azimuthal angle.

It is convenient to define the following terms.

$$\beta = (mC_\alpha/8)(c/r) \qquad (15)$$

$$\delta = C_D/C_\alpha \qquad (16)$$

Hence,

$$a = \beta x (\cos \phi \cos^2 \theta + 2 \, \delta \sin^2 \theta/\cos \phi) \quad (17)$$

For a blade element at any azimthal angle, the local normal force coefficient is:

$$C_n = C_L \cos_\alpha + C_D \sin_\alpha = (S_L + C_D)\alpha .$$

The normal force on the element in the direction of the radius is, therefore,

$$dN_\theta = (S_L + C_D) [x(r'/r) - (1-a) \sin \theta]$$

$$(1-a) \cos^2 \phi \, \cos \theta \, c \frac{\rho V^2}{2} db \qquad (18)$$

The normal force on the blade perpendicular to the axis of rotation is obtained by integrating from $\phi = 0$ to ϕ_m for the curved section and then along the blade for $\phi = \phi_m = $ const. The results of this integration are the functions

$$n_c = \gamma\{[J_2 - \beta x(J_3 \cos^2 \theta + 2\delta J_1 \sin^2 \theta)]$$

$$- \sin \theta[C_2 - 2\beta x(C_3 \cos^2 \theta + 2\delta C_1 \sin^2 \theta)$$

$$+ (\beta x)^2(C_4 \cos^4 \theta + 4\delta C_2 \sin^2 \theta \cos^2 \theta$$

$$+ 4\delta^2 C_0 \sin^4 \theta)]\} \qquad (19)$$

$$n_s = (r_m/r)[1 - \beta x(\cos \phi_m \, \cos^2 \theta + 2\delta \sin^2 \theta/\cos \phi_m)]$$

$$(\cos^2 \phi_m/\sin \phi_m)\{(x/2)(r_m/r) - \sin \theta[1 -$$

$$\beta x (\cos \phi_m \cos^2 \theta + 2\delta \sin^2 \theta/\cos \phi_m)]\} \qquad (20)$$

and

$$C_{N\theta} = (16/m) \beta \cos \theta \, (n_c + n_s)/(A/r^2) \qquad (21)$$

In writing Eqs. 19 thru 21, the following notation is used to define the various integrals and terms

$$C_i = \int_o^{\phi_m} \cos^i \phi d\phi \qquad (22)$$

$$J_i = (1 - \gamma) \, C_i + \gamma \, C_{i+1} \qquad (23)$$

The local tangential force coefficient for small angles is:

$$C_t = C_L \sin \alpha - C_D \cos \alpha = S_L \alpha^2 - C_D \qquad (24)$$

The power output by a pair of blade elements at any aximuthal angle is, therefore,

$$dP_\theta = dP_{\theta U} + dP_{\theta U} = \omega r' (dT_{\theta U} + dT_{\theta D})$$

This simplifies to

$$dP_\theta = 2 x [(r'/r)(S_L \cos^2 \phi \cos^2 \theta - C_D \sin^2 \theta)$$

$$(1-a)^2 - C_D x^2 (r'/r)^3](\rho V^3/2)cdb \qquad (25)$$

Equation 25 is integrated along the blade in the same manner as Eq. 18. The result of this integration is the following relationships.

$$p_c = \gamma\{(1-\delta)\cos^2 \theta \, [J_2 - 2\beta x(J_3 \cos^2 \theta + 2\delta J_1 \sin^2 \theta)$$

$$+ (\beta x)^2(J_4 \cos^4 \theta + 4\delta \sin^2 \theta \cos^2 \theta J_2 + 4\delta^2 \sin^4 \theta J_0)]$$

$$- \delta \sin^2 \theta[J_0 - 2\beta x(J_1 \cos^2 \theta + 2\delta \sin^2 \theta J_{-1})$$

$$+ (\beta x)^2(J_2 \cos^4 \theta + 4\delta \sin^2 \theta \cos^2 \theta J_0$$

$$+ 4\delta^2 \sin^4 \theta J_{-2})] - \delta x^2 I_3\} \qquad (26)$$

$$p_s = \frac{(r_m/r)^2}{2\sin\phi_m} \{[(1-\delta)\cos^2\theta\cos\phi_m - \delta\sin^2\theta]$$

$$\left[1 - \beta x \left(\cos\phi_m\cos^2\theta + \frac{2\delta\sin^2\theta}{\cos\phi_m}\right)\right]^2$$

$$-\frac{\delta x^2}{2}(r_m/r)^2\} \tag{27}$$

and

$$C_{p\theta} = 16\ \beta x\ (p_c + p_s)/(A/r^2) \tag{28}$$

The term I_3 is defined as

$$I_3 = (r'/r)^3\ d\phi = (1-\gamma)\ J_0$$

$$+ 2\gamma(1-\gamma)J_1 + \delta^2 J_2 \tag{29}$$

The terms $C_{N\theta}$ and $C_{p\theta}$ are functions of the azimuthal angle and the tip speed ratio.

The average power coefficient for the rotor is:

$$C_p = \frac{1}{\pi}\int C_{p\theta}d\theta$$

$$= \frac{16\beta x}{(A/r^2)}\left[\int\frac{p_c d\theta}{\pi} + \int\frac{p_s d\theta}{\pi}\right] \tag{30}$$

The integration of the power functions p_c and p_s involved a number of $\cos_m\theta\sin_n\theta$ terms. When the integrals are evaluated between the limits (0 to π) the following characteristic constants are obtained.

$$G_0 = \gamma J_2/2\ ,$$

$$G_1 = \gamma(J_3 + 2\delta J_1)\ ,$$

$$G_2 = \gamma(5J_4 + 4\delta J_2 + 4\delta^2 J_0)\ ,$$

$$G_3 = I_3\gamma\ ,$$

$$S_0 = (r_m/r)^2/4\sin\phi_m\ ,$$

$$S_1 = 2S_0(\cos\phi_m + 6\delta\sec\phi_m)\ ,$$

$$S_2 = 2S_0(\cos^2\phi_m + 4\delta + 20\delta^2\sec^2\phi_m)\ ,$$

$$H_0 = \gamma J_0/2\ ,$$

$$H_1 = \gamma(J_1 + 6\delta J_{-1})\ ,$$

$$H_2 = \gamma(J_2 + 4\delta J_0 + 20\delta^2 J_{-2})\ ,$$

$$R_0 = S_0\cos^2\phi_m\ ,$$

$$R_1 = 2S_0\cos^2\phi_m(3\cos\phi_m + 2\delta\sec\phi_m)\ ,$$

$$R_2 = 2S_0 cps^2\phi_m(5\cos\phi_m + 4\delta + 4\delta^2\sec^2\phi_m)\ ,$$

$$R_3 = S_0(r_m/r)^2$$

$$(31a - 31k)$$

It is convenient to define the following general constant:

$$K_i = (G_i + R_i) - \delta(G_i + H_i + R_i + S_i) \tag{32}$$

The relationship for the average power coefficient is, therefore,

$$C_p\ [16\beta x/(A/r^2)][K_0 - (\beta x/4)K_1 +$$

$$(\beta x/4)2K_2 - \delta x^2(G_3 + R_3)] \tag{33}$$

The variable in Eq. 33 is the tip speed ratio.

4 USE OF THE MODEL

To use the model to determine the performance factors, one first selects the aerodynamic characteristics C_{Do} and S_L. These characteristics can be obtained for the selected airfoil from a reference such as Ref. 7.

The rotor geometry is determined by selecting the chord ratio (c/r) and the height ratio (h/r). The geometric parameters for the rotor can be determined from Eqs. 3 to 5.

For the aerodynamic characteristics and geometric parameters, the blade parameters and the various integrals can be evaluated. The blade parameters are determined from Eqs. 15 and 16. The value of the integrals are obtained from Eqs. 22, 23, 29, 31, and 32. If the airfoil and the rotor planform are not changed, these values will remain constant.

The blade normal force coefficient power coefficient can be determined at any tip speed ratio and azimuthal angle from Eq. 21 and Eq. 28 respectively. The average power coefficient C_p is obtained from Eq. 33 as a function of tip speed ratio.

5 COMPARISON WITH EXPERIMENTAL RESULTS

Wind tunnel test data for a Darrieus rotor is presented in Fig. 3. This data is from Sharpe, Ref. 5. The planform for the Sharpe rotor is a troposkin with two blades and chord ratio of 0.1185.

The average power coefficient, which was predicted from the model, is presented in Fig. 3 for a NACA 0012 and a NACA 0021. The aerodynamic characteristics used to generate Fig. 3 are:

NACA	0012	0021
C_{Do}	0.0078	0.0096
S_L	6.068	5.626

These values were obtained from Ref. 7 for a half rough airfoil.

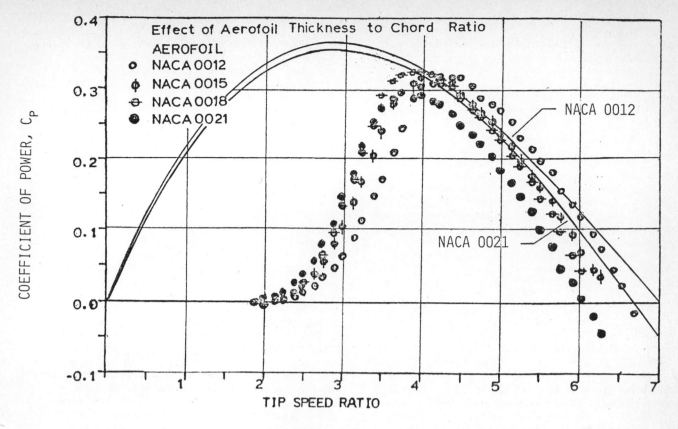

Fig 3   Comparison of model and wind tunnel data

One can see from Fig. 3 that the model agrees quite well with the test data for the NACA 0012 up to the point of maximum C_p. At this point the airfoil is stalled and the model is not valid beyond this point. The model does predict a decrease in performance for the NACA 0021. The decrease is not as much as observed in the wind tunnel.

The average power coefficient at tip speeds greater than the value for maximum C_p, which is the region of validity, is sensitive to the value of profile drag. The coefficient of drag for a rotor is difficult to assess other than by experimental means. The rotor drag includes more than the airfoil drag.

The effect of profile drag on performance is presented in Fig. 4. The airfoil is a nominal NACA 0015. The profile drag values are representative of a smooth, half rough and rough airfoil.

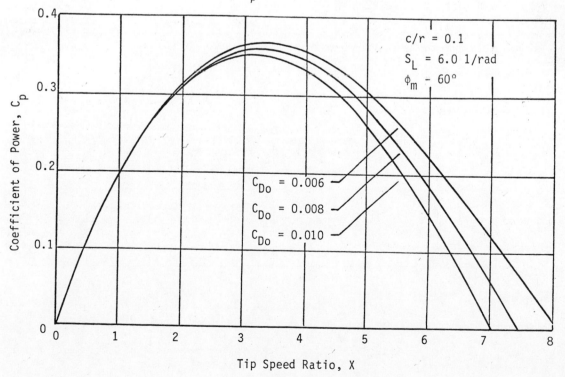

Fig 4   Effect of profile drag

6 NUMERICAL EXAMPLES

The advantage of the model is that the effect of the rotor design parameters upon performance can readily be assessed. Fig. 3 is an example of this. One can see from Fig. 3 that the maximum power coefficient and the cut-in tip speed ratio ($C_p = 0$) decrease with increasing airfoil thickness. This is due to an increase in profile drag with thickness.

The effect of planform shape is illustrated in Fig. 5 and 6. In Fig. 5, the C_p-x relation is presented for various maximum blade angles. Blade angles of 45° and 90° describe a triangular rotor planform and a circular arc planform, respectively, which are the limiting cases. Figure 5 indicates that performance is not a strong function of the rotor planform. This conclusion supports that made by Parashiviou and Delclaux (6).

The maximum C_p, the tip speed ratio at maximum C_p and the cut-in tip speed ratio are presented in Fig. 6. The point of maximum power coefficient can be obtained by differentiation, Eq. 33. The tip speed ratios x_m and x_o both decrease with increasing blade angle. Figure 6 shows that the maximum power coefficient attain a minimum at a blade angle of 56°. This blade angle basically describes a troposkien. This effect is due to a change in sweep area.

The effect of solidity or chord ratio is presented in Fig. 7. As the chord ratio increases, so does the interference factor, see Eq. 14, and the maximum power coefficient. The tip speed ratios for cut-in and for C_{pm} both decrease. This occurs because the β parameter is nearly constant for both conditions.

The cyclic variation for the blade normal force coefficient and the power coefficient are illustrated in Fig. 8. The normal force is positive in the radially outward direction. A negative value for power is due to blade drag.

7 CONCLUSIONS

A model consisting of closed-form equations has been developed for the performance of Darrieus wind turbines. The analysis, which led to the model, is based on the assumptions of constant aerodynamic coefficients and the trigonometry of small angles. The analysis is a combination of disc impulse and blade element theories and results in an interference factor which varies with the azimuthal angle.

The model predicts performance which agrees with wind tunnel data up to the point of rotor stall. This is as expected since the model uses a non-stalling airfoil. The region of validity for the model is from the point of maximum power coefficient to cut-in speed. This is the region of general interest for wind turbines.

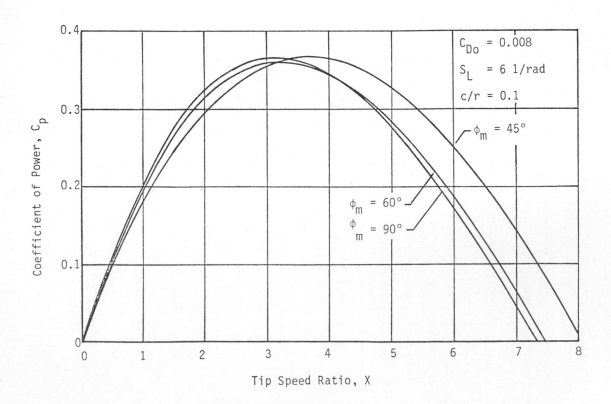

Fig 5   Effect of blade planform

The model is useful for parametric design studies. Both average and cyclic performance can be assessed. The model predicts a minimum in the maximum power coefficient as a function of blade planform. This minimum occurs at a blade angle of 56° which is basically the troposkien. The minimum is only 2% less than the maximum value. Therefore, the performance is relatively independent of planform and that the planform may be selected based on other considerations such as strength.

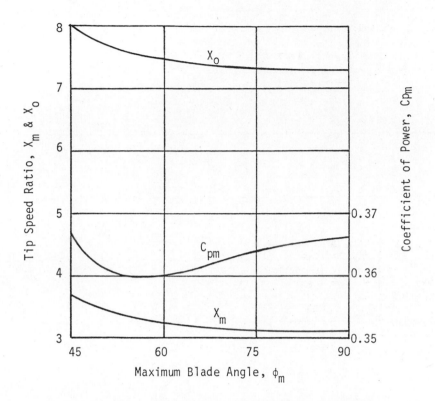

Fig 6   Effect of blade planform on maximum performance

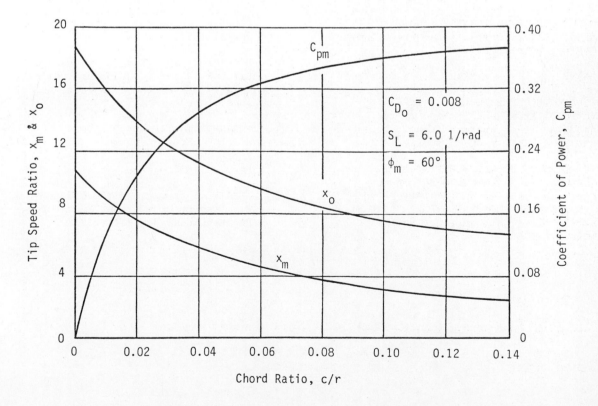

Fig 7   Effect of chord ratio on maximum performance

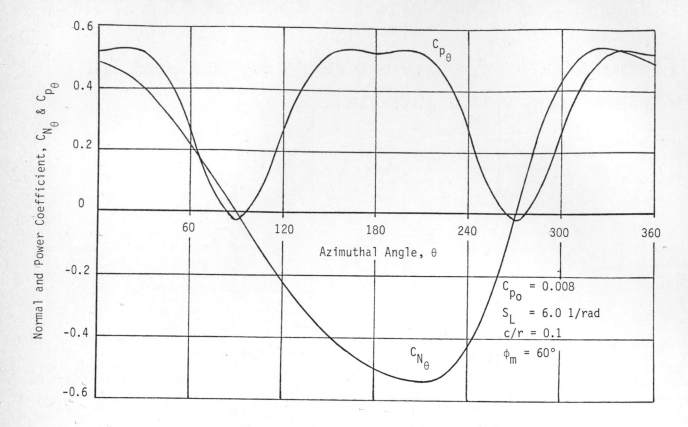

Fig 8   Azimuthal performance

8 REFERENCES

(1) STRICKLAND, J. H. "The Darrieus Turbine,
     A Performance Prediction Method Using
     Multiple Stream Tubes", SAND 75-0431,
     Sandia Laboratories, Albuquerque, New
     Mexico, October 1975.

(2) SHARPE, D. J. and READ, S. "A Critical
     Analysis of the Extended Multiple Stream
     Tube Theory for the Aerodynamics of a
     Vertical Axis Wind Turbine", Proc. 1982
     Wind and Solar Energy Technology Confer-
     ence, University of Missouri, 1982.

(3) STRICKLAND, J. H., WEBSTER, B. T. and
     NGUYGEN, T. "A Vortex Model of the
     Darrieus Turbine: An Analytical and
     Experimental Study", SAND 79-7058, Sandia
     Laboratories, Albuquerque, New Mexico,
     February 1980.

(4) BEANS, E. W. "Approximate Aerodynamic
     Analysis for Horizontal Axis Wind
     Turbines", Journal of Energy, Vol. 7,
     No. 3, 1983, pg. 243.

(5) SHARPE, D. J. "Wind Tunnel Data for
     Darrieus Rotor", Queen Mary College,
     University of London, London, England,
     unpublished.

(6) PARASHIVOIU, I. and DELACLAUX, F. "Double
     Multiple Streamtube Model with Recent
     Improvement", Journal of Energy, Vol. 7,
     No. 3, 1983, p. 250.

(7) ABBOTT, Ira. H., and VAN DOENHOFF, Albert
     E., "Theory of Wing Sections", Dover
     Publications, New York, 1958.

Turbulence – a frequency domain analysis for vertical axis wind turbines

M B ANDERSON
Sir Robert McAlpine & Sons Ltd, Hemel Hempstead

SYNOPSIS A technique is presented for analysing the effects of turbulence on the aerodynamic loads of a vertical axis wind turbine. A frequency domain approach is adopted in order that a parametric analysis can easily be undertaken.   The atmospheric turbulence is assumed to be a stationary Gaussian process which is both homogeneous and isotropic thus allowing the von Karman correlation tensor to be used.   Results are presented which indicate that as a result of rotational sampling a significant amount of turbulent energy is concentrated at multiples of the rotational speed.

1   INTRODUCTION

The effects of turbulence on loads, dynamic response, fatigue life and energy capture have been shown to be important for horizontal axis wind turbines, e.g. Anderson (1982), Powles and Anderson (1984). However, little work has been undertaken, to date, on the effects of turbulence on any of the above aspects for vertical axis wind turbines. Veer's (1984) research, at Sandia Laboratories, deals with the effects of turbulence on the aerodynamic loads, for a 17m Darrieus turbine, using a time domain approach, but no account was taken of dynamic effects. The effects of turbulence becomes more important as the rotor size becomes comparable with turbulent length scales and as a result turbulence will :-

a)   excite all modes of the turbine;
b)   concentrate significant amounts of energy at and around integer multiples of the rotational frequency;
c)   modify the peak aerodynamic loads;
d)   reduce the amount of energy which the rotor can extract from the turbulent component.

A general methodology for analysing the effects of turbulence is shown in Table 1. Apart from the step designated "Rotational Sampling" standard techniques can be used (e.g. Bendat and Piersol 1980). A comparison, between a horizontal and a vertical axis wind turbine, of the physical and statistical properties and assumptions generally made for each step is shown in Table 2 and is exemplified below :

Turbulence
For both types of turbine it is generally assumed that atmospheric turbulence is a stationary Gaussian process which is homogeneous and isotropic. However, for a vertical axis wind turbine, it is necessary to consider both the longitudinal and lateral turbulent velocity components unlike a horizontal axis wind turbine where only the longitudinal component is of importance (Jensen and Frandsen 1978). In practice turbulence is neither homogeneous or isotropic, but Fordham and Anderson (1982) have shown that these limitations do not affect the results significantly.

Rotational Sampling
Owing to the turbines' rotational motion it is necessary to convert the turbulence, as seen by a stationary observer, into a rotating frame of reference. This technique is usually called "Rotational Sampling" and in principle is the same for both types of turbine. For a horizontal axis wind turbine the main feature of this transformation is that it shifts the turbulent energy to higher frequencies and particularly to the rotor speed and its harmonics. The transformed spectrum is however still stationary, i.e. independent of rotor position. As will be shown, in this paper, for a vertical axis wind turbine this transformation causes not only a redistribution of energy but, perhaps more importantly, the spectrum becomes non-stationary with respect to frequency. This effect can be explained by considering an observer rotating with the blade. As the observer approaches the oncoming wind the turbulent spectrum will be shifted to higher frequencies whereas when he is travelling in the same direction as the oncoming wind the converse will occur, the spectrum will be shifted to lower frequencies. This effect is very similar to the well known Doppler effect.

Aerodynamic Loads
For both types of turbine the formulation of the aerodynamic loads is basically the same except that for a vertical axis wind turbine they are periodic. As a result of this the effect of turbulence is dependent on rotor position.   In addition, both turbulent components have to be considered.

Structural Response
Considering only the rotor, the formulation of the equations of motion are fundamentally the same. However, when the supporting structure is included the equations become periodic

unless, for a rotor consisting of more than three blades, a multiblade co-ordinate system is adopted. For a vertical axis wind turbine the equations of motion may be simplified considerably if it is assumed that the supporting structure co-rotates, i.e is isotropic relative to a fixed co-ordinate system.

Statistics

Similar techniques for both types of turbine can be used except with a vertical axis wind turbine the deterministic periodic aerodynamic loads must be combined with the stochastic turbulent components.

This paper presents results specifically for a straight bladed wind turbine but the techniques used are equally applicable for a Darrieus type turbine.

2. STRUCTURAL MODEL

The problem is approached by assuming a modal representation of the turbine. Given the normal mode shapes and frequencies, the response of a blade at a point 'l' and time 't' is given by

$$y(i,t) = \sum_i \phi_i(i) q_i(t) \qquad (1)$$

where ϕ_i and q_i are the mode shape and co-ordinate of the 'i'th mode. The co-ordinate 'y' can represent either lagging, flapping or torsion of the blades assuming that they are mutually uncoupled by the effects of rotation or aerodynamics. The generalised co-ordinate of each mode is assumed to satisfy the following relationships :-

$$M_i \ddot{q}_i + C_i \dot{q}_i + K_i q_i = F_i \qquad (2)$$

where M_i, C_i and K_i are the generalised structural properties and F_i is the generalised force.

From the results of previous work it was found that the applied load is relatively insensitive to the displacement or its derivatives and hence the power spectrum of $y(l,t)$ is given by

$$G_{yy}(\omega) = \sum_i \sum_j \phi_i \phi_j H_i^* H_j \iint \phi_i \phi_j \cdot$$
$$G_{P_1 P_2}(l_1, l_2, \omega) dl_1 dl_2 \qquad (3)$$

where G_{P1P2} is the cross-spectral density matrix of the applied load for all points on the blade and

$$H_i(\omega) = \left\{ K_i \left[1 - (\omega/\omega_i)^2 + j 2 \zeta_i \omega/\omega_i \right] \right\}^{-1} \qquad (4)$$

where ω_i and ζ_i are the undamped natural frequency and damping ratio of the ith mode.

3. STATISTICAL ANALYSIS

For a Gaussian process spectral moments may conveniently be used to characterise a spectral distribution just as probability moments are used to characterise a probability distribution. In general the n-th spectral moment is defined as

$$m_n = \int_0^\infty \omega_n G_{yy}(\omega) d\omega \qquad (5)$$

For an arbitary spectrum Cartwright and Longuet-Higgins (1956) have shown, based on the work of Rice (1944), that it is possible to determine the probability of peak values in terms of a spectral width parameter

$$\varepsilon = \left[1 - m_2^2 / m_0 m_4 \right]^{1/2} \qquad (6)$$

which takes values between 0 (narrow band spectrum) and 1 (wide band spectrum). The probability distribution of maxima (i.e. peak values), denoted by h, is given by

$$p(\gamma) = (2\pi)^{-1/2} \varepsilon \exp\left(-\frac{1}{2} \gamma^2/\varepsilon^2\right) - \frac{1}{2}(1-\varepsilon^2)^{1/2} \cdot$$
$$\gamma \exp\left(-\frac{1}{2}\gamma^2\right) \cdot$$
$$\left[1 + erf\left(\gamma(1-\varepsilon^2)^{1/2}\right)/\varepsilon \cdot \sqrt{2} \right] \qquad (7)$$

where $erf(\ )$ is the error function and $\gamma = h/m_0^{\frac{1}{2}}$. From this distribution the mean of h, denoted $\bar{h}$, the root mean square value h_{rms} are given respectively as

$$\bar{h}/m_0^{1/2} = (\pi/2)^{1/2}(1-\varepsilon^2)^{1/2} \qquad (8)$$

$$\bar{h}_{rms}/m_0^{1/2} = (2-\varepsilon^2)^{1/2} \qquad (9)$$

The expected value of h_{max}, that occurs within a specified time, T is given by

$$E(h_{max})/m_0^{1/2} = \left[2\ln(\tau)\right]^{1/2} + \frac{1}{2} 0.5772 \left[\ln(\tau)\right]^{-1/2}$$

where $\tau = T(m_2/m_0)^{\frac{1}{2}}$. $\qquad (10)$

4. AERODYNAMIC LOADS

Neglecting the steady centrifugal loads the applied load due to the fluctuating aerodynamic forces can be considered to comprise three components. The first arises from the blades periodic motion through a steady flow, the second from turbulence and the third from aeroelastic effects.

For reasons which will become apparent the formulation of the aerodynamic forces, with the effects of turbulence included, are easier if derived in the rotating frame of reference. If the longitudinal and lateral turbulent velocity components, in the stationary frame, are respectively u and w then the velocity components relative to the rotating blade are

$$V_x' = \bar{U}\cos\Theta + u'$$

$$V_y' = \bar{U}\sin\Theta + \Omega r - w' \tag{11}$$

Where u' and w' are the turbulent velocity components in the rotating frame of reference (Figure 1). The blade aerodynamic normal and tangential forces are defined as

$$\begin{Bmatrix} dF_N \\ dF_t \end{Bmatrix} = \frac{1}{2}\rho c W^2 dL \begin{bmatrix} \cos\alpha & \sin\alpha \\ \sin\alpha & -\cos\alpha \end{bmatrix} \begin{Bmatrix} C_L \\ C_D \end{Bmatrix} \tag{12}$$

where the usual symbol convention is applied. As a frequency domain solution is wanted it is necessary to linearise this equation through making the following assumptions :-

$$C_L\cos\alpha \gg C_D\sin\alpha \qquad W = \Omega r$$

$$C_L = C_{L_\alpha}\sin\alpha$$

$$C_D = C_{D_0}$$

which results in

$$\begin{Bmatrix} dF_N \\ dF_t \end{Bmatrix} = \frac{1}{2}\rho c\, dL \begin{Bmatrix} V_x' V_y' C_{L_\alpha} \\ V_x'^2 C_{L_\alpha} - \Omega r V_y' C_{D_0} \end{Bmatrix} \tag{13}$$

Substituting equation (11) into (13) gives for the normal force

$$dF_N = \frac{1}{2}\rho c\, C_{L_\alpha} \{ \bar{U}^2\cos\Theta\sin\Theta$$
$$+ \bar{U}\cos\Theta\,\Omega r - \bar{U}\cos\Theta w'$$
$$+ \bar{U}\sin\Theta u' + u'\Omega r + u'w'\} \tag{14}$$

and assuming that u'w', u' . $\bar{U}\sin\Theta$ and w'$\bar{U}\cos\Theta$ are small compared with the other terms leads to the following linear equation :

$$dF_N = \frac{1}{2}\rho c\, C_{L_\alpha} dL \{ \bar{U}^2\cos\Theta\sin\Theta$$
$$+ \bar{U}\cos\Theta\,\Omega r$$
$$+ u'\Omega r\} \tag{15}$$

As a result of transforming the turbulent velocities into the rotating frame of reference it can be seen from equation (15) that the effect of turbulence is independent of azimuth angle. In addition the only component which effects the normal force is that component which is normal to the blade and hence if the turbulence is Gaussian then the loads will also be Gaussian.

Following a similar analysis, for the tangential force, results in the following equation :

$$dF_t = \frac{1}{2}\rho c\, dL \{ C_{L_\alpha}(\bar{U}^2\cos^2\Theta$$
$$+ 2\bar{U}u'\cos\Theta)$$
$$- C_{D_0}(\Omega r)^2\} \tag{16}$$

Again only the turbulent component normal to the blade is important but unlike the normal force the effects of turbulence are now dependent on the azimuth angle. As a result of this it has not been possible to obtain a solution in the frequency domain without considerably more effort possibly involving Markov processes and the Fokker-Planck equation. However, as the normal force is usually the design load this limitation may not be significant.

5. TURBULENCE

In order to determine $G_{P1P2}(w)$ and hence $G_{vv}(w)$ it is necessary to determine the cross-spectral density function of the turbulent velocity fluctuations.

Assuming that the turbulence is both isotropic and homogeneous then a similar method to that used by Rosenbrock (1955) and more recently by Anderson (1981) can be used. The correlation coefficient between two velocity components 'p' and 'q' at points P and Q in space (see Figure 2) can be shown (Karman (1937)) to be

$$R_{pq} = \langle u^2 \rangle [f(r,t)\cos\alpha\cos\beta$$
$$+ g(r,t)\sin\alpha\sin\beta]\sin\gamma \tag{17}$$

where $\langle u^2 \rangle$ is the variance of the turbulence and the angle brackets denote an ensemble average. The functions f(r,t) and g(r,t) are the correlation coefficients between velocity components parallel and at right angles to the line joining P and Q. In general 'f' and 'g' are functions of separation, i.e. 'r' and also time, however at the frequencies of interest atmospheric turbulence can be considered to be a stationary process and hence time invariant. The condition of fluid continuity provides a relationship between these two functions

$$g(r) = f(r) + \frac{r}{2}\frac{d}{dr}f(r) \tag{18}$$

so that the specification of f(r) implies g(r). The form of f(r) will be discussed later.

In a time interval γ, a point on a blade rotating at a constant angular velocity Ω will move, relative to an observer translating with the mean flow, between two points separated by a vector $\underline{r}$, (see Figure 1) given by :

$$\underline{r} = [R\cos\Theta_0 - R\cos(\Omega\tau + \Theta_0) + \bar{U},$$
$$R\sin\Theta_0 - R\sin(\Omega\tau + \Theta_0),$$
$$Z]^T \tag{19}$$

where $[\ ]^T$ is the transpose, R the radius of the turbine, Z the vertical separation between the two points on the blade.

253

Even though we have assumed that f(r) and g(r), for a given r, are stationary the correlation coefficient R_{pq} will not be stationary as r is a function of not only time lag τ but also Θ_o hence t (absolute time). It is fairly obvious from an analysis of equations (17) through (19) that R_{pq} is periodic, viz :-

$$R_{p\gamma}(t+\tau,t) = R_{p\gamma}(t+\tau+nT, t+nT)$$

(20)

where $\tau = \Theta_o/\Omega$ and skew symmetric viz:-

$$R_{pq}(t+\tau,t) = R_{pq}(-t-\tau,-t)$$

(21)

where T is the period of rotation and 'n' is an integer. An example of the single point (p=q) correlation coefficient for the case of the normal velocity component on a blade at a radius of 50m rotating at 0.7 rad/sec in a 10 m/s wind is shown in Figure 3. It can be observed from this figure that correlation coefficient is dependent on blade azimuth angle. Only half of the function has been plotted as the other will be skew symmetric according to equation (21). As the spectral density of stationary data can be defined by the Fourier transform of a stationary correlation function, so the spectral density of a non-stationary process can be defined by the double Fourier transform (Papoulis 1965) as

$$\Gamma(\omega_1,\omega_2) = \iint_{-\infty}^{\infty} R(\tau_1,\tau_2) e^{-j(\omega_1\tau_1 - \omega_2\tau_2)} d\tau_1 d\tau_2$$

(22)

For the correlation function defined as in equation (20) we have

$$\Delta(\omega_1,\omega_2) = \iint_{-\infty}^{\infty} R_{p\gamma}(t+\tau,t) e^{-j(\omega_1\tau - \omega_2 t)} d\tau dt$$

$$= \Gamma(\omega_1, \omega_1 + \omega_2)$$

(23)

Now from equations (20) and (23) it is possible to show that

$$\Delta(\omega_1,\omega_2) = e^{jnT\omega_2} \Delta(\omega_1,\omega_2)$$

(24)

for all w_2. This equality is possible only if the function $\Delta(w_1,w_2)$ equals zero everywhere except at the points $(\tilde{w}_1, w_2)$ such that

$$e^{jnT\omega_2} = 1$$

(25)

for any integer 'n'. For this to be the case, we must have

$$nT\omega_2 = 2\pi k \qquad \text{k:integer}$$

(26)

This represents a family of straight lines in (w_1, w_2) space which are parallel to the w_1 axis. As a result of this we can define the average $\tilde{R}(\tau)$ correlation coefficient and hence its transform $\tilde{G}(\omega)$, the average cross-spectral power density function :-

$$\tilde{R}_{p\gamma}(\tau) = \frac{1}{T} \int_0^T R_{p\gamma}(t+\tau,t) dt$$

(27)

$$\tilde{G}_{p\gamma}(\omega) = \frac{1}{2\pi} \int_{-\infty}^{\infty} \tilde{R}_{p\gamma}(\tau) \cos\omega\tau \, d\tau$$

(28)

In addition the correlation coefficient is also ergodic. It remains to describe the functional form of f(r), which is not determined by the symmetry properties leading to equation (18). A number of different functional forms were examined :

a) simple exponential,
b) von Karman model,
c) Fourier inversion of the spectrum.

Based on the work of Fordham and Anderson (1982) and Anderson et. al. (1984) the von Karman model was adopted as a compromise between accuracy and computational speed.

The angles α and β in equation (17) are determined by the velocity component of interest, in the case of the normal force it is the component normal to the blade. The angle γ is a function of the vertical separtion between the points p_1 and p_2. If the points lie in the same horizontal plane the angle will be zero. If the points p_1 and p_2 are not on the same blade, but on blades separated by Θ_s, equation (23) is then

$$r = [R\cos\Theta_o - R\cos(\Omega\tau+\Theta_o+\Theta_s)+\bar{U}\tau,$$
$$R\sin\Theta_o - R\sin(\Omega\tau+\Theta_o+\Theta_s),$$
$$z]^T$$

(29)

6. RESULTS

To gain an understanding of the effects of turbulence and also substantiate the assumptions made regarding the removal of non-stationary effects it was decided to neglect the modal response and concentrate on integrated normal aerodynamic blade forces; this in effect assumes that the blade behaves like a rigid body. This assumption also allows a certain amount of verification of the results, with a time domain simulation model. (Anderson et al. 1985).

The power spectrum of the turbulent component only for the integrated half blade normal aerodynamic force is

$$G_N(\omega) = A_N^2 \iint_{LL} G_{u_1u_2}(l_1,l_2,\omega) dl_1 dl_2$$

(30)

where

$$A_N = \frac{1}{2}\rho c C_{L_\alpha} \Omega R$$

(31)

From equation (30) it can be observed that the variance of the normal force is directly proportional to the variance of the turbulence and the statistics will be Gaussian.

The following parameters will be used to describe a straight bladed 100m diameter vertical axis wind turbine :-

Radius	50 m
Chord	5 m
Rotational Speed	0.7 rad/sec
Blade Length (tip to tip)	72 m
Mean Wind Speed	10 m/s
Turbulence Intensity	15%

Equation (30) has been evaluated for the above parameters for a range of turbulent length scales and the results are presented in Figures 4 and 5. The spectra are shown plotted against the reduced frequency (i.e. non-dimensional with respect to the rotational frequency). For all the spectra it can be observed that a significant amount of energy is concentrated around 1P but not at the higher harmonics. Perhaps surprisingly there appears to be a removal of energy at 2P compared with adjacent frequencies. Increasing the length scale from 50m to 150m causes the 1P peak to become larger. A statistical analysis of these spectra, based on equations (6) to (9) is presented in Table 3. From this table a number of observations can be made :

i)   The variance increases with increasing length scale.

ii) The spectral width, an indication of the distribution of energy, does not appear to have any obvious relationship with the length scale. This is not surprising as the calculation of this parameter is prone to numerical error.

iii) The mean and root-mean-square peak values, which are functions of the spectral width, are approximately 1.05 and 1.30 respectively.

iv) The maximum likely peak value decreases with increasing length scale.

v)   The zero crossing rate decreases with increasing length scale.

Presented in Figures 6 to 8 are results obtained from Anderson et. al 1985 using a time domain simulation technique for the turbulent component only; a direct comparison with results obtained from this analysis can be made. A turbulent length scale and intensity of 85m and 15% respectively was used in both computations. From Figure 9 it can be seen that the two spectra are in reasonable agreement especially in the prediction of the 1P peak and the trough at 2.5P. The spectra from the simulation is considerably more 'noisy', at higher frequencies, as a result of statistical errors. Both spectra also predict the existence of a peak at 3.5P even though the structure was considered to be rigid in both cases. The probability and level crossing distributions, for the turbulent normal force component from the simulation, are shown in Figures 10 and 11 respectively. Also shown are the predicted maximum peak value and the zero crossing frequency from this analysis. It can be clearly observed that good agreement exists and that the distributions are approximately Gaussian.

7. CONCLUSIONS

An analytical model has been developed, in the frequency domain, for predicting the modal response of a vertical axis wind turbine to turbulence. Unfortunately owing to non-stationary terms, in the tangential force equation which could not be removed, the model is limited to the normal force only. As discussed this is not a severe limitation as the normal force is usually the design load.

It has been shown that as a result of rotational sampling a significant amount of turbulent energy is concentrated at 1P and and to a lesser extent at 2.25P and 3.5P. These latter two peaks are not thought to be real but the result of the numerical analysis. To a large extent it has been shown that the statistical properties of the turbulent induced normal forces can be assumed to be independent of the turbulent length scale. It has been shown that the maximum peak and root mean square peak values are 3.4 and 1.35 respectively times the standard deviation. Confidence in the assumptions made during the development of this model has been gained by comparing the results with a time domain simulation model.

8. ACKNOWLEDGEMENTS

M.B. Anderson acknowledges Sir Robert McAlpine & Sons Ltd. and the U.K. Department of Energy, who financed the work, for permission to publish this paper.

9. REFERENCES

Anderson, M.B., 1981 'An Experimental and Theoretical Study of Horizontal Axis Wind Turbines'. Ph.D. dissertation, University of Cambridge.

Anderson, M.B., 1982. 'The Interaction of Turbulence with a Horizontal Axis Wind Turbine'. Proc. of the 4th BWEA Conference, Cranfield.

Anderson, M.B., Garrad, A.D. and Hassan, U., 1984. 'Teeter Excursions of a Two-Bladed Horizontal Axis Wind Turbine Rotor in a Turbulent Velocity Field'. J. of Wind Eng. and Ind. Aero., Vol. 17, pp. 71-88.

Anderson, M.B., Bossanyi, E. and Powles, S.J.R., 1985. 'The Response of a Vertical Axis Wind Turbine to Fluctuating Aerodynamic Loads'. Proc of the 7th BWEA Conference, Oxford. Pub. M.E.P.

Bendat, J.S. and Piersol, A.G., 1980. 'Engineering Applications of Correlation and Spectral Analysis'. John Wiley & Sons Inc.

Cartwright, D.E. and Longuit-Higgins, M.S., 1956. 'The Statistical Distribution of the Maxima of a Random Function'. Proc. Roy. Soc., Ser. A, Vol. 237, pp. 212-232.

Fordham, E.J., 1984. 'Atmospheric Turbulence in Relation to Horizontal Axis Wind Turbines and the Construction of Two Research Machines'. Ph.D. dissertation, University of Cambridge.

Jensen, N.O. and Frandsen, S., 1978. 'Atmospheric Turbulence in Relation to Wind Generator Design'. Proc. 2nd Int. Symp. on Wind Energy Systems, Amsterdam.

Papoulis, A., 1965. 'Probability : Random Variables and Stochastic Processes'. New York: McGraw Hill.

Rosenbrock, H.H., 1955. 'Vibration and Stability Problems in Large Wind Turbines having Hinged Blades'. Electr. Res. Assoc., Rep., C/T 113.

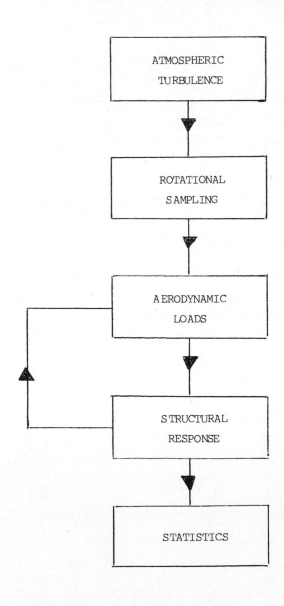

Table 1 General Method of Analysis

TURBINE TYPE

COMPONENT	HORIZONTAL	VERTICAL
Atmospheric Turbulence	Gaussian distribution. Stationary. Only longitudinal component usually of interest. Isotropic and homogeneous. Wind shear neglected.	Gaussian distribution Stationary. Lateral and longitudinal components required. Isotropic and homogeneous. Wind shear neglected.
Rotational Sampling	gaussian distribution preserved. Stationary.	Gaussian distribution preserved. Non-stationary with respect to frequency.
Aerodynamic Loads	Gaussian distribution preserved. Stationary. Independent of blade position	Non-Gaussian distribution Non-stationary. Dependent on blade. position.
Structural Response	Linear structural model. Gaussian distribution preserved. Independent of blade posit/n except when tower is included for a two-bladed turbine.	Linear structural model. Non-Gaussian distribution. Non-stationary. Dependent on blade position.

Table 2 Comparison of Basic Assumptions

257

Parameter	LENGTH SCALE (^xL_u) (m)					
	50	75	100	125	150	Infinity
Variance (m_o) (N^2)	9.07×10^8	9.94×10^8	10.04×10^8	10.07×10^8	10.97×10^8	12.17×10^8
Spectral Width (ε)	0.600	0.349	0.482	0.542	0.585	0.456
Mean Peak ($\bar{h}/m_o^{\frac{1}{2}}$)	1.00	1.17	1.09	1.05	1.01	1.11
R.M.S. Peak ($h_{rms}/m_o^{\frac{1}{2}}$)	1.28	1.37	1.32	1.30	1.28	1.33
Zero Crossing Rate (Hz)	0.735	0.622	0.556	0.511	0.479	0.432
Maximum Peak ($h_{max}/m_o^{\frac{1}{2}}$)	3.48	3.44	3.41	3.40	3.36	3.33

Table 5.3   Variation of the Statistical Properties of the Aerodynamic Normal Force with Length Scale. (Turbulent Component Only)

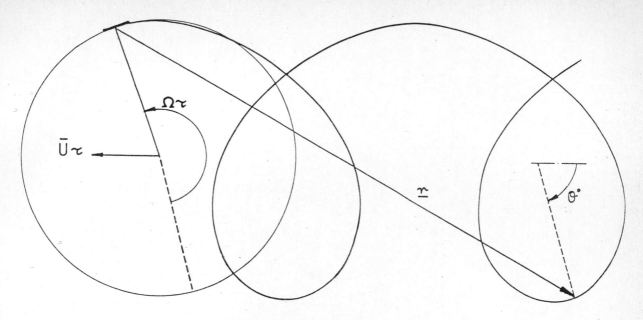

Fig 1   Trajectory of a point on the blade relative to an
observer translating with the mean flow

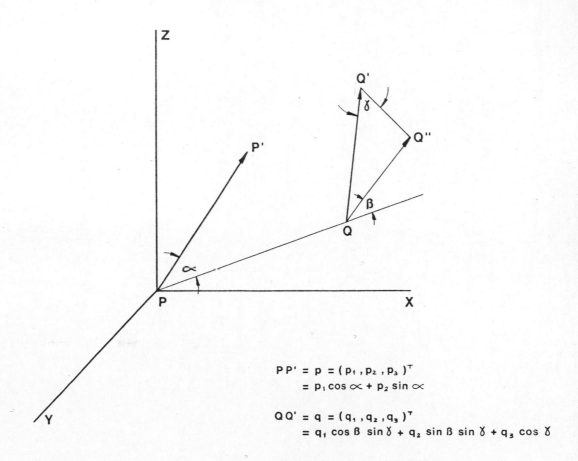

$$PP' = p = (p_1, p_2, p_3)^T$$
$$= p_1 \cos \alpha + p_2 \sin \alpha$$

$$QQ' = q = (q_1, q_2, q_3)^T$$
$$= q_1 \cos \beta \sin \gamma + q_2 \sin \beta \sin \gamma + q_3 \cos \gamma$$

Fig 2   Correlation

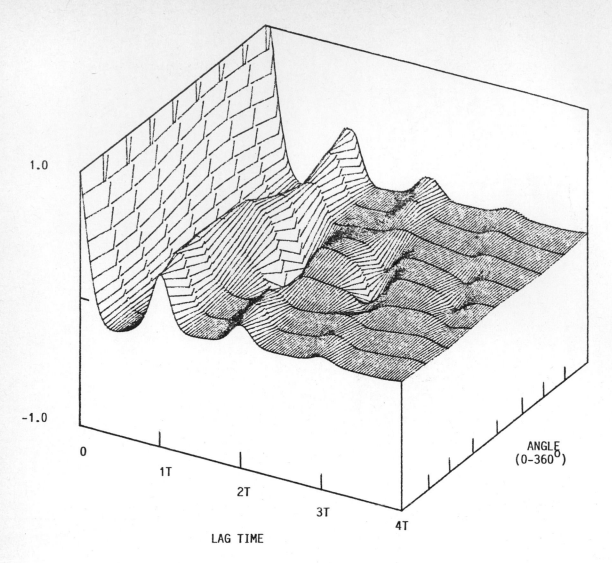

1.0

−1.0

0

1T

2T

3T

4T

ANGLE
(0–360°)

LAG TIME

Fig 3   Correlation coefficient

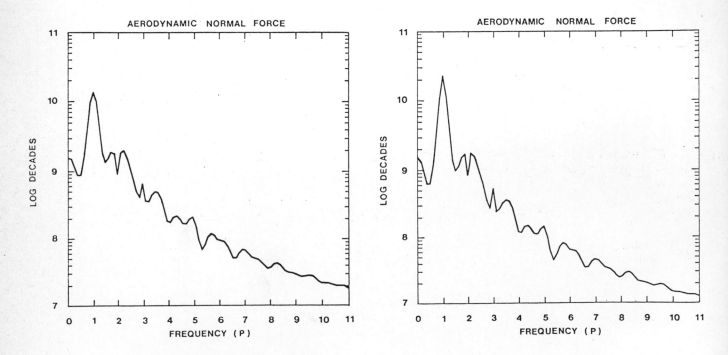

AERODYNAMIC NORMAL FORCE

LOG DECADES

FREQUENCY (P)

Fig 4   Power spectral density function of normal blade
force (turbulence only). Length scale = 100m

AERODYNAMIC NORMAL FORCE

LOG DECADES

FREQUENCY (P)

Fig 5   Power spectral density function of normal blade
force (turbulence only). Length scale = 150m

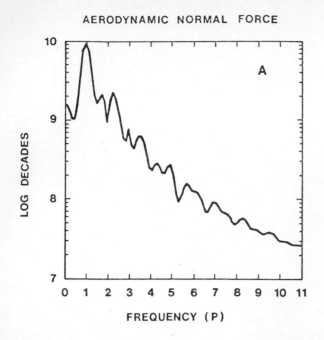

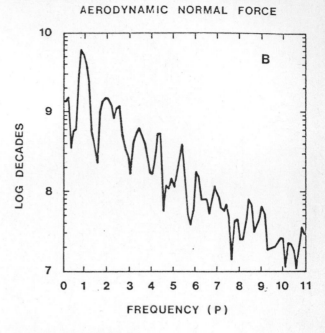

Fig 6   Comparison of the predicted power spectral density
functions of the normal force from a direct frequency
domain analysis (A) and indirectly via the time
domain (B)

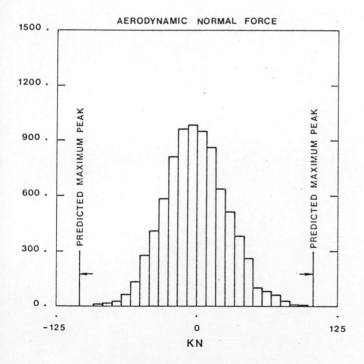

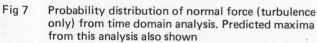

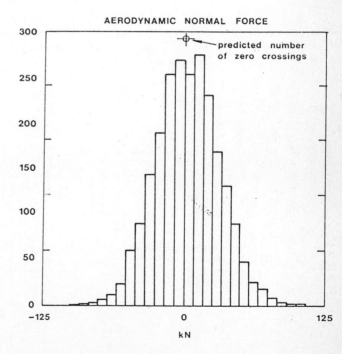

Fig 7   Probability distribution of normal force (turbulence
only) from time domain analysis. Predicted maxima
from this analysis also shown

Fig 8   Level crossings of normal force (turbulence only)
from time domain analysis. Predicted number of
zero crossings from this analysis also shown

Methods of predicting dynamic stall

R A McD GALBRAITH, BSc, PhD, CEng, MRAeS
Department of Aeronautics and Fluid Mechanics, University of Glasgow
M VEZZA, BSc
James Howden & Company, Glasgow

This report contains a brief discussion on the catagories and capabilities of current methods of predicting dynamic stall. The catagories considered are similar to those of previous surveys[1,2,3], whilst the quality and capabilities of each model considered, are assessed on the basis of the currently accepted salient features of the stall[1,4], and other influential factors. The subjective assessments of the model details are concisely presented in tabular form which covers the major portion of the work. The table is intended as a quick reference guide to the models. Finally, the report concludes with a speculative discussion on future prospects.

1)   Introduction

During the forward flight of a helicopter, the rotor is subjected to the cyclic pitching about a mean angle of incidence dictated by the collective pitch applied. The cyclic pitch is necessary to balance the lift over the rotor disc and, on the retreating side, where the local dynamic pressure is low, the rotor blade incidence often exceeds the value at which "static stall" occurs. Provided the incidence is reduced before the onset of dynamic stall, this can have a beneficial effect and extend the flight envelope of the aircraft. The occurrence of rotor dynamic stall, however, is a most serious event and imposes unacceptable loads on the control linkages.

Figures 1 and 2 illustrate the typical extreme loadings that can occur during dynamic stall, and these may be compared to their "static" counterparts. Included in fig.2. is a diagramatic presentation of the accepted physical description of dynamic stall the salient features of which are:

a)   A large overshoot of C_n and C_m, over the static counterparts up to the onset of stall.

b)   After stall onset, a large vortex appears and is convected down the chord resulting in a divergence of the normal force and pitching moment coefficients, followed by a complete stall and loss of lift.

c)   As the geometric incidence of the aerofoil decreases, there is subsequent boundary layer re-attachment and a return to fully attached flow.

The importance of dynamic stall on the performance of a helicopter is well understood[5] and, thus, the need to possess a predictive capability for it is evident[1-5,6]. In contrast to this, it is only recently that dynamic stall has been given due consideration for wind turbine applications. This is hardly surprising for a technology with such a short modern history and rapid development, albeit wind turbines have existed for several centuries. Dynamic stall is of particular relevance to the VAWT whose blades experience a similar once per cycle oscillatory variation in effective incidence. Whilst the details of the effective incidence variation are not truly sinusoidal, the effective cyclic pitch is directly analogous to that of the helicopter rotor. Even the HAWT, when operating in yaw, experiences cyclic changes of incidence. As our understanding and predictive capabilities are at present, these minor differences hardly matter; although it is a source of some controversy[27]. Indeed, the helicopter rotor experiences several other motions such as flapping, lag and twist, all of which are coupled and, in addition, the flow field is highly three-dimensional. The authors are therefore of the opinion, based on the research carried out for the helicopter industry, that the dynamic stall, at least for VAWT, is a very important consideration for fatigue life assessement and overall performance.

As may be expected, many predictive schemes for dynamic stall have been proposed and are still under continuous development within the helicopter industry. These have been surveyed and a broad classification adopted whereby the different methods are grouped according to the mathematical technique, or otherwise. To date, however, no quick comparative summary of the relevant methods has been carried out. Such a summary forms the basis of this report and, inevitably, the survey will be subjective and should be treated so.

It is not the purpose of the discussion herein, to set one model against another, but simply to highlight, and roughly grade, the salient features of each and succinctly present them. The result (tab. 1) is, it is hoped,

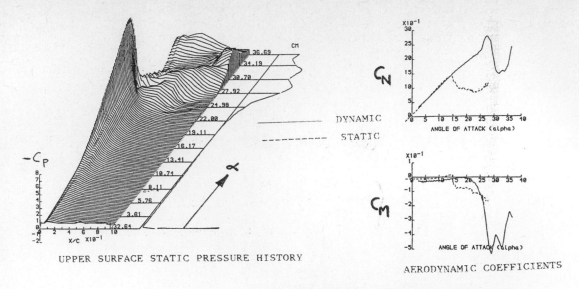

UPPER SURFACE STATIC PRESSURE HISTORY

AERODYNAMIC COEFFICIENTS

Fig 1   Dynamic stall of a NACA 23012 aerofoil for constant pitch rate
displacement (Ref. 31) ($\alpha \simeq 167°$/s)

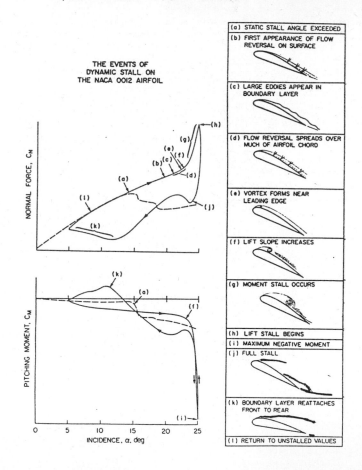

THE EVENTS OF
DYNAMIC STALL ON
THE NACA 0012 AIRFOIL

(a) STATIC STALL ANGLE EXCEEDED

(b) FIRST APPEARANCE OF FLOW
    REVERSAL ON SURFACE

(c) LARGE EDDIES APPEAR IN
    BOUNDARY LAYER

(d) FLOW REVERSAL SPREADS OVER
    MUCH OF AIRFOIL CHORD

(e) VORTEX FORMS NEAR
    LEADING EDGE

(f) LIFT SLOPE INCREASES

(g) MOMENT STALL OCCURS

(h) LIFT STALL BEGINS

(i) MAXIMUM NEGATIVE MOMENT

(j) FULL STALL

(k) BOUNDARY LAYER REATTACHES
    FRONT TO REAR

(l) RETURN TO UNSTALLED VALUES

Fig 2   General development of dynamic stall on an oscillating
aerofoil (Ref. 30)

fair to all and can be usefully enhanced as the
models and criticisms develop. Each of the
models has been developed to fulfill a specific
requirement, and any relative grading of the
models (not done here) must reflect this need.

The manner in which the models will develop
is inextricably linked to both the physical
understanding of the stall process and the
computational power and speed of the available
computers. To date, there is no obvious single
way to proceed and the reasons for this are

discussed.

2)   Survey of current methods

2.1   The model categories and salient stall
features

The various models to be considered have
been grouped under similar headings to those
used in other surveys[1,2,3]. In this case,
however, a concise version of the survey is

METHOD FOR PREDICTIONS		REF	TO STALL ONSET					STALL ONSET				POST STALL						MOTION				OTHER FACTORS					
PREDICTIVE CAPABILITY			UNSTEADY PRESS. DIST	UNSTEADY BOUNDARY LAYER	LAM.–TURB. TRANSITION	FLOW REVERSAL	COMPRESSIBILITY	L.E. CRITERION	TRAILING EDGE SEPARATION	SHOCK-WAVE INTERACTION	ACOUSTIC WAVES	VORTICITY BUILD-UP/SHEDDING	VORTICITY TRANSPORT	WAKE MODELLING	INDUCED EFFECTS OF WAKE	SUBSEQUENT VORTEX SHEDDING	RE-ATTACHMENT	SINUSOIDAL	RAMP	PLUNGE	OTHER	SWEEP / 3-D	REYNOLDS Nº	BLADE VORTEX INTERACTION	ROUGHNESS ETC.	FREE STREAM TURBULENCE	NOISE
NAVIER STOKES	METHA	7	✳	✳	—	✳	—	✳	✳	—	—	✳	✳	✳	✳	✳	✳	✳	—	—	—	—	+	—	—	—	—
	SHAMROTH	8	✳	✳	✳	✳	✳	✳	✳	—	—	✳	✳	✳	✳	✳	✳	—	✳	—	—	—	—	—	—	—	—
DISCRETE VORTEX	HAM	9	+	—	—	—	—	O	—	—	—	✳	✳	✳	✳	—	+	✳	—	—	✳	—	—	—	—	—	—
	BAUDU	10	✳	—	—	—	—	+	—	—	—	✳	✳	✳	✳	✳	—	✳	—	—	—	—	—	—	—	—	—
	VEZZA	11	✳	—	—	—	—	—	O	—	—	✳	✳	✳	✳	✳	O	✳	✳	Δ	✳	—	—	—	—	—	—
	SPALART	12	✳	Δ	+	Δ	—	+	+	—	—	✳	✳	✳	✳	✳	+	✳	—	—	✳	—	—	—	—	—	—
	LEWIS	13	✳	—	—	—	—	+	+	—	—	✳	✳	✳	✳	✳	—	Δ	—	—	O	—	O	—	—	—	—
	ONO	14	✳	—	—	—	—	O	O	—	—	+	✳	✳	✳	+	+	✳	—	✳	—	—	—	—	—	—	—
	KATZ	15	O	—	—	—	—	O	O	—	—	✳	✳	✳	✳	✳	.	—	—	—	+	—	.	—	—	—	—
ZONAL	RAO	16	O	O	✳	—	—	—	✳	—	—	—	—	O	O	—	+	✳	—	—	—	—	—	—	—	—	—
	SCRUGGS	17	✳	✳	—	✳	—	—	✳	—	—	—	—	—	O	—	—	—	✳	—	—	—	—	O	—	—	—
	CRIMI	18	+	O	✳	—	—	✳	✳	—	—	—	—	O	O	—	✳	✳	✳	✳	✳	—	—	O	—	—	—
PREDOMINANTLY EMPIRICAL	BEDDOES	19	+	—	—	—	✳	✳	✳	✳	✳	✳	✳	—	—	—	✳	✳	✳	—	—	—	—	—	—	—	—
	GANGWANI	20	—	—	—	—	✳	✳	✳	—	—	✳	✳	—	—	—	✳	✳	—	✳	✳	✳	O	—	—	—	—
	TRAN	21	—	—	—	—	✳	O	O	—	—	O	O	—	—	—	✳	✳	—	+	—	—	—	—	—	—	—
	ERICSSON	22	—	—	—	—	O	O	O	—	—	—	—	—	—	—	+	✳	✳	—	—	—	—	—	—	—	—
	GORMONT	23	—	—	—	—	✳	O	—	—	—	—	O	—	—	—	+	✳	—	—	—	—	—	—	—	—	—
	JOHNSON	24	—	—	—	—	—	O	—	—	—	✳	—	—	—	O	—	✳	—	—	—	—	—	—	—	—	—

✳ GOOD CONSIDERATION   + APPROXIMATE   O VERY APPROXIMATE   Δ BEING DEVELOPED   — NOT MODELLED.

TABLE 1

presented in tabular form (table 1) from which the strengths and weaknesses of the models can be seen more readily. The model categories are as follows:

a) Navier-Stokes methods
b) Discrete vortex methods
c) Zonal methods
d) Predominantly empirical methods

Navier-Stokes solutions, of which there are two notable methods, attempt to solve the relevant equations in their fundamental form by numerical techniques. In contrast to this the "Discrete Vortex" approach normally ignores the viscous terms in the basic equations and assumes potential flow outwith the boundary layer. The viscous nature of the flow is modelled or taken account of, by the generation and subsequent induced transport of discrete combined vortices. The manner and location of their generation is normally obtained empirically or via appropriate boundary layer calculations. A further and related simplification of the Navier-Stokes methods is used in the "Zonal" models in which the predicted separation or viscous "zone" is taken as the boundary for the external potential flow. In the numerical procedure of the model, the regions interact in an iterative manner.

In the above three categories, there is a distinct attempt to invoke the basic equations of fluid motion and around this philosophy the flow model is constructed. In contrast to this, the last and very broad category, considers all models in which little or no direct account is taken of the basic equations. These models rely on good quality empirical data matched to a model of the gross features of the stall process; they commonly form part of a rotor performance program.

From table 1, it can be seen that various categories have also been employed for the assessment of the predictive capabilities of

the methods. The headings cover most of the relevant features of dynamic stall[4] and the main ones are five in number:

a) To stall onset
b) At stall onset
c) Post stall
d) Types of motion considered
e) Other factors.

Within the above, the details considered relevant are listed (tab.1) and include such events as transition, induced effects of the wake and re-attachment etc. It is assumed that all models should be capable of giving C_N and C_M predictions and so these have been omitted from the table.

If a model considered any of the listed features either explicitly or implicitly, an appropriate symbol was placed in the relevent location of table 1. These symbols were:

✳	good consideration of the phenomenon
+	approximate
0	very approximate
Δ	being developed
—	not modelled.

The allocation of these symbols was based on the relevant published work and should not be considered as being an exact process. Nevertheless, it is thought that the tabular presentation provides an easily digestable means of assessing the present state of the art in numerical studies of dynamic stall, as well as the future research needs.

2.2 Fundamental Navier-Stokes methods

Here the two most commonly quoted works have been considered. The first, Metha[7], only considers the laminar state, but the resulting flow patterns clearly showed the characteristics which are normally associated

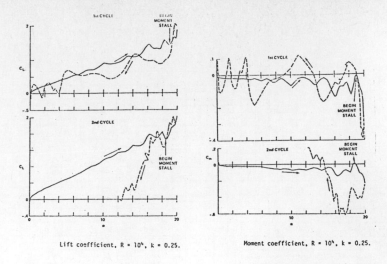

Lift coefficient, R = 10⁴, k = 0.25.   Moment coefficient, R = 10⁴, k = 0.25.

Fig 3   Navier-Stokes method (Laminar flow, Metha. Ref. 7)

with dynamic stall such as vortex build up, its
convection and eventual shedding from the
trailing edge. Figure 3 illustrates the C_L, C_M
and α cycles obtained. As a consequence of the
fundamental nature of the process it is
accepted that a good consideration of the
listed factors is as indicated, but due to the
restriction of laminar flow, the Reynolds
number variation is assessed as approximate.
At the stall onset, both leading and trailing
edge criteria are implicitly modelled.

The second such solution, Shamroth[8], takes
account of turbulent flow and, in doing so,
requires the inclusion of a suitable turbulent
model or closure hypothesis. On the quality
and appropriatness of the turbulence model and
necessary empirical inputs, will the accuracy
of the predictions depend, once the numerical
problems have been minimised. Unfortunately,
no overall C_N and C_M data were given, but the
pressure distributions obtained from ramp like
motions, fig. 4, are encouraging.

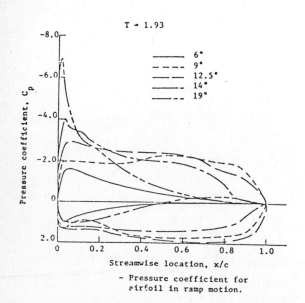

Fig 4   Navier-Stokes method (Turbulent flow, Shamroth
Ref. 8)

2.3   Discrete vortex methods

Only three of the seven listed methods
(tab. 1) will be considered in detail and these
are those of refs. 11, 12 and 13. The first of
these, Vezza and Galbraith[11], represents the
aerofoil surface by a vortex sheet from which,
at the separation points, two further sheets
emerge and, in time, these are replaced by
discrete vortices (fig. 5). At present, the
method is only applicable to trailing edge
separation. So far only ramp and impulsive
motions in pitch have been considered, but the
oscillatory case is at an advanced stage of
development and the necessary separation and
re-attachment criteria are based, on the data
of ref. 17.

In contrast to this, the model of Spalart
et al[12] is highly developed and the technique
employed envelopes the aerofoil in a line of
discrete vortices positioned a small distance
from the aerofoil surface. On this surface the
zero velocity condition is invoked. The
vortices may be continuously created or
absorbed, and convected in the induced flow
field. Initially the convection was
unrestricted, but subsequently linked to a
separation point obtained via a suitable
boundary layer calculation. As may be seen in
fig. 6, the predictions are most encouraging
and consideration of an unsteady boundary layer
calculation with flow reversal has been made.

The method of Lewis[13] is, as yet, only
applicable to fixed incidence aerofoils,
consideration is being given to the unsteady
case and although it is similar to that of ref.
12, it contains an interesting "random walk"
technique to account for viscous diffusion.
The method is highly developed and widely
applicable. For example, it is capable of
giving a good reproduction of the Blasius
profile for the zero pressure gradient boundary
layer (see fig. 7).

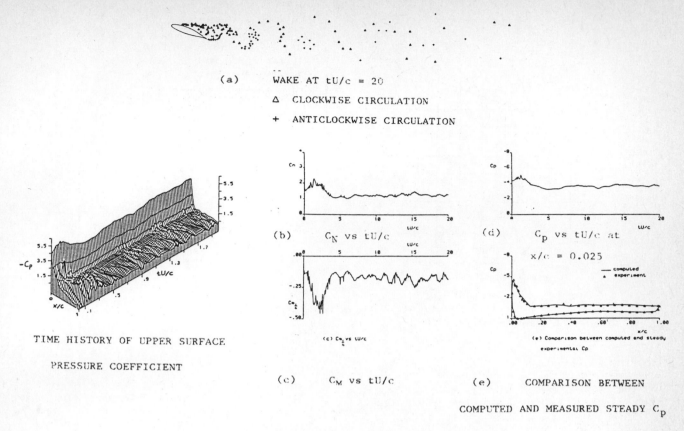

(a) WAKE AT tU/c = 20

Δ   CLOCKWISE CIRCULATION

+   ANTICLOCKWISE CIRCULATION

TIME HISTORY OF UPPER SURFACE

PRESSURE COEFFICIENT

(b) C_N vs tU/c

(c) C_M vs tU/c

(d) C_p vs tU/c at

x/c = 0.025

(e)   COMPARISON BETWEEN

COMPUTED AND MEASURED STEADY C_p

Fig 5   Results obtained following a step change in incidence from
0 to 21.14° using the GA(W)−1 aerofoil (Ref. 11)

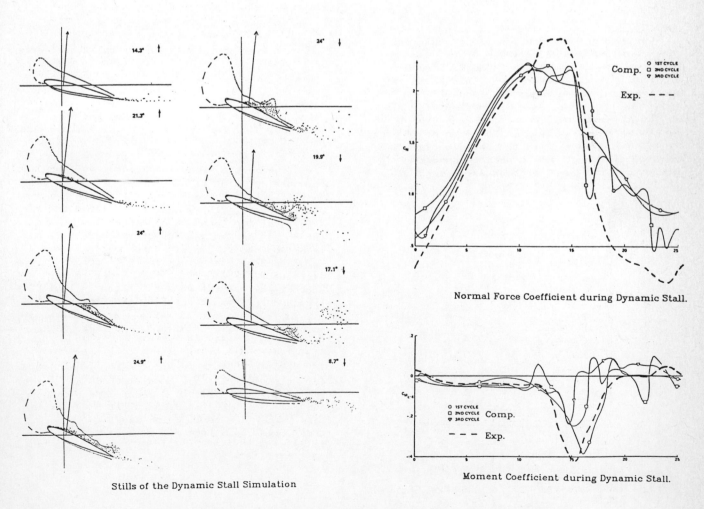

Stills of the Dynamic Stall Simulation

Normal Force Coefficient during Dynamic Stall.

Moment Coefficient during Dynamic Stall.

Fig 6(a) Discrete vortex method, Spalart *et al*, Ref. 12 (NACA 0012 aerofoil)

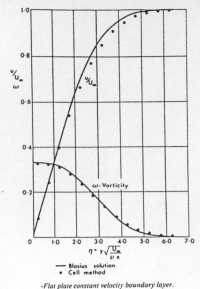

— Blasius solution

• Cell method

-Flat plate constant velocity boundary layer.

Fig 7   Lewis and Porthouse, Ref. 13

2.4   Zonal Methods

The three methods considered in the survey include that of Scruggs et al[17] which is not a pure dynamic stall method. It is, however, an important contribution, in that the effects of pitch rate on flow reversal within the viscous zone/boundary layer are clearly expressed. Their results have been used, for example, by Vezza and Galbraith[11] and Beddoes[19].

The model of Rao et al[16] is a quasi-steady method which models trailing edge separation. The shape and induced effects of the wake are included by enclosing the "dead air" region within two vortex sheets lying along streamlines of the flow. The dynamic stall overshoot in C_N etc. and the hysterisis loop associated with oscillatory cases are predicted as may be seen in fig. 8. The method has been developed further[29] and the new technique is a highly developed, hybrid between discrete vortex and the zonal methods as defined under the catagorisation.

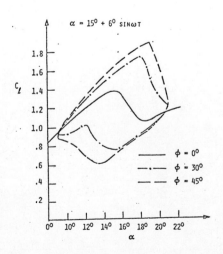

Fig 8   Zonal method, Rao *et al*, Ref. 16

The model of Crimi and Reeves[18] is similar to that of Rao et al, but is more general in that it includes bubble growth and bursting criteria which may be invoked depending on the boundary layer state predicted. Figure 9 shows typical results and it may be seen that, as with ref 16, several of the gross effects of dynamic stall are predicted, but the vortex induced lift is not.

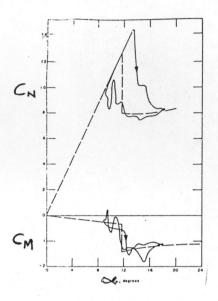

Fig 9   Zonal method, Crimi and Reeves, Ref. 18

2.5   Predominantly empirical

These methods, like most of the above, are well considered in refs. 1, 2 and 3. The category is, however, very large as may be realised from the very different approaches of Beddoes[19] and Gangwani[20], which are respectively, a time delay method and an apparently highly empirical and accurate curve fit.

Although the method of Beddoes is dependent on empirical time delays, it is based on, and models, accepted gross features of the dynamic stall process. It is therefore biased towards a physical interpretation of the flow, where as the method of Gangwani (synthesised approach) is predominantly empirical and borders on a pure and highly accurate curve fitting exercise. In principle, the latter could account for all the effects listed in table 1, provided sufficient good quality data were available. Both these methods are remarkably successful as may be judged from figs. 10 and 11.

A significant distinguishing feature of this catagory is, that these methods are mainly those currently used in the helicopter industry and have been developed to fulfil a particular task. They are still being developed and their predictive capability may be observed in figs. 10 to 13.

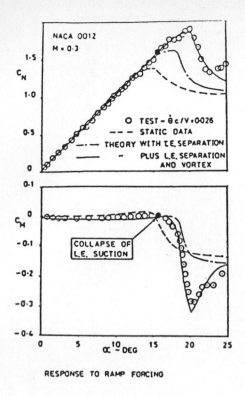

RESPONSE TO RAMP FORCING

Fig 10   Beddoes, Ref. 19

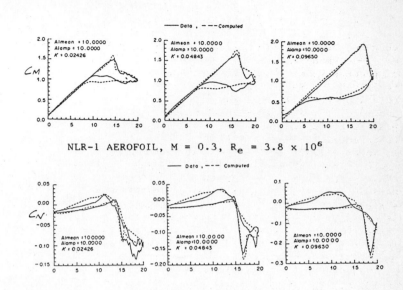

NLR-1 AEROFOIL, M = 0.3, $R_e = 3.8 \times 10^6$

Fig 11   Gangwani, Ref. 20

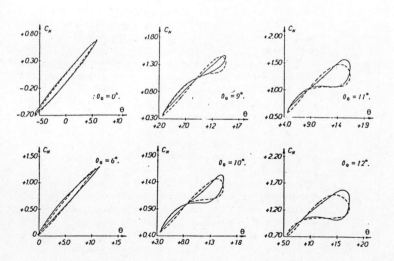

Fig 12   Tran and Petot, Ref. 21

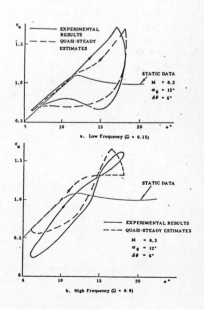

Fig 13   Erricson and Reding, Ref. 22

3) General Discussion

When considering the summary table, it should be recalled that it relates to the phenomena modelled and not the overall performance in predicting the time histories of C_N and C_M. If the table is used to assess predicted C_M and C_N histories, one would probably conclude that the Navier-Stokes methods, because of their comprehensive nature, would be among the best performers. Their comprehensive nature follows from the philosophy of modelling the flow via the basic equations and assuming that, in doing so, the observed flow phenomena will automatically be predicted. In this respect they are fundamentally different from all other models in which individual flow phenomena of the stall process are considered to a greater or lesser extent. Whilst, for the laminar case this may be justified, and very impressive predictions are obtained, this is primarily a consequence of having a well understood and accurate stress-strain relationship. Such accuracy is not possessed, however, by current turbulence models, and on the inginuity and applicability of the chosen hypothesis will the quality of the predicted flow development depend. To permit flow development from the basic equations employing a very approximate turbulence model, will likely result in very approximate predictions, albeit the salient features of the stall are implicitly modelled.

For the discrete vortex methods, it may be seen that, whilst the unsteady pressure distribution is normally predicted up to the stall onset, the detailed boundary layer effects are not. Spalart et al, however, did include such a calculation, but with limited success and work on the inclusion of an unsteady boundary layer method continues. The incentive for such work is, of course, to provide a suitable stall criterion. If no boundary layer assessment is included, the stall onset criterion is simply an empirical input, and thus the methods are on a par, for this salient feature, with predominantly empirical procedures. Once the stall is initiated, however, the subsequent development is well considered up to the process of re-attachment as indicated in tab. 1.

The predominantly empirical methods do not, at first sight, appear particularly favourable, but a closer inspection of tab. 1 and figs. 11 and 12 may reveal that the important stall onset and re-attachment are well considered and may even account for the effects of sweep.

From all the data presented herein, it is clear that there is a lack of comparisons between the empirical and predicted C_N and C_M histories (less so in the predominantly empirical methods) and each modeller tends to present his own personal test cases. This makes a detailed comparison difficult and suggests that there is a distinct requirement for a set of accepted and pertinent tests cases such as exists for boundary layer work (first used in 1968 at the Stanford Conference[25]). Even fewer comparisons are given for the unsteady pressure distributions, where appropriate, and such information would be useful when assessing predictive capability in detail. A set of test cases could, as it develops, cover a range of motions and the "other factors" which are little considered by all the methods.

The proposed data base implies future development of the methods; but which ones? It is suggested that this is an unanswerable question. The purist may prefer continued strong effort in the fundamental Navier-Stokes equations because, once the numerical algorithms are developed, all that remains is to improve the turbulence model. If the finite difference grid were such that the explicit modelling of the large scale turbulence could be eliminated, then the problem of the turbulence model would be simplified; but the small scale would remain. This, however, requires massive computing power and, at present, such power is unavailable and future prospects of attaining it are debatable[26,27].

In contrast to this, the predominantly empirical methods require little computer resources but numerous criteria to individually model the relevant manifestations of the stall. The more comprehensive this is, the greater the requirement for good quality empirical data, the collection of which is expensive and time consuming. In principle, however, all the factors noted in tab. 1 may be accounted for by suitable correlations to provide more general models. These would use limited computer resources and yield economic predictions when used in rotor performance codes.

There is, thus, a dilema on how to proceed. Navier-Stokes solutions require, in essence, a single closure hypothesis but large computer resources, whilst the empirical models need only limited computational power but large and expensive data bases. Unfortunately, the problem is further complicated by the prospect of a "rational" computer which will use both large computing capacity and the accumulated data bases. Such methods could possess a heuristic quality that is at present unavailable, and the desire for effective "rational" algorithms has provided the incentive to develop the necessary computers labled, 5th Generation[26].

If the above suggested codes are realised, they could be incorporated in an "integrated" package for aerofoil performance etc (eg ref. 28), and will require access to appropriate data banks and prediction procedures. In fact, all the hard earned data collected to date, would find a new use and as more was accumulated and made available they, in effect, would add to the "rational" program's "experience". There is thus the possibility that, in pursuing the further development of empirically dominated methods, the necessary collected data would have alternative uses.

4) Conclusions

It is concluded that,

1) There is a definite need for accepted test

cases by which the performance of dynamic stall methods may be assessed.

2) There is a currently unresolved dilema as to which basic prediction philosophy should be adopted. Using the appropriate form of the Navier-Stokes equations in finite difference form requires large and expensive computing power, whilst the predominantly empirical procedures need little of such power but much expensive data.

3) The summary table provides a quick reference to the methods and this should be developed and modified where necessary.

Acknowledgements

The authors wish to express their thanks to the U.S.A.R.D.S.G. - U.K. and SERC who kindly funded the summary under contract number DAJA45-84-M-0446 and 82801965 respectively.

References

1 McCROSKEY, W.J.
   The Phenomenon of Dynamic Stall.
   Lecture Series 1981-4 in Unsteady Airloads and Aeroelastic Problems in Separated and transonic Flow. von Karman Institute for Fluid Dynamics, March 1981.

2 BEDDOES, T.S.
   Short Course in Unsteady Aerodynamics (Rotor Applications).
   von Karman Institute, March 1980.

3 PHILIPPE, J.J.
   Dynamic Stall: An example of Strong Interaction between Viscous and Inviscid Flows.
   AGARD CP-227, 1977.

4 YOUNG, W.H.Jr.
   Fluid Mechanics Mechanism in the Stall Process for Helicopters.
   USAAVRADCOM TR 81-B-1, NASA.TM.81956, 1981.

5 BYHAM, G.M. and BEDDOES, T.S.
   The Importance of Unsteady Aerodynamics in Rotor Calculations.
   AGARD CP 277, September 1977.

6 WHITE, R.P.
   Periodic Aerodynamic loadings: the problem, what is being done and what needs to be done.
   Jo. of Sound Vib. (1966) 4, (3), 282-304. 1966.

7 MEHTA, U.B.
   Dynamic Stall of an Oscillating Airfoil.
   AGARD CP 277 September 1977, AGARD Meeting on Unsteady Aerodynamics, Ottawa, Paper 23.

8 SHAMROTH, S.J. and GIBELING, H.J.
   Analysis of turbulent flow about an oscillating aerofoil using a time dependent Navier-Stokes procedure.
   AGARD CP No. 296, 1981.

9 HAM, N.D.
   Aerodynamic Loading on a Two-Dimensional Airfoil during Dynamic Stall.
   AIAA Journal, Volume 6, No. 10, October 1968.

10 BAUDU, N., SAGNER, M. and SOUQUET, J.
    Modelisation du rechrochage dynamique d'un profile oscillant.
    AAAF 10th Colloque d'Aeronautique Applique, Lille, France, 1973.

11 VEZZA, M. and GALBRAITH, R.A.McD.
    Modelling of unsteady, incompressible separation on an aerofoil using an inviscid flow algorithm.
    G.U. Aero. Rept. 8412, 1984.

12 SPALART, P.R., LEONARD, A. and BAGANOFF, D.
    Numerical simulation of separated flows.
    NASA TM 84328, 1983.

13 LEWIS, R.I. and PORTHOUSE, D.T.C.
    Recent advances in the theoretical simulation of real fluid flows.
    Gen. Meeting of N.E. Coast Inst. of Eng. and Shipbuilders, U.K., 1983.

14 ONO, K., KUWAHARA, K. and OSHIMA, K.
    Numerical analysis of dynamic stall phenomenon of an oscillating aerofoil by the discrete vortex approximation.
    Paper No. 8, 7th International Conference on Numerical Methods in Fluid Dynamics, Stanford, 1980.

15 KATZ, J.
    A discrete vortex method for the non-steady separared flow over an aerofoil.
    J.F.M. Volume 102, pp 315-328, 1981.

16 RAO, B.M., MASKEW, B. and DVORAK, F.A.
    Theoretical Prediction of Dynamic Stall on Oscillating Airfoils.
    Presented at the 34th Annual Natural Forum of the Americal Helicopter Society, May 1978.

17 SCRUGGS, R.M., NASH, J.F. and SINGLETON, R.E.
    Analysis of Flow-Reversal Delay for a Pitching Foil.
    AIAA 12th Aerospace Sciences Meeting, AIAA Paper No. 74-183, February 1974.

18 CRIMI, P. and REEVES, B.L.
    A Method for Analysing Dynamic Stall.
    AIAA 10th Aerospace Sciences Meeting, AIAA Paper No. 72-87, 1972.

19 BEDDOES, T.S.
    Representatives of airfoil behaviour.
    AGARD-CCP-334, 1982.

20 GANGWANI, S.T.
    Synthesized airfoil data method for prediction of dynamic stall and unsteady airloads.
    VERTICA Volume 8, No. 2, pp 93-118, 1984.

21 TRAN, C.T. and PETOT, D.
    Semi-empirical model for the dynamic stall of airfoils in view of the applications to the calculation of responses of a helicopter blade in forward flight.
    VERTICA, Volume 5, pp 33-53, 1981.

22 ERICSON, L.E. and REDING, J.P.
    Dynamic stall of helicopter blades.
    Jo. of Am. Helicopter Society, Volume 17,
    No. 1, 1972.

23 GORMONT, R.E.
    A mathematical model of unsteady
    aerodynamics and radial flow for
    applications to helicopter rotors.
    US Army AMRDL-EUSTIS Directorate Report
    TR-72-51, 1972.

24 JOHNSON, W.
    The effect of dynamic stall on the response
    and airloading of helicopter rotor blades.
    J.A.H.S. Volume 14, No. 2, pp 68-79, 1969.

25 COLES, D.E. and HIRST, E.A. (Editors)
    Proceedings of the Stanford Conference on
    turbulent boundary layer prediction.
    Volume 2, AFOSR-IFP, University Press,
    Stanford, California, 1968.

26 FEIGENBAUN and McCORDUCK.
    The Fifth Generation; Pan Books, 1984.

27 International Conference on Supercomputers.
    Technology Training Corporation,
    Paris/London, 1984.

28 KERR, A.W. and STEPHENS, W.B.
    The development of a system for
    interdisciplinary analysis of rotorcraft
    flight characteristics.
    AGARD-CPP-334, 1982.

29 MASKEW, B. and DVORAK, F.A.
    Prediction of dynamic separation
    characteristics using a time-stepping
    viscid/inviscid approach.
    3rd Symposium on Numerical and Physical
    Aspects of Aerodynamic Flows, California
    State University, Long Beach, C.A., 1985.

30 CARR, L.W., McALISTER, K.W., McCROSKEY, W.J.
    Analysis of the development airfoil
    experiments.
    NASA TN-D 8382, 1977. (Also AIAA J. Vol.
    14, No. 1, January 1976).

31 SETO, L.Y. and GALBRAITH, R.A.McD.
    The collected data for ramp function tests
    on a NACA 23012 aerofoil. Volume 1:
    Description and pressure data.
    G.U. Report No. 8413, November 1984.

Application of tangential wall-jets for high lift generation

H V RAO and **G E L PERERA**
Department of Mechanical & Production Engineering, The Polytechnic, Huddersfield

SYNOPSIS

The principle of high lift generation by a tangential wall-jet of air is established from the relevant aerodynamic theory and is also verified through wind-tunnel studies. The feasibility for utilising this principle for improving the performance of conventional machines or for application in a new concept of Vertical Axis Wind Turbine Generator (VAWTG) is examined. Preliminary theoretical studies indicate several advantages for the proposed new concept of VAWTG consisting of polygonal cylinders. Comparison is made with conventional wind turbine generators based on computer simulated optimum operation. Further work is required through prototype development and testing.

NOTATION

b	height of wall-jet at entrance
B	number of cylinders/blades
C_D	drag coefficient
C_L	lift coefficient
C_m	$\equiv 0.5\ (b/r)(V/W_r)^2$, wall-jet momentum coefficient
C_p	$\equiv P/\rho\ HRW^3$, Power coefficient
ρ	density of atmospheric air
r	radius of the cylinder/the circle enclosing a regular polygon
H	effective height of VAWTG
N	radian angular velocity of VAWTG
P	power developed by VAWTG
R	radius of the orbiting cylinders of VAWTG measured at their axes
s	$\equiv rB/R$, solidity
V	wall-jet air velocity at entrance
V_m	maximum value of V
W	wind velocity
W_r	wind velocity relative to the orbiting cylinder/blade

INTRODUCTION

The power extracted by any wind machine directly depends on the lift/drag generated on the working surfaces for the prevailing relative flow conditions. In the case of conventional wind turbines the attainment of high lift coefficient through the use of appropriately shaped working surfaces leads to a compact and efficient machine. Apart from the influence of the geometry of the surface, modifications of the boundary layer flow also have a significant effect on the lift generation.

In principle it is possible to achieve a high lift coefficient by - (i) movement of the solid boundary surface as in the well-known Magnus effect and (ii) the injection/suction of fluid to/from the boundary layer. In this paper the possibility for exploiting the second of the above categories of high lift generation through the use of air injection in the form of a tang-

ential wall-jet is considered to obtain an improved performance of conventional wind turbine generator and to apply in a new concept of VAWTG having cylinders of polygonal cross-section as working surfaces. The wall-jet principle may also be utilised in the Madaras rotor concept in order to avoid the axial rotation of the cylinders. Initial theoretical investigation shows that as much as 30% increase in power output may be achieved after accounting for the power required for air jets. However, the Madaras rotor concept is not considered here, as it will not be possible to achieve the optimum utilisation of the wall-jet principle and also such a system has many other difficulties in practical implementation. This is brought out clearly in the report by Whiteford, et al. (1)

1 WALL-JET ACTION ON A CYLINDER

A high lift coefficient may be achieved for a cylinder in a cross-flow by the introduction of a wall-jet of air at an appropriate location as shown in Fig.1. Furuya and Yoshino (2) and Yoshino et al. (3) considered the aerodynamics of wall-jet action on cylinders. The lift and drag coefficients were experimentally determined by Waka et al. (4) and their data may be represented by the following equations:

$$C_L = 16\ C_m + 1.4 \qquad (0<C_m<0.5) \ \ldots\ldots(1)$$

$$C_D = 1.2\ C_m + 0.73 \qquad (0<C_m<0.5) \ \ldots\ldots(2)$$

The above equations are valid for the case of a single wall-jet located at 135° from the upstream stagnation point on the circular cylinder. The data is strictly applicable only for a two-dimensional flow but may be used with little error for aspect ratios (H/r) greater than 6.0 and especially where end plates are provided.

2 APPLICATION FOR A WIND MACHINE

The principle of high lift generation may be utilised to enhance the performance of a conventional wind machine by introducing tangential wall-jets on to the turbine blades. A preliminary study indicated that the complexity of the mechanical arrangement, and the power required in forming the wall-jets will render this technique impractical in general. However, the tangential wall-jet may be applied with several advantages in a new concept of VAWTG consisting of two or more polygonal cylinders acting as turbine blades as shown in Fig.2. The performance is optimised by introducing the wall-jets at the appropriate locations dependent on the relative wind direction, as the cylinders orbit about the axis of the machine. Both boundary layer considerations and wind-tunnel studies indicate that it is adequate to use only one or two wall-jets located at an angular position of about 135° measured from the upstream stagnation point with respect to relative wind velocity.

3 EXPERIMENTAL STUDIES

In order to establish the phenomenon of lift generation by the effect of tangential wall-jet for geometries other than that of a circular cylinder, wind tunnel tests were conducted on a NACA 0018 aerofoil section and on an octagonal cylinder having sides of length 42 mm and height 600 mm. The actual arrangement for injecting the wall-jets is indicated in Fig.3. The lift force was qualitatively observed in the case of the aerofoil by mounting it on suitable ball-bearings. These experiments were carried out in a wind tunnel measuring 1.5 x 1.0 m in cross-section. A maximum free stream air velocity of 15 m/s and wall-jet velocities up to 45 m/s are used. These tests clearly establish the principle of lift generation by the effect of wall-jet on an aerofoil section. It is to be noted that all the experiments on the aerofoil are conducted with zero angle of attack and hence any observed lift produced is entirely due to the modification of the boundary layer flow, rather than due to the effect of delaying the boundary layer separation brought about by the wall-jet. The lift and drag forces on the octagonal cylinder are planned to be evaluated by the measurement of pressure distribution on the surface. A sensitive pressure transducer in conjunction with a digital data-logger is employed in these experiments. Experimental data is currently being processed.

4 MATHEMATICAL MODELLING OF THE FLUID FLOW

The successful application and optimisation of the wall-jet principle to a wind machine requires lift and drag data obtained from wind tunnel tests to be corrected for size effect. The mathematical models of fluid flow will enable these corrections to be determined. A flat plate in a free stream of air subjected to tangential wall-jet and oriented with various angles of attack as shown in Fig.4, is initially considered for developing the mathematical model. The model is based on a two dimensional ideal fluid flow outside the boundary layer which is to be matched with the edge of boundary layer flow.

The flow within the boundary layer is modelled by the 'momentum integral' differential equation. The velocity field is determined by the Runge-Kutta method of integration of the above equation with boundary condition dictated by the flow outside the boundary layer. The computed pressure distribution and the wall shear stress provide data for calculating the lift and drag forces on the flat plate. The above computer model has been successfully developed and is now being extended for the case of an octagonal cylinder in the presence of wall-jet/s for various orientations within the free stream of air. Comparison of the computer predictions with the wind-tunnel test data enables the calculation of the size effect parameters required for prototype design.

5 THE POWER COEFFICIENT

A computer simulation is done for evaluating the optimum power coefficient for various values of blade-tip/wind speed ratio (RN/W). The results of this computation are shown in Figures (5) and (6). In this computer simulation, the lift and drag coefficients quoted in equations (1) and (2) are used and two polygonal cylinders are considered (B=2). Further, a magnitude of 0.1, for the jet height to radius ratio (b/r) is used. The latter ratio influences the magnitude of the jet momentum coefficient (C_m). The power coefficient (C_p) will increase with an increase in C_m, in general. However, the magnitude of b/r is limited by the practical arrangement for introducing the two-dimensional wall-jets. It is also to be expected that the drag coefficient will increase with greater values of (b/r), due to the increased projected area of the polygonal cylinder. It may be noted that the calculated power coefficient (C_p) is based on the net power produced after accounting for the power required for the formation of jet air flow. Further, the computed value is a pessimistic estimate as the maximum value of the lift coefficient was restricted to 8.0, although it could attain values higher than this according to equation (1). There is also a further improvement possible by optimising the two stitch-over periods for the wall-jets, while each of the cylinders completes a full orbit around the axis of the machine.

6 ADDITIONAL FEATURES OF THE PROPOSED VAWTG

The following are the main additional advantages of the proposed VAWTG utilising the wall-jet principle for high lift generation:

(i)   The wall-jets may be used for aero-dynamic braking and speed control.
(ii)   The flexible nature of air jets makes it simple to alter the operating conditions with changes in wind velocity, maintaining optimum performance throughout.
(iii)   Less liable for 'stalling' as wall-jets stabilise the boundary air flow.
(iv)   The structural stability of polygonal cylinders will result in material savings as critical stress conditions are avoided.
(v)   The system is self-starting unlike some of the VAWTG.

The following are the major disadvantages which require careful consideration in the prototype development:

    (i)   Fluctuating forces on the cylinders and the possible vibrations.

    (ii)   The complexity of supplying and controlling the wall-jet air flow.

7 CONCLUSIONS

Theoretical studies indicate that it is feasible to utilise the principle of high lift generation by the tangential wall-jets on a polygonal cylinder in a new concept VAWTG. Such a machine will have an optimum value of power coefficient for a blade-tip/wind speed ratio of about 1.0. The proposed VAWTG will have some advantages over the conventional machines. Further work is required through prototype development and testing.

REFERENCES

(1)   Whiteford, D F; Minandi, J; West, B S; Dominic R J. Analysis of the Madaras Power Plant. An alternative method for extracting power from the wind. Technical Report, University of Dayton Research Institute, Ohio, June 1978.

(2)   Furuya, Y; Yoshino, F. Aerodynamic forces acting on a circular cylinder with tangential injection of air. B. JSME v18, n123, September 1978.

(3)   Yoshino, F; Waka, R; Iwasa, T; Hayashi, T. The effect of a side-wall on aerodynamic characteristics of a circular cylinder with tangential blowing. B. JSME v24, n192. June 1981.

(4)   Waka, R; Yoshino, F; Hayashi, T; Iwasa, T. The aerodynamic characteristics at the mid span of a circular cylinder with tangential blowing. B.JSME v26 n 215. May 1983.

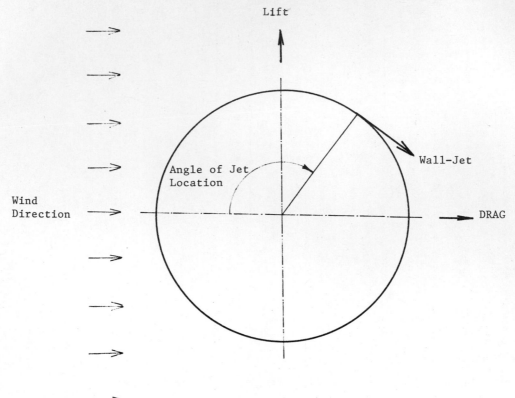

Fig 1   Cylinder with wall-jet

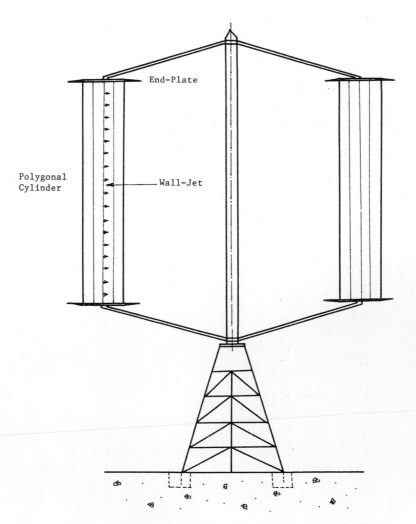

Fig 2   Concept of VAWTG using polygonal cylinders with
wall-jets

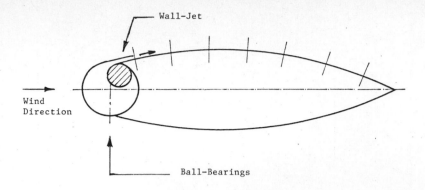

Wall-Jet

Wind
Direction

Ball-Bearings

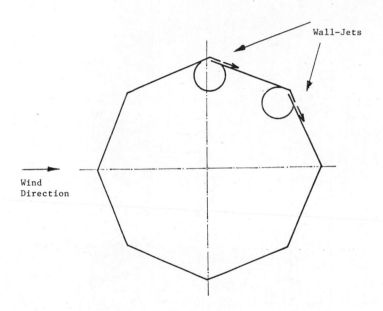

Wall-Jets

Wind
Direction

Fig 3   Arrangement of wall-jets for aerofoil and octagonal
cylinders

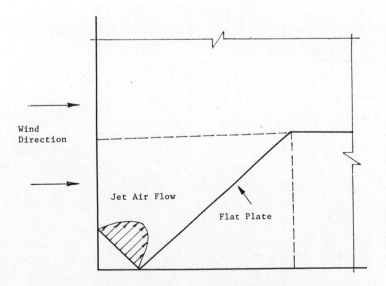

Wind
Direction

Jet Air Flow

Flat Plate

Fig 4   Two-dimensional flow model for a flat-plate with
wall-jet

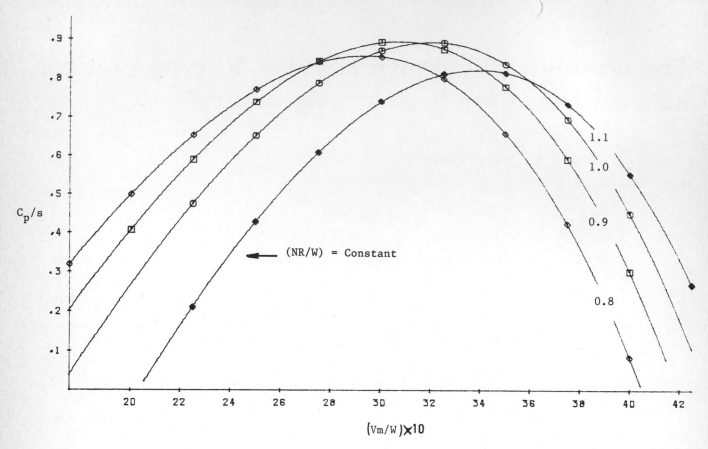

$(V_m/W) \times 10$

Fig 5   Variation of power coefficient for the proposed
VAWTG with (V_m/W) and $(R.N/W)$

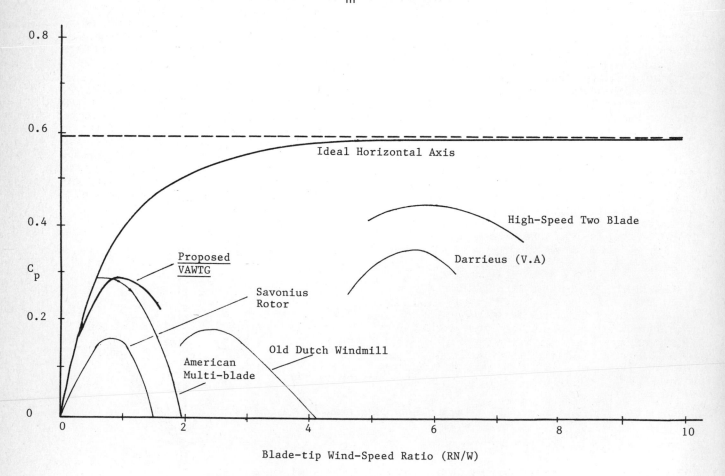

Blade-tip Wind-Speed Ratio (RN/W)

Fig 6   Comparison of optimum power coefficient of the
proposed VAWTG

The aerodynamic control of the 'V' type vertical axis wind turbine by blade tip control

A J ROBOTHAM
Appropriate Technology Group, The Open University, Walton Hall, Milton Keynes
D J SHARPE
Dept of Aeronautical Engineering, Queen Mary College, London

SYNOPSIS

This paper presents results of wind tunnel tests carried out on a small V-type Vertical Axis Wind Turbine (V-VAWT). This model wind turbine was specifically designed to study the effect of tip area and tip pitch on the performance of this novel vawt. The results have demonstrated that tip pitch is highly suitable for overall control of the V-VAWT and that tip areas as little as 5% of the total blade area could provide both power regulation and overspeed control. While it has not been possible to match the predicted performance data directly to these tests results, the computer model VAWTTAY6 can be used with some confidence to predict the performance of larger sized V-VAWTs.

1 INTRODUCTION

The V-type Vertical Axis Wind Turbine, conceived by Derek Taylor at The Open University, has been described at a number of international wind energy and solar energy conferences [1-4].

The development of the V-VAWT concept has concentrated on the enhancement of Sharpe's aerodynamic performance prediction model VAWTTAY6, the verification of this model with data from wind tunnel tests of small V-VAWTs and the design and construction of a 5kW free-air wind turbine. The 5kW V-VAWT has recently been erected on the Appropriate Technology Group's field test facility at The Open University, but only preliminary evaluation of the performance of this machine has been conducted.

This paper will consider the aerodynamic control methods that have been perceived as suitable for rotor power control and overspeed protection of a medium sized V-VAWT, and will compare the predicted performance and measured performance of a small V-VAWT with pitchable blade tips, Fig 1.

2 AERODYNAMIC PERFORMANCE PREDICTION

The aerodynamic performance of the V-VAWT is modelled using the computer program VAWTTAY6, which embodies Sharpe's extended multiple streamtube theory [5]. The predictions using VAWTTAY6 have been verified by wind tunnel tests with model V-VAWTs undertaken at Queen Mary College, London [2].

3 POWER REGULATION AND OVERSPEED CONTROL

All wind turbines, whether they be of horizontal axis or vertical axis configuration, require some form of power regulation and overspeed protection, either actively or passively activated and the V-VAWT is no exception.

Blade tip control has been extensively used on horizontal axis wind turbines of all sizes and has proven to provide power regulation and overspeed control. However the only vertical axis machine known to the authors employing such control is the recently developed Westwind 75kW wind turbine; an 'H'-type VAWT with two blades of fixed pitch but with moveable tips at both ends of each blade.

Blade tip control is also considered highly suitable for the V-VAWT, but the tip device known as the T-brake [4] has a number of structural advantages and because of its simplicity is favoured for the 5kW free-air V-VAWT. However, in order to assess the potential of tip pitch control, a wind tunnel sized V-VAWT with moveable tip portions was constructed and its performance evaluated at Queen Mary College.

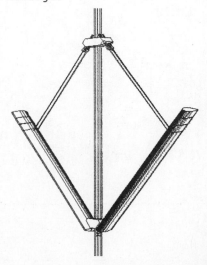

Fig 1   General view of model V-VAWT with pitching tips

3.1 Wind tunnel test model and testing technique

A two-bladed model V-VAWT of tip diameter 940mm was constructed to assess the potential of tip pitch control. Each blade was 615mm in length, with a uniform chord of 80mm. For strength the blades were of a NACA0025 aerofoil cross section and were made from English Ash encapsulating a high tensile aluminium spar for rigidity. Each blade was held to the hub at the 30% chord, inclined at 45 degrees to the vertical and supported by a pair of cables attached 115mm from the tip.

Additionally, each blade had three moveable tip portions, the area of each measuring 5% of the total blade area. The pitch angle of each tip portion was adjustable and could be pre-set with either positive, 'nose-in' pitch or negative, 'nose-out' pitch. The position of each tip portion was locked by two grubscrews that were accessible through the leading edge. During tests, the access holes were filled with plasticene and covered with vinyl tape to restore the leading edge profile. Using this model, the effect of variation of tip pitch on overall performance for tip areas of 5%, 10% and 15% of the total blade area on overall turbine performance has been studied.

The model V-VAWT was tested in the blowdown wind tunnel at Queen Mary College using the acceleration method [6], which is a simple and quick method for determining the complete Cp-lamda characteristic of the turbine.

The test results have been corrected for cable drag and bearing friction, both of which were measured separately. Cable drag is the most significant of these parasitic losses and was measured in a manner previously described [2]. Corrections for wind tunnel blockage have not been included because the experiments were conducted in an open jet tunnel.

The rotor would only accelerate to a rotational speed where the torque being developed by the turbine was equal in magnitude to the parasitic drag losses from the cables and the bearings. In order to obtain performance data for rotational speeds greater than this equilibrium speed, it was necessary to drive the turbine via a friction contact with an electric motor. When the motive power was released, the deceleration of the turbine was measured. This technique allowed measurement of rotor torque over a greater range of tip speed ratios and tip pitch settings.

All tests were conducted at an average Reynold's Number of approximately 2-300000 which is much lower than the operating Reynold's Numbers of larger-sized, free-air machines. The consequence of testing at such low Reynold's Numbers is discussed below.

3.2 Wind tunnel test results

Some of the results from this series of tests are presented in figures 2-6. Figures 2-4 show

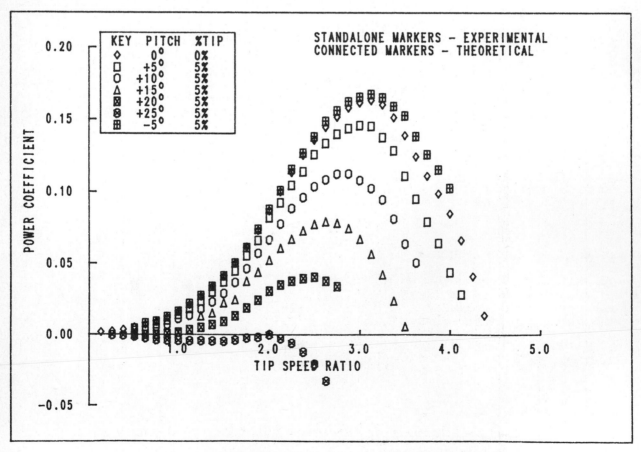

Fig 2   Measured effect of tip pitch on power coefficient for
         model V-VAWT with 5 per cent tip area

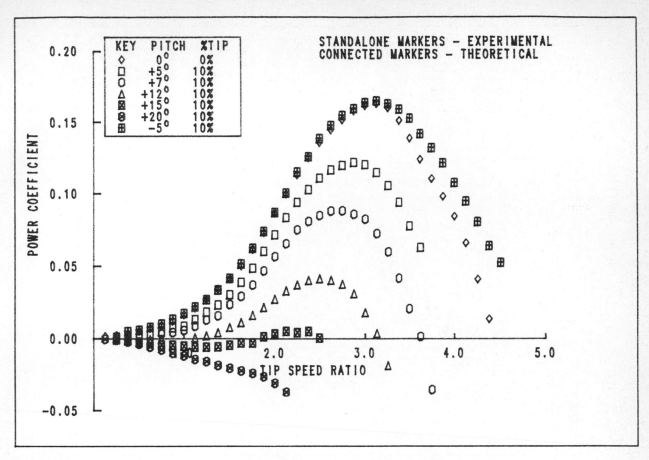

Fig 3   Measured effect of tip pitch on power coefficient for
model V-VAWT with 10 per cent tip area

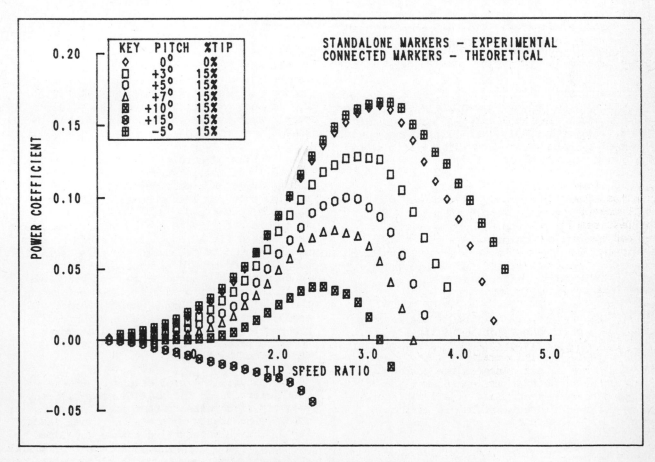

Fig 4   Measured effect of tip pitch on power coefficient for
model V-VAWT with 15 per cent tip area

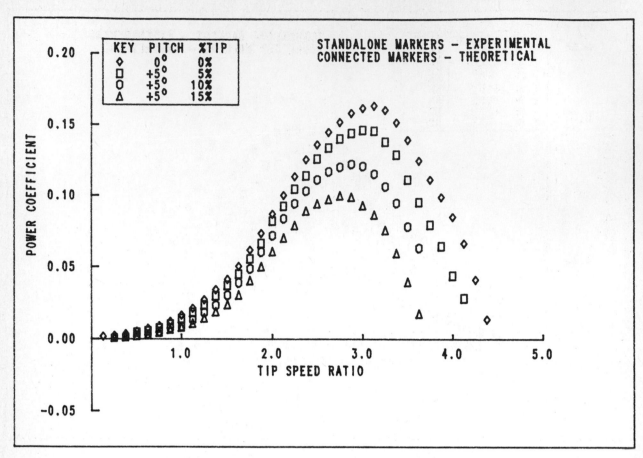

Fig 5   Measured effect of tip area on power coefficient for
model V-VAWT for +5° tip pitch angle

how Power Coefficient is affected by changes in tip pitch angle for tip areas of 5%, 10% & 15% of the total blade area. Figure 5 compares the effect of tip area for a pitch setting of +5 degrees.

These results illustrate the effectiveness of tip pitch control as a means of power modulation and overspeed protection. Even a 5% tip area, at pitch angles in excess of +25 degrees, can effectively 'kill' all the power developed by this model wind turbine at all but the lowest tip speed ratios.

Looking again at figures 2-4, it is seen that Power Coefficient is enhanced with small negative (nose-out) pitch angles. From the test results available, a pitch angle of -5 degrees appears to optimise this enhancement. Enhancement of power with small, full-span pitch offsets was noted by Stacey and Musgrove [7] for the 'H'-VAWT, so a similar effect was not unexpected with the V-VAWT.

Figure 6 shows how Torque Coefficient varies with pitch angle for a 5% tip area. It shows the high starting torque that is a feature of the V-VAWT, and illustrates how a small negative pitch offset enhances the turbine performance. Note that the enhancement of developed torque is apparent at all tip speed ratios for a -5 degree pitch angle even at starting.

3.3 Predictions of tip pitch effects

The computer program VAWTTAY6 can be used to predict the effect of tip pitch on the performance of the model V-VAWT. However, it is not possible for these predictions to be matched directly with the wind tunnel results because at present VAWTTAY6 uses NACA0012 static aerofoil data, whereas the wind tunnel model was constructed using the thicker NACA0025 section.

Published aerofoil data for the NACA0025 aerofoil section is scarce, though Sandia Laboratories have published data for this section that has been generated using the Eppler computer code [8]. Despite the fact that this data covers angles of incidence upto 180 degrees for a range of Reynold's Numbers, none of the data presented for this section has been verified by wind tunnel tests. Consequently the authors have not used this aerofoil data for predicting the performance of the model V-VAWT. Low Reynold's Number static aerofoil tests of a NACA0025 section are planned for later this year so that more accurate predictions of performance of can be made using VAWTTAY6.

However, the predictions of Power Coefficient using the NACA0012 aerofoil data show tip pitch effects that are similar to those demonstrated by the wind tunnel tests, Fig 7. The high starting torque developed by the model V-VAWT and the enhancement of torque with small negative, nose-out pitch positions is also predicted by the theory, Fig 8.

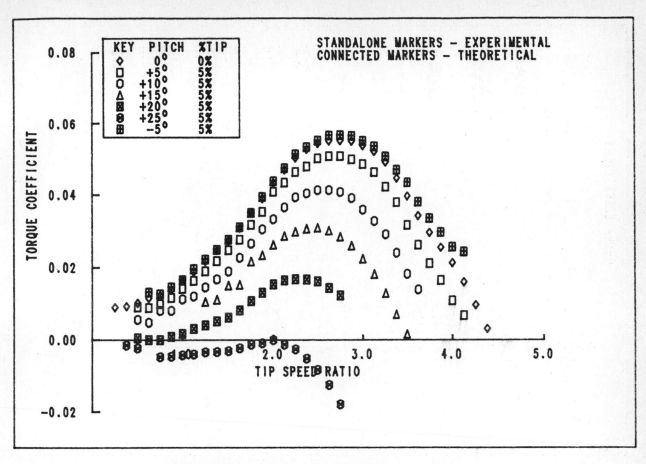

Fig 6   Measured effect of tip pitch on torque coefficient for
model V-VAWT with 5 per cent tip area

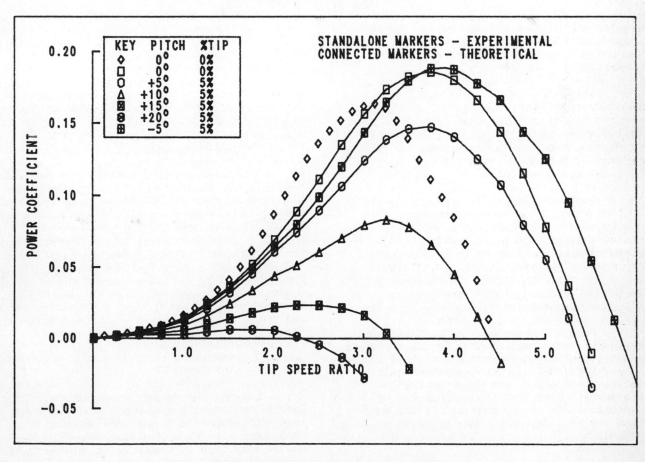

Fig 7   Predicted effect of tip pitch on power coefficient for
model V-VAWT for 5 per cent tip area

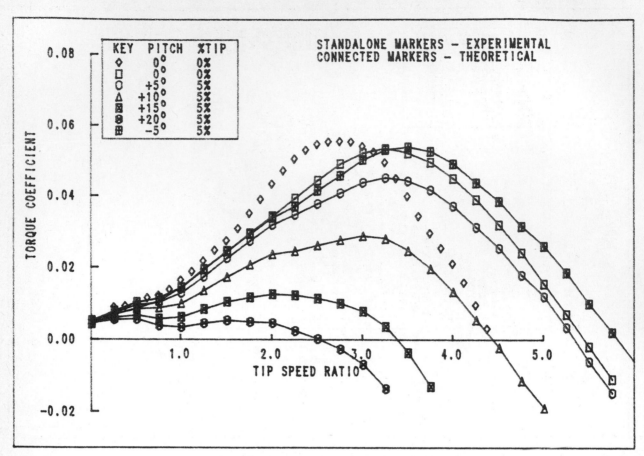

KEY	PITCH	%TIP
◇	0°	0%
□	0°	0%
○	+5°	5%
△	+10°	5%
⊞	+15°	5%
⊗	+20°	5%
⊟	−5°	5%

Fig 8   Predicted effect of tip pitch on torque coefficient for
model V-VAWT with 5 per cent tip area

As a result of these tests, the performance of larger V-VAWTs using tip pitch control can be predicted with some confidence. Figure 9 shows the predicted performance of the 5kW 3-bladed free-air V-VAWT, that has been described previously [2-4], and the predicted effect of tip pitch for a 5% tip area.

These predictions have been made for operational Reynold's Number of approximately 1500000, and show how the developed power can be destroyed for pitch angles as little as +30 degrees. The effectiveness of a blade tip deployed at a large pitch angle is largely Reynold's Number independent, though the lift, and hence torque, generated by the fixed portion of the blade will generally increase with Reynold's Number. Consequently as the operating Reynold's Number increases, the tip must be deployed at a larger pitch angle to obtain the same net braking effect. The choice of tip size and the pitch angle range through which it is deployed will be crucially dependent upon the operating conditions and the control criterion specified for the machine. At this stage of the V-VAWT development however, blade tip areas of approximately 5% of the total blade area seem appropriate for both power regulation and overspeed control.

3.4 The T-Brake as a control device

The T-Brake offers a number of advantages over pitching tip control: it can be simply mounted at the blade tip without creating any discontinuity in the blade structure; actuation can be of a push-pull nature as opposed to pitching, and the swept area of the turbine is increased by the additional aerofoil surfaces. However, the effect on performance of the T-Brake control surface is very similar to that of the pitching the blade tip and can be predicted with some confidence using VAWTTAY6, Fig 10.

These predictions should be treated with caution though, since they do not account for any flow interaction between the T-Brake and the fixed blade. In its pitched position the trailing edge of the T-Brake will spoil the flow over the outer portion of the fixed blade. It is therefore likely that these predictions may underestimate the braking effect of the T-Brake, but at present it is not possible to quantify this underestimation. The T-Brake should also act as an end plate thus reducing the blade tip losses and so increasing the lift developed by the fixed blade.

4. CONCLUSIONS

Recent wind tunnel tests of a model V-VAWT have shown that the aerodynamic performance of this wind turbine can be modulated with tip pitch control. These tests have demonstrated that both overspeed control and power regulation can be achieved with tip areas as little as 5% of the total blade area. Some enhancement of torque at all operating speeds has been demonstrated by small negative pitch settings.

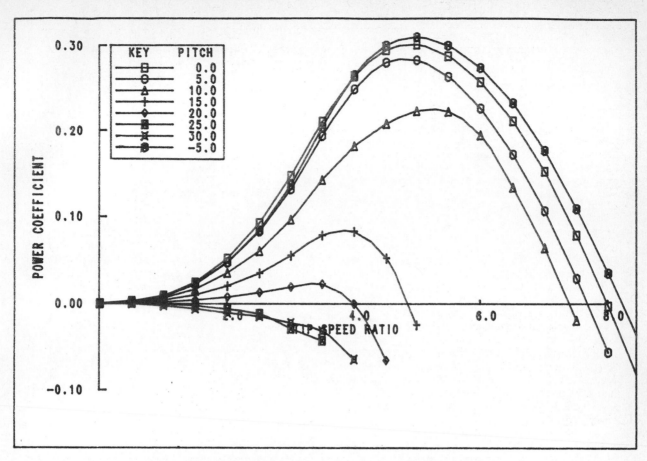

Fig 9   Power coefficient vs tip speed ratio: VAWTTAY6 predictions
         3-bladed, 5KW V-VAWT with 5 per cent blade tip control surface

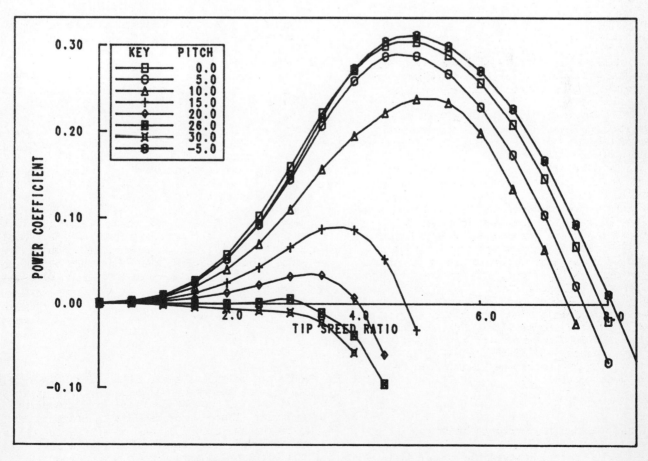

Fig 10   Power coefficient vs tip speed ratio: VAWTTAY6 predictions
          3-bladed, 5KW V-VAWT with 5 per cent T-brake control surface

While it has not been possible to match the results exactly to predicted data, the use of the aerodynamic prediction model VAWTTAY6 for development of larger sized V-VAWTs with tip control seems reasonable.

ACKNOWLEDGEMENTS

The authors wish to express thanks to Derek Taylor and Godfrey Boyle of the Open University, Scott Forrest who made the wind tunnel model and the technical staff at Queen Mary College for assistance during testing. Tony Robotham is sponsored by the Science and Engineering Research Council.

REFERENCES

1. D. J. Sharpe and D. A. Taylor, Preliminary investigations into an innovative vertical axis wind turbine. Proc. World Solar Energy Congress, Perth, Australia, 1983.

2. D. J. Sharpe and D. A. Taylor, The aerodynamic performance of the vee type vertical axis wind turbine. Proc. Seventh BWEA Wind Energy Conference, Oxford, 1985.

3. A. J. Robotham, D. J. Sharpe, D. A. Taylor and G. A. Boyle, Further developments in the Taylor 'V' type VAWT concept. Proc. Intersol 85 World Solar Energy Congress, Montreal, Canada, June 1985.

4. G. A. Boyle, A. J. Robotham, D. J. Sharpe and D. A. Taylor, The Taylor 'V' type vertical axis wind turbine: current status. Proc. Windpower 85 Conference, San Francisco, U.S.A., August 1985.

5. D. J. Sharpe, Refinements and developments of the multiple streamtube theory for the aerodynamic performance of vertical axis wind turbines. Proc., Sixth BWEA Wind Energy Conference, Reading, 1984.

6. D. J. Sharpe, A theoretical and experimental study of the Darrieus vertical axis wind turbine. Research report, School of Mech., Aero. and Prod. Eng., Kingston Polytechnic, October, 1977.

7. G. Stacey and P. J. Musgrove, The effect of fixed pitch offset on a high solidity vertical axis windmill. Proc., Third BWEA Wind Energy Conference, Cranfield, 1981.

8. R. E. Sheldahl and P. C. Klimas, Aerodynamic characteristics of seven symmetrical airfoil sections through 180-degree angle of attack for use in aerodynamic analysis of vertical axis wind turbines. Sandia Laboratories Report, SAND80-2114, 1981.

The use of spoilers to control a horizontal axis wind turbine

S M B WILMSHURST
The Cavendish Laboratory, Cambridge

1. ABSTRACT

Experimental results are reported from field tests of spoilers mounted on the blades of a HAWT. Three different spoilers were tested: they were of square cross-section and placed near the leading edge on the suction surface of the blades. The results are analysed and discussed with the aim of establishing the viability of spoilers as a control system for larger machines.

2. INTRODUCTION

The use of spoilers is one possible means of control for wind turbines. They act by decreasing the lift and increasing the drag along part of the blade, both effects tending to reduce or reverse the tangential force on the blade which provides the driving torque. A helpful comparison of spoilers with other types of air-brakes has been made by Pedersen [1]. Spoilers may be used as a primary control mechanism to keep the machine running at the desired speed; or simply as a braking system, in either a primary or back-up role.

In actual operation, spoilers would obviously have to be retractable, either being withdrawn into the blades or folded flat against them, as appropriate. On the scale of the Cambridge machines, however - 5m diameter and with the narrow blades characteristic of a high design tipspeed ratio [2] - this would be difficult to arrange. In addition, one aim was to experiment with varying the position of the spoilers, and this necessitates temporary attachment.

Before this work was begun, wind tunnel tests to investigate the effects of spoilers had been carried out at small scale [3]. A 3cm chord dural model blade, with the same aerofoil section (NACA 4415) as those of the Cambridge turbines, was used: Reynolds Numbers up to 20000 were covered. The results are not reported in detail here, because they turned out to be very different from those obtained in the field, and hence not a useful guide to full-scale conditions.

This paper does not address the subject of deployment, although some light is shed on the need for thoroughly effective retraction of the spoilers when not in action. The problem of transient effects caused by deployment is clearly an important one, but this is best examined at larger scale.

3. THE ACTION OF SPOILERS

Spoilers take the form of a ridge, fence or other obstacle generally placed on the suction surface of the blade. Their action is critically dependent on the blade angle of attack, and hence on the tipspeed ratio (in the case of a fixed-pitch machine).

Assuming the Reynolds Number is high enough, spoilers will give a clean separation of the flow without reattachment: the blade is thus effectively in a stalled condition with resulting loss of lift and greatly increased drag. It is convenient to approximate the change in lift by the expression from Kirchhoff flow for a flat plate at incidence [4]:

$$C_L = 2\pi\alpha \left(\tfrac{1}{2} + \tfrac{1}{2} f^{1/2}\right)^2$$

where f is the fractional chord at which flow separation occurs and α is the blade angle of attack. With spoilers in position, f is known accurately.

In fact it is the increase in drag which dominates the change in torque when spoilers are fitted: drag may be increased to many times the clean blade value.

Jacobs [5] gives a series of experimental results for the drag of protuberances on blades. The smaller ones did not protrude beyond the boundary layer and did not result in separation of the flow. Data from the largest protuberances tested give sufficient support to the approximation for C_L, quoted above, to justify its use in the analysis below. Results for the drag indicate that C_D effects will also increase dramatically as the protuberance is moved towards the leading edge.

4. FIELD EXPERIMENTS

4.1 Choice of Spoilers

The selection of spoilers was controlled by three requirements: simplicity, ease of attachment and removal, and reproducibility. These were met by using lengths of balsa wood of square cross-section. A double coat of varnish was given, to provide weather protection. Attachment to the blades was achieved with narrow strips of double-sided sticky tape. This turned out to be very successful.

The spoilers investigated were as listed in Table 1.

Thus spoilers 1 and 2 were identical except for their positioning: spoiler 3 was twice the height of 1 and 2 and half the length. All the spoilers were centred 2m from the hub (0.8 radius). An attempt was made to run the machine with a 5mm spoiler the same length as the smaller ones, but it proved impossible to reach operating speed (the Cambridge machines have electrical starters which cut out at 100 RPM, which is beyond the negative torque region with clean blades, but evidently not with large spoilers on - see below).

Fig.1 shows a view of the blade with spoiler 2 in position. Fig.2 gives a close-up view of the arrangement.

From the work carried out at smaller scale in the wind tunnel, a number of lift and drag curves were available for different sizes and positions of spoiler. In spite of the differences in regime, it was decided to select spoiler positions used in these tests as a basis for the field work.

4.2 Power Measurements

The turbine was run in constant RPM mode to obtain the power curve: two different methods of datalogging were used. The need to maintain constant RPM is not as great as might be supposed: the blades of the Cambridge machines are extremely light, so the corrections for angular acceleration would be small. In these measurements, therefore, the requirement that RPM be controlled at a fixed value has been somewhat relaxed.

Early data were processed using a Fortran program written for the purpose. One-minute averages were taken of windspeed and power, with corrections for time-of-flight and anemometer proximity effects. Corrections were also made for electrical power losses (which have been measured in the laboratory to high accuracy for the generator). This method, however, was too time-consuming in that all the data processing had to be carried out after the run was complete: the logging system involved is optimised for very high speed data collection such as the following of blade behaviour. This speed is not required for power measurements, so a simpler system was used for most of the measurements reported here. In this system, a different method was used in which the data was averaged and corrected during the run itself. A simple BASIC program was written to do this: the logging here is controlled by a BBC microcomputer. The data are printed out each minute: tipspeed ratio, power coefficient, mean RPM, mean windspeed, mean power generated, and mean load duty cycle. Datasets were combined to give adequate coverage of the accessible range of tipspeed ratio: the power curves were then drawn up in the usual way, with power coefficients binned by tipspeed ratio.

5. RESULTS

Fig.3 shows the clean blade power curve with the spoiler curves to the same scale: Fig.4 displays the results from the three spoilers on an expanded scale. Error bars show standard errors. One dataset was also obtained when the blade was nominally 'clean', but in fact small pieces of tape used to secure the spoilers in place were still adhering. The peak power coefficient was reduced by about 20%.

Figs.5 and 6 show the drag required to produce these power curves, derived from a performance prediction program written by Anderson [6]: in Fig.5 this is the 'whole-blade drag coefficient', with the clean blade data shown for comparison; and in Fig.6 the additional drag is reduced to a 'spoiler drag coefficient' based on the frontal area of the spoiler itself. In this case the clean blade drag coefficient has been subtracted from the total drag; and for both Figs.5 and 6 allowance was made for lift reduced in accordance with the expression above.

There is a small systematic error in the drag values for the spoilers, due to the fact that the performance program solves for the blade behaviour at 20 radial stations. Taking spoilers 1 and 2 as occupying four of these stations, and spoiler 3 as occupying two, resulted in effective spoiler lengths of 0.5m and 0.25m respectively, instead of the true values of 0.45m and 0.225m. The substantive conclusions are unaffected by this, however. Fig.7 shows approximate machine runaway speeds for all configurations. Although the windspeeds on this graph are low, extrapolation to higher values is possible, because the runaway tipspeed ratio converges to a constant value above about 5 m/s (below this speed the rotor losses have a major effect).

6. DISCUSSION

In studying these results, it must be remembered that the length of spoilers was deliberately restricted in order to make it possible to test them. The results can be extrapolated for longer or larger spoilers, which will reduce the torque still further. A further run of the performance prediction program was made to test this: this shows that a spoiler 1.5 times the length of spoiler 2, and centred in the same place, would give negative torque for all tipspeed ratios.

The run carried out with tape still adhering presumably represents the case where the flow is 'tripped' rather than separated, thus producing a turbulent boundary layer with lower lift. The effects of tripping on the NACA 4415 aerofoil section are discussed by Ostowari and Naik [7].

The field tests were conducted at Reynolds Numbers in the range 150000 to 400000, and as would be expected there is some degree of variation in behaviour between these extremes. The low tipspeed ratio end of the power curve for spoiler 3, for example, showed power about 15% higher at 250 RPM than at 150 RPM. Nevertheless the Reynolds Numbers should be high enough for valid extrapolation to larger sizes of machine.

With reference to the power curves, it is clear that the shorter but higher spoiler, 3, is less efficient at reducing power than spoiler 2 which was at the same position. This may be partly because of circulation around the ends of the spoilers which would tend to reduce the effectiveness of shorter ones, but can be entirely accounted for by the fact that with 3 there is a greater length of clean blade, giving higher total lift. This is borne out by the very similar spoiler drag coefficients for these two cases. It is also clear that the spoiler positioned further forward (2) is more effective than the one further back (1), as would be expected from the expression for lift quoted above; and indeed, when the effective drag coefficients are compared (Fig.6) spoilers 1 and 2 show very similar behaviour, indicating that differences must be due to the lift characteristics.

Where data are available for high angles of incidence (spoiler 2) the spoiler drag coefficient is seen to fall off from an angle of about 13. This is because the blade itself is entering stall here. It is in this region that operation of spoilers is most problematical: in a separated flow their effectiveness is much reduced. For many machines, however, including those at Cambridge, the torque coefficient will anyway be negative by the time the whole blade is stalled (and it is of course the tip region, where the spoilers are positioned, which enters stall last). Note, however, that in the absence of the spoiler the blade alone would _not_ be in stall at the equivalent tipspeed ratio (about 3.5): the presence of the spoiler, by reducing the lift, also reduces the axial interference coefficient and hence increases the angle of attack. Without the spoilers this part of the blade enters stall at a tipspeed ratio of about 2.5.

The graph of runaway speeds (Fig.7) shows similar behaviour for all three spoilers. The runaway speed slope is determined largely by the tipspeed ratio at which the power coefficient drops to zero. Clearly, for operation as brakes, spoilers must stop the machine completely and not just reduce the runaway speed: again it should be said that a simple scaling-up of the spoilers will achieve this.

7. CONCLUSIONS

The operation of blade-mounted spoilers has been successfully demonstrated. The experiments are in agreement with the theory in showing stronger effects with spoilers near the leading edge. The drag caused by the spoilers appears to be almost independent of position, at least at low angles of attack.

It is striking that such a small spoiler can effectively wreck the power output of a wind turbine, as shown by the work reported here. Further work should examine the behaviour at very low tipspeed ratios, when the blade tip itself will enter stall.

An important point is demonstrated by the result from the blades affected by residual sticky tape. This is that for use as brakes, spoilers must be truly retractable. The least protuberance remaining when they are not in use is likely to have a severe impact on the performance.

8. ACKNOWLEDGEMENTS

SMBW acknowledges the support of an SERC CASE award with the Central Electricity Generating Board. Thanks are also due to Mr G.T. Atkinson, for the BBC microcomputer program.

9. REFERENCES

1. Pedersen, T.F. 1984. Air Brakes on Stall-Regulated Windmills, Proc. 6th BWEA Conference, Reading, U.K., CUP.

2. Powles, S.J.R. 1984. Horizontal-Axis Wind Turbines, Ph.D. Thesis, University of Cambridge.

3. Bullinger, D. 1985. Experiment to Investigate the Effect of Boundary Layer Tripping on an Aerofoil at Low Reynolds Number, Internal Report, Cavendish Laboratory.

4. Thwaites, B. 1960. Incompressible Aerodynamics, OUP.

5. Jacobs, E. 1932. Airfoil Section Characteristics as Affected by Protuberances, NACA Report 446.

6. Anderson, M.B. 1981. An Experimental and Theoretical Study of Horizontal-Axis Wind Turbines, Ph.D. Thesis, University of Cambridge.

7. Ostowari, C. and Naik, D. 1984. Post Stall Studies of Untwisted Varying Aspect Ratio Blades with a NACA 4415 Airfoil Section, Wind Engineering, Vol.8. No.3.

Table 1

	Height (mm)	Height (chord)	Length (m)	Length (radius)	Position (from leading edge) (mm)	Position (from leading edge) (chord)
1	2.5	0.033	0.45	0.18	17	0.194
2	2.5	0.033	0.45	0.18	7	0.076
3	5.0	0.066	0.225	0.09	7	0.076

Fig 1   Spoiler on blade in position No. 2

Fig 2   Close-up with spoiler in position No. 2

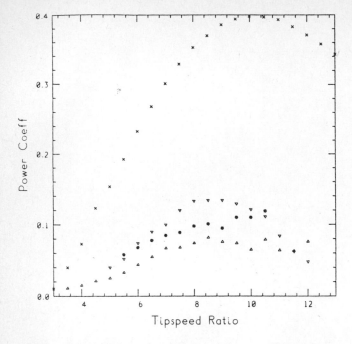

Fig 3   Power curves: clean blade and three spoilers, drawn
to same scale
  ✕ Clean blade   △ Spoiler 2
  ✳ Spoiler 1   ▽ Spoiler 3

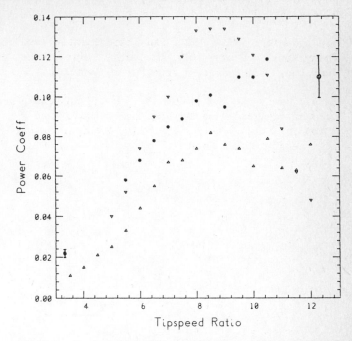

Fig 4   Power curves: three spoilers. Expanded scale
  ✕ Clean blade   △ Spoiler 2
  ✳ Spoiler 1   ▽ Spoiler 3
        ╪ Approximate errors

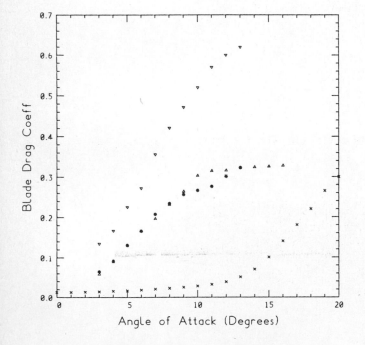

Fig 5   Calculated drag curves (whole blade)
  ✕ Clean blade   △ Spoiler 2
  ✳ Spoiler 1   ▽ Spoiler 3

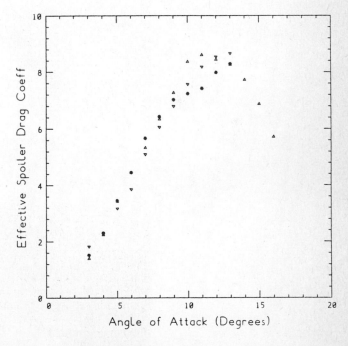

Fig 6   Calculated drag referred to spoiler frontal area (see
text for explanation)
  ✕ Clean blade   △ Spoiler 2
  ✳ Spoiler 1   ▽ Spoiler 3

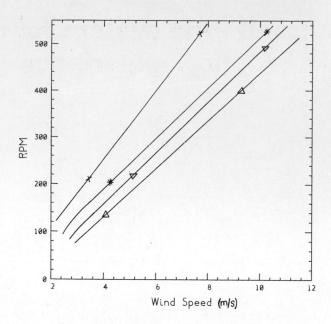

Fig 7   Approximate machine runaway speeds
        ✕ Clean blade   △ Spoiler 2
        ✳ Spoiler 1   ▽ Spoiler 3

A control and monitoring system for a wind turbine – a practical experience

C C HORN, CEng, MIEE and **A P WITTEN**
CAP Industry Ltd, Reading, Berks

SYNOPSIS This paper traces the successful development by CAP of a complex control and monitoring system for a 25m Vertical Axis Wind Turbine. It examines some of the problems encountered and shows how they were successfully overcome. The paper concludes by looking into the future at CAPs continued involvement with the wind energy programme.

1 INTRODUCTION

In November 1983 CAP won a major contract to supply a fixed price, turnkey Control and Monitoring system for the 25m Variable Geometry Vertical Axis Wind Turbine (VGVAWT). This was being constructed in South Wales by the VAWT consortium, headed by Sir Robert McAlpine, and included NEI Cranes.

The contract was for the supply of control, monitoring and analysis computer systems and software, together with an integrated data acquisition and control system.

This paper traces the successful development of the system, and highlights some of the unique problems which were encountered while developing equipment for this application.

The paper concludes with a report on the current status of the project and a look into the future of CAP's involvement with the wind power programme.

2 PROJECT REQUIREMENTS

The 25m VAWT being constructed at Carmarthen Bay in South Wales is predominantly a research machine, and consequently is heavily instrumented, the total number of sensors being in the order of two hundred. The control and monitoring system therefore has two primary roles. Firstly it must control the turbine, ensuring safe operation, and secondly it must provide the means of capturing and analysing the data from the large number of sensors for research purposes.

2.1 Description of the wind turbine.

The turbine itself comprises a crossarm and blade assembly mounted on top of a concrete tower. The two blades are hinged in their centres and a reefing mechanism, located at the centre of the crossarm, allows the blades to be folded down in an arrow-head configuration to control the power output.

The blade and crossarm assembly is attached to a hollow pintle shaft which drives a gearbox in the top of the tower. This in turn drives a high speed shaft running down inside the tower and offset to one side. At the base of the tower, the transmission is turned through 90 degrees by another gearbox and drives horizontally mounted synchronous and induction generators. Two generators are included to enable evaluation of different configurations.

All wiring to sensors on the crossarm and blades is brought down the centre of the pintle shaft and is picked up by an electronics unit mounted on the lower end of the shaft immediately below the gearbox.

To start the turbine (which is not self-starting), the generators are used as motors. In addition, a small motor is included to allow the transmission to be inched round slowly for maintenance purposes. Control of this rather complex electrical system necessitates a significant amount of switchgear and this, together with sensors mounted on the generators, is connected to the computer systems via an electronics unit housed in the generator room.

Two meteorological masts are located some 50m away from the turbine. These sense wind speed and direction at 5 different heights providing data to the computer for controlling the turbine.

2.2 Equipment location.

Figure 1 shows the location of the various parts of the control and monitoring system.

The computers and their associated peripherals are housed in an air conditioned computer room at the base of the tower. The central part of the data acquisition system is also located in the computer room and is actually built into the pedestal of the main control desk.

Data is collected (and control commands issued) from four remote data acquisition units (DAUs). Two of the units are capable of output as well as input.

The rotor DAU is mounted on the rotor assembly (attached to the lower end of the pintle shaft) and captures data from about 100 blade and crossarm mounted sensors. Outputs from it also control the reefing and unreefing of the blades and the operation of the top brake.

The generator room DAU is housed with the generators and switchgear in the generator room. It acquires data from the switchgear, the generators and from sensors mounted in the tower itself. It also provides quite a large number of relay control outputs to the switchgear to enable the computers to start up and shut down the turbine.

Two smaller DAUs are housed in weatherproof cabinets and are attached to each of the met masts to collect data from the anemometer units. Each has 10 input signals, 5 wind speeds and 5 wind directions.

All four DAUs are connected to the computer room equipment by means of high speed fibre optic links.

2.3 Computer systems.

Three prodiminantly independent computer systems are provided. Control of the machine is carried out by a Digital LSI 11/23 minicomputer. A second LSI 11/23 provides a logging function and is mainly there to provide a short term record of events leading up to and following the occurance of some incident. It performs a function similar to an aircraft "black box" flight recorder.

The third computer is a somewhat larger PDP 11/44 machine and is used for data analysis. This provides the tools needed to use the turbine for research. Various analysis packages are provided on this computer to allow sophisticated acquisition, processing and presentation of data. In addition, remote access is provided to this machine via a British Telecom line from McAlpines offices in Hemel Hempstead.

2.4 Safety requirements.

The 25m VGVAWT machine represents a significant investment in terms of development time and costs. It is imperative therefore that the greatest possible precautions are taken to protect the machine from damage.

For the control and monitoring system, this implies two forms of protection. Firstly the control and monitoring system itself must be protected from damage, the most likely cause of damage being lightning strike. Secondly, the turbine must be protected from a failure of the control and monitoring system. Computer systems can fail and so a "watchdog" system has been included which will insure that the turbine will shut down automatically should a fault occur in any part of the control and monitoring system.

In addition, an entirely independent system monitors a few critical parameters and if all else fails can override the computers and initiate an emergency shut down. This system comprises a number of trip amplifiers monitoring accelerometers in the crossarm.

3 SPECIFIC PROBLEM AREAS

The supply of a control and monitoring system posed some unique problems associated with this particular application. These arose either from the mechanical characteristic of the wind turbine or from the somewhat harsh environment in which the system had to survive.

3.1 Rotor sensors.

Approximately 100 sensors are mounted on the blade and crossarm assembly. Many of these are pressure gauges with outputs in the order of a few millivolts. To bring the analogue signals from all these sensors off the rotor would have required several hundred slip-rings, and was clearly impractical. Consequently, one of the data acquisition units had to be mounted on the rotating part of the machine to acquire and digitise the data into a single digital data stream before passing it through slip-rings and down the tower.

This DAU had to include full signal conditioning amplifiers for all the low level signal. Multi-pole low pass anti-aliasing filters were also needed on all channels since because of the large number of channels and bandwidth limitations, sampling rates had to be kept low with respect to signal frequency.

3.2 Lightning protection.

Perhaps the most severe of all the requirements was the need for the equipment to survive lightning strike. The wind turbine is of course a tall structure, and is of necessity located in an exposed position. The probability of a direct lightning strike is high. Whereas it is probably impossible to protect the actual sensors from a direct strike, it is imperative that damaged caused by a strike is kept to a minimum.

Elaborate precautions have been taken to protect the structure of the turbine by means of lightning conductors and earthing. In addition, the design of the control and monitoring has emphasised protection from high voltages.

Firstly, all sensor inputs are passed through lightning protection modules. These are multi-stage surge protection devices designed to divert high voltages safely to ground. For the two main DAUs, on the rotor and in the generator room, these units are housed in a separate enclosure from the electronic so that high voltages are kept well away from the delicate electronics.

Secondly, all DAUs are well screened to prevent a strike inducing voltages within the electronics.

Thirdly, the DAUs are isolated as far as possible from the computer room equipment. This is mainly achieved by using fibre-optics for the communications links.

In the case of the rotor DAU, the use of fibre-optic slip rings was considered, but they proved excessively expensive. The solution finally adopted was to use conventional slip-rings and then convert to fibre-optics immediately below the slip-rings in a small unit mounted below the rotating DAU.

Another area of concern was the supply of power to the met. mast DAUs. Battery power was considered to provide total electrical isolation of the met. masts, but proved impractical. The solution adopted was to supply power at a low DC level which could be thoroughly protected by means of surge protectors.

3.3 Communications.

Several aspects of the communications requirements proved challenging.

The main characteristics were for a system providing high speed, high integrity and good isolation between units. Fibre-optics were therfore the obvious choice.

A second characteristic was the need to collect data from a number of remote locations (actually four) and combine the data into a single frame for use by the computers. Using conventional techniques such as pulse code modulation (PCM), this proved very difficult since 4 asynchronous high speed data streams would have had to be combined. This would have required complex synchronising and buffering electronics.

In addition, there was a need for data to flow in the other direction, (out from the computers) to perform the control functions. A convenient solution which satisfied both the above requirements was to devise a synchronous bi-directional system. In essence, the remote DAUs respond to commands issued by a central control unit and it is this unit which defines the order and frequency at which channels are sampled. The commands issued from the central unit may be "read a channel" or may be "set an output". The system can therefore perform output as easily as input.

By careful design, a synchronous system of this nature can operate as fast as an asynchronous system, since the command for one transaction can overlap the data from the previous one.

Having settled on a synchronous system, a number of additional benefits could be realised. In particular, the system could be continuously self-checking and easily made fail-safe. The failure of a communication link can be immediately detected by both the central unit, (which fails to get a responce to a command), and by the remote unit, (which times out a watchdog timer). The design therefore is inherently secure.

4 DATA ACQUISITION SYSTEM

The data acquisition system required for the VAWT project falls into the category of "medium speed". For some reason there appears to be a shortage of proprietary systems available within this range.

At the top end of the market are the high speed PCM systems. These are predominantly built for military applications and are therefore to full military equipment specifications. They carry an appropriate price tag.

At the bottom end of the market are a wide range of data loggers. These are cheap, but have very limited capabilities. Their performance was well below that required for VAWT.

Given the general lack of suitable proprietary equipment, coupled with the particular requirements for fibre-optics and a bi-directional synchronous system, it was decided to develop a custom data acquisition system for this application. This system we called CAPDAS.

4.1 CAPDAS

The approach taken in developing CAPDAS was to make maximum use of proprietary board level components, in particular input/output boards. Many different types of interfaces were required for VAWT including both analogue and digital inputs and outputs, and so a search was made to select a range of I/O boards which would provide all the functions required. The range finally selected was based on the STD bus.

The STD bus is one of the standard 8 bit wide microprocessor busses. It is widely used in the US for industrial applications. The cards are fairly small (approx 6" x 4"), low cost and sourced by a large number of manufacturers. There are well over 200 different cards to choose from, and all the required I/O configurations were available as well as functions such as memory cards and a real-time clock. It was therefore decided to design CAPDAS around the STD bus.

Figure 2 is a block diagram of the control and monitoring system, which indicates how CAPDAS is used to connect together the computers with their remote sensors.

4.2 Remote units.

Although the STD bus was designed as a microprocessor bus, the speed requirements for CAPDAS were too high to allow a micro to control the remote DAUs. A "hardware" bus controller was therfore designed. In practice, this was implimented using a bit-slice processor and was specifically designed to handle high speed data communications.

The functions required of the remote units were very limited and basically involved receiving commands from the control unit, acting upon the commands by controlling the STD bus, and returning the results of the operation back to the control unit.

The maximum speed of CAPDAS in its present form is one operation every 40 microseconds, or a throughput of 25kHz.

4.3 Control unit.

The control unit was constructed in the same way as the remote units. It again used the STD bus and the CAPDAS bus controller. However, the software was different to perform the control unit functions rather than the remote unit functions.

The control unit uses a sequence table held in EPROM (Erasable Read Only Memory) to control the sequence and sample rates for individual channels. By repeating some channels many times within the table, they can be sampled at a frequency which is a multiple of the basic frame rate.

As data is received from the remote units it is passed via a high speed parallel interface to DMA (Direct Memory Access) input ports on the three main computer systems. Synchronising pulses are also fed to the computers to indicate the start of the data frame.

4.4 Data protection.

Data transmitted over the high speed serial links is protected by parity checks and a sumcheck. The system will tolerate two successive frames containing an error. However, if three or more occur, the computers will initiate a shut down of the turbine. This provides a very high level of system integrity.

Facilities are included which allow the cause of a fault to be rapidly traced to a particular data link or remote DAU.

4.5 Output operations.

As already mentioned, the system is capable of output as well as input operations. To accomodate this some of the entries in the sequence table are specified as output "slots". A small microprocessor is co-resident on the STD bus with the CAPDAS controller and acts as an interface to the computers for control purposes. Control messages generated by the computers are sent via RS232 links to the microprocessor which assembles a command to the respective remote DAU. This is then passed to the CAPDAS controller for transmission during the next "output" time slot allocated within the sequence table.

4.6 The control console.

A small control console is driven by the microprocessor. This contains an alphanumeric display and keypad and provides a means for the operator to monitor the health of the data acquisition system. By using the keypad, any input to the system can be directly viewed on the display.

Any output can also be controlled from the keypad. This is a very valuable feature for use during commissioning and testing, but is obviously not desireable in an operational environment. It means, for instance that all safety monitoring and interlocking can be overridden. To account for this, a keyswitch on the console allows all control functions to be disabled during normal operation.

4.7 Watchdog timers.

Watchdog timers are used in the remote DAUs to guard against loss of communications with the computers, or failure of the computers themselves. These are small units with a relay output. Provided a continuous stream of pulses is received, the relay will be kept energised. If the pulses fail in either the on or off state the relay will release, breaking an external circuit and stopping the turbine.

The pulses are generated from the control computer software, and so a failure of the computer hardware, the software, the communication links, the CAPDAS control unit or the remote unit will all cause the watchdog relay to release.

The watchdog unit in the rotor DAU is connected in series with a number of trip amplifiers. These provide an entirely independent backup system capable of stopping the wind turbine if acceleration forces on the rotor exceed safe limits. This only operates as a last resource since under normal circumstances the computer system should shut down the turbine long before these limits are reached.

4.8 Uninterruptable power supplies.

As part of the general safety features of the control and monitoring system, the critical parts of the system are protected against failure of the mains power. Short term battery backup operation is provided for all the remote DAUs, the CAPDAS control unit and the control and logging computer systems. The analysis computer system is not protected since it does not contribute to the safe operation of the wind turbine.

5 COMPUTERS

The system includes three computers for control, logging and analysis respectively.

5.1 The control computer.

The control computer is responsible for ensuring the safe operation of the turbine. It uses information from the met. masts to determine whether or not wind conditions are favourable for running the turbine, and will initiate a start up or shut down sequence accordingly. Once running, the reef angle will be continuously adjusted to control output power.

A number of parameters are also continuously checked by the control computer against preset safety limits, and if any exceeds its limit, the turbine is stopped. An out of limits situation counts as an "incident" for the logging computer which freezes its recording of data.

5.2 The logging computer.

The logging computer has two main functions. Firstly, it maintains a continuous recording of a number of sensors so that the cause of an event such as an emergency shutdown can be determined. A circular buffer is continuously maintained containing three minutes worth of data. This is frozen when an event occurs. The event may be either a shut down initiated by the control computer, or the loss of a watchdog signal from the control computer which may indicate that the computer itself has failed.

The second function of the logging computer is to acquire and record long term statistical data on the turbines operation. This data can then be made available to the analysis computer for subsequent processing.

5.3 The analysis computer.

The analysis computer is a significantly more powerful computer than the other two, and provides the analysis facilities required to use the turbine as a research tool. A number of sophisticated analysis software packages are provided on this machine for use by researchers.

This computer system includes a high resolution graphics display which can be used to generate real-time plots of selected parameters.

6 THE CURRENT POSITION

The development of the control and monitoring system has been successfully completed and the system has passed its factory acceptance.

It was delivered to site shortly after Christmas and partly installed. We are now awaiting installation of the switchgear in the generator room to be completed before we can return to site and complete the majority of the installation. This should take place within the next few weeks, and at the end of this installation phase, all the computer systems and most of the data acquisition system will be installed and tested. The only remaining work will be the installation of the rotor DAU which will take place immediately after the rotor assembly itself is installed in May.

Subsequent to installation, there will be a period of commissioning during which CAP will provide support to McAlpine during the commissioning of the turbine. It is anticipated that we will be working closly with McAlpine during this period and will most probably have a continuous presence on site.

7 THE FUTURE

The development of the control and monitoring system for the 25m VAWT has provided CAP with an interesting and challenging project. It has also meant that we have had to face up to, and successfully overcome a number of problems which are specific to the field of wind energy. As a result, CAP is now in the unique position of being the only systems integrator able to offer specific expertise to the wind energy industry. We therefore look forward to many years working with this industry providing control and monitoring systems for all types of wind turbines.

CAPDAS, the data acquisition system developed for the 25m VAWT is a very flexible system capable of adapting to many applications. It provides at reasonable cost a very high performance system suitable for applications where large numbers of sensors are required, and it is anticipated that it will play a major role in many research type projects.

For production turbines, CAPDAS is clearly too sophisticated and expensive, and CAP is already developing a much smaller and cheaper system. This will be a microprocessor based system capable of acquiring data from up to 32 analogue channels. For control only, a single board system will be installed. If monitoring is also required, a second system will be installed together with a bubble memory cassette unit for recording the data. Both units will be housed in weatherproof cabinets.

The first units will be used on the 17m VAWT machine currently being developed for the Scilly Isles and the control unit, and optionally the monitoring unit, will be fitted to subsequent production machines.

With this new product, CAP can now offer systems to the wind energy industry to cover both research and production requirements. We look forward in the immediate future to the commissioning of the 25m and 17m VAWT machines, and in the longer term to many years involvment with, and contribution to the wind energy programme. We hope that this will continue through the research programme and on to subsequent provision of commercial turbines.

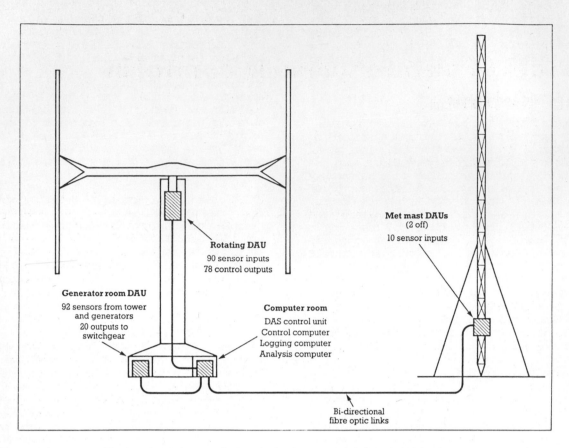

Fig 1   VAWT control and monitoring system. Equipment location

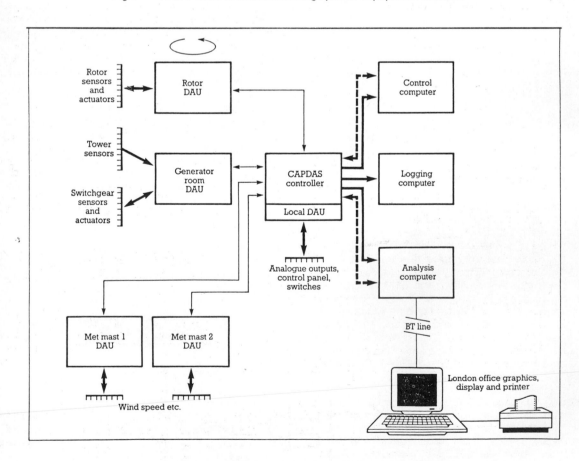

Fig 2   VAWT control and monitoring system. Block diagram

The role of electrohydraulic control in wind turbines

A C MORLEY
Vickers Systems Limited, Havant, Hants

1 INTRODUCTION

Wind turbine generators are established as a viable energy source into the next century. A significant number of machines have been successfully implemented around the world and many more are in embryonic stages.

This population of working installations, many concentrated in the U.S.A., has supplied data which has impacted the world-wide trends in commercial wind turbine design. One such trend is the use of medium to large (100-500kW) wind turbines for commercial utility applications. This interest in machines utilising long aerofoils generating large forces has created a need for high performance control systems to modulate their input power. Electrohydraulic control systems find applications here and in many other turbines of all sizes for the control of auxiliary functions.

2 STATE OF THE ART ELECTROHYDRAULICS

In recent years, the dramatic increase in the population of industrial robots and fully automated systems has brought about a change in the classic concept of industrial hydraulics. Servo valves with their design roots in the Aerospace and Defence industries found more and more applications in the production environment.

The marriage between high technology servo products and conventional hydraulics was initially somewhat strained by differing requirements in system design and environmental conditions. The advent of proportional valves in the last decade is the result of intense design effort by industrial hydraulics suppliers to meet the challenges posed by ever increasing specifications of productivity, accuracy and efficiency.

In order to support the new developments in electrohydraulic interfaces, a comprehensive range of electronic accessories is manufactured by several hydraulic equipment suppliers. The importance attached to this technology has brought about a new name – Electrohydraulics.

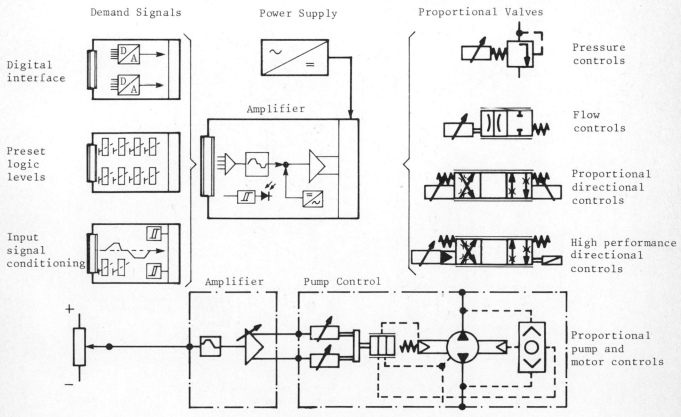

Fig 1   Electrohydraulic hardware combinations

Besides the application of new technology, existing hydraulic products have evolved to handle higher pressures and flow rates in smaller envelopes. Cartridge insert valve assemblies in machined manifold blocks have integrated these features with considerable design flexibility. Proportional electrically modulated cartridge valves have contributed a powerful combination which has been successfully applied to dedicated control systems for lifts, presses, injection moulding machines and other similar applications. Electrohydraulics manufacturers who build customised hydraulic power units and control assemblies are finding a significant proportion of these systems are required with integral electrical or electronic features. These systems can be tailor made to suit customers control requirements and can include intelligence in the form of programmable controllers or microprocessors.

Manufacturers of hydraulic actuators have also advanced with the technology and an interesting development in transducer technology has allowed position and velocity transducers to be integrated inside the envelope of a cylinder. This protects one of the most vulnerable elements of a control system and allows the designer to position the actuator in confined or aggressive environments.

3 APPLICATIONS OF ELECTROHYDRAULICS ON WIND TURBINES

3.1 Hydraulic Brake Circuits

Hydraulic brakes, both spring applied and pressure applied, offer large retarding forces in a small envelope. The hydraulic power sources for these systems are compact, inexpensive and energy efficient.

Products of particular interest here are direct acting pressure controls and poppet type directional valves. These seating valve designs offer leak free and reliable systems.

Energy storage to maintain brake status over long periods can be achieved by the use of gas pre-charged hydraulic accumulators.

Such brake systems have found many applications on the main rotor and yawing controls of wind turbines.

3.2 Yawing and other rotation systems

The nature of the design of wind turbines often requires a compact system to control the wind seeking ability of horizontal axis machines. Electrohydraulic systems have been successfully applied to 'Free yawing' or 'Self-seeking' machines to control rotational speed under varying load conditions.

Small to medium size turbines have used high torque, low speed hydraulic motors with a variety of load and speed control systems. Large machines have used lift and rotate systems which reduce the effort required to rotate while providing a rigid support in the static condition.

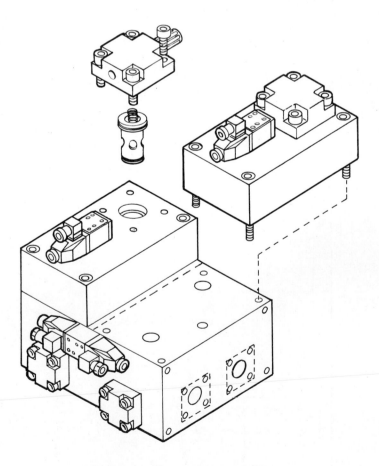

Fig 2   Cartridge manifold assembly

3.3 Blade attitude control systems

The majority of medium to large wind turbines require a compact and powerful method of adjusting blade attitude to the wind. These systems often require to be modulated so that power output can be precisely controlled over a wide range of wind conditions.

The ability of these systems to react quickly to gusting winds is important and the ability of hydraulics to supply instant power at the right place at the right time is unrivalled by any other power transmission system.

Several types of control system are possible, each offering a different degree of controllability.

(a) On/Off systems; otherwise known as 'Bang-Bang' systems give fixed speeds and pressures selected by remote control.

(b) Pulsed proportional systems; are a refinement of 'Bang-Bang' systems where solenoids are subject to frequent short pulses under the control of some intelligence built into the turbine management system.

(c) Hydraulic feedback systems; where preset controls react to changes in load (sensed as pressures in service lines) and control them within prescribed limits.

(d) Open loop electrically modulated systems; these systems give programmable control of actuator output force, velocity and rates of change within limits of accuracy determined by the hydraulics and the nature of the load.

(e) Closed loop electrically modulated systems; offer the best accuracy with the load effects cancelled by the use of electrical feedback devices signalling an analog or digital control system.

(f) Hybrid systems; these may offer desirable combinations of the above mentioned control regimes.

3.4 Transmissions

The flexibilty of hydraulic transmissions is unrivalled and systems capable of handling shaft horsepowers into hundreds of kilowatts are possible. The ability to provide a comprehensive array of torque and speed controls while maintaining high efficiencies makes them an attractive proposition for wind turbine applications.

3.5 Pressure sources

The hydraulic pressure for these control systems is commonly supplied by an electrically driven power unit. In some cases the hydraulic pump may be driven from a high speed shaft geared to the main rotor or any other prime mover.

Hydraulic power packages can be optimised to meet customer's installation requirements and apart from the hydraulic functions may also include equipment related to cooling, electrical control and management systems.

CONCLUSIONS

Modern Electrohydraulics are widely applicable to solving the specialised control problems associated with wind turbines. Their high power to weight ratio, fast response, flexible controls and long term reliability make them a designers choice that is hard to beat.

WECS condition monitoring – the remedy for long life?

A J GARSIDE and **J H WEBB**
Cranfield Institute of Technology, Cranfield

SYNOPSIS

Long operational life is essential for a WECS to maximise the return from investment of the high capital cost.   A concern is whether this requirement places WECS out of step with known experience with other mechanical systems and whether procedures are available to ensure trouble free, safe operation.   Condition monitoring, the application of NDT procedures and performance evaluation for example, is used successfully in the aeronautical field to manage the operation of a turbo machine or gear box for example, allowing intervention for remedial action to be planned and permitting service lives to be extended.   Possible application of some of the techniques is discussed along with suggestions of their implementation.   Examples are given from helicopter components, aircraft operation and turbo machinery, and possible hazards discussed.   The tools are available, but will need some adaption for wind turbine application.   Their use will enhance the maturing of wind turbine operations and provide a firm basis for long life operation.

1.   INTRODUCTION

The target life considered for large wind turbines is currently approaching thirty years, exceptional considering that gas pumping turbines may be operational for five years, cars ten years, locomotives perhaps fifteen years, ships twenty years and utility power stations twenty five years.   Of these only the former and latter listed are called to give 24 hour service.   Is a wind turbine being asked to make a 'quantum leap', beyond it's peers or can inspection and monitoring provide the confidence to furnish information on component health and give the expectation of planned, long term, trouble free running. A British Airways aero engine is brought back to base frequently for detailed inspection. It may have completed 3 long haul flights, with relatively small number of flight cycles, or be on shuttle duty with 10 or more flights incurred over equal hours, but including a larger number of cycles and increased fatigue damage.   This may indicate the need for frequent yet different approaches to inspection being taken in the wind turbine case.

A compatriot, operating in the earth's boundary layer, and requiring continual control variation to maintain safe operation is the helicopter.   These after half a century of development have begun to demonstrate reasonable lives in crucial rotor and gear box components.   Many of these mechanisms have been subject to forms of condition monitoring which has prolonged safely, operating times. Can the wind turbine be monitored such that it exhibits mature behaviour earlier in it's development?

The wind turbine is exceptional considering the unsteady nature of the wind energy resource in that certain components may have to endure over 10^7 fatigue cycles each year.   There are few comparable structures which endure this. Marine structures are subject to varying sea loadings and may provide some experience to draw on.   However, the attraction of siting WECS in exposed estuaries and shallow sea regions, to take advantage of unhindered wind conditions may take a wind turbine even beyond marine experience, as corrosion fatigue effects have to be faced.

2.   WHAT IS CONDITION MONITORING?

Removal of a unit from operation for regular maintenance has been practice in many industries, this causes a break in production and a consequent loss of revenue.   Such action often proves unnecessary since nothing is found to be malfunctioning, oil perhaps only needing to be changed, yet features have been adjusted, replaced, etc., without the real need.   Recalling the advice of motoring organisations concerning servicing prior to holidays - allow two weeks before you go to give time for readjustments/corrections to be made - the hand of man can sometimes do harm.

A result in the automotive field is that some cars are now fitted with diagnostic plugs for checking a variety of items and settings. This facility reduces intervention, speeds up checking and gives the facility for regular checking to monitor changes.   At the other end of the scale, the modern jet engine has thousands of components, each interacting with the other.   Plainly the removal of an engine, strip, check and rebuild would stop airline operation as we know it, but the careful logging of operational parameters - temperatures, pressures, rotational speeds and vibration levels can, when analysed, provide an accurate guide to the elapsed life of the engine, the need for minor adjustments, or a guide to the level of correction or renewal at the next service period.

Aspects of health monitoring for an aero engine, helicopter or aircraft may include:-

Performance
Mechanical operation
Chemical composition of lubricants and coolants
Component inspection

Such a check list could apply to a wind turbine.

Methods of monitoring health for a wind turbine component or assembly, assist in building confidence in running for longer periods, allow changes to be monitored to establish guide lines for yet further life development in revised design or revised operational procedures.

For a wind turbine the monitoring may include:-

Power output and wind speed
Frequency responses
Particle examination within lubricants
Visual inspection of accessible parts
Crack inspection using penetrants
Ultrasonic crack detection in hollow components
Operating temperature
Pressure leakage from blade main spars to indicate cracks

Many of these methods can be enacted without shutting down the turbine. These and other methods will be discussed in detail along with their application. Some are more appropriate for small components, but for rotor blades up to 50m length and a confined hub including bearings and bolts particular methods will be required.

3   PERFORMANCE MONITORING

Unlike the aero gas turbine, where the inlet and internal conditions can be clearly established with limited instrumentation at critical areas and the related areas are well known, the wind turbine has a varying wind stream entering it's sphere of influence and the extent of the influence can change moment by moment of the instantaneous operating characteristic of the blades, which will be continually changing especially for a constant speed machine.

The net power developed may be derived easily for electrical generators. A single cup, propeller or Dines type anemometer is used commonly as a control for a turbine. This provides only single point data for estimating the energy available but can provide little data relating to the whole swept area, and will be unlikely to provide a useful response to substantial short term variations in the wind. Use of ten minute averaging will provide a guide level of energy in the relevant wind stream; this can be compared with the net output power over ten minutes. Continuous bining of these data will allow a power wind speed curve to be constructed. (It is unlikely this will be of use to get absolute measurement of machine performance because of the intervening losses and the inadequate wind data).

Comparison of successive power/speed curves (whether complete curves or segments thereoff) could highlight discrepancies from installed performance; this will not be a speedy check because of the wide scatter involved. Differences may arise because of mal settings of blade control surfaces (tips and ailerons for example), giving unexpected kinks in the curves or a general deterioration in the output power because of blade profile damage or insect build up (the latter is commonly encountered in the USA).

It may be possible to isolate the particular offending component by examining the frequency content of the output electrical signals. There will be a strong rotor related content to the power as blades pass through the different layer of the incoming wind stream. The resulting peaks would be equal in a uniform stream, the wind stream is anything but uniform. However, successive logging of the characteristics will isolate differences between blades and perhaps signal some remedial action. This presupposes that an isolating coupling between the power input shaft and the generator is not so good as to mask completely variations. The techniques discussed in Chapter 4 will be applicable.

For more detailed evaluation additional instrumentation is needed, means of measuring torque through the shafts, strain in the blade root. These could be used to isolate the portions of the drive train and the detailed root loads of individual blades. The addition of slip rings or telemetry is necessary.

The orientation of a HAWT turbine must form part of a performance quality package in view of some of the results published on the Mod-O experiments (1).

Performance parameters may be useful in identifying degrees of icing in addition to special instrumentation which may be added to affected areas. Icing conditions will affect the blades and the wind speed sensor. Icing conditions experienced in the recent UK cold period (February 1986) and reported in the media (2) show the effects of glazed ice on anemometers and on electric power cables, leading to failure of the latter with the unpleasant task of repair. Concerning rotor blades of HAWT and VAWT an alteration in characteristics occurs. Mortimer (3) points out that the acretion of rime ice causing a smooth increase in chord can bring about an improvement in lift, more likely for HAWT configurations, so that the assumption that the rotor will steadily reduce in power output may be erroneous. (The VAWT rotor with the varying incidence airflow is less likely to have a smooth aerodynamic acretion. The concern is the icing of the control wind sensor. Icing does not always occur in calmer conditions as was noted in the UK in February 1986, wind speeds can be significant. An icing cup or propeller anemometer may appear to register acceptable wind speeds, whereas the rotor is enduring excessive wind speeds in an incorrect configuration - it may even need to be stopped (Chappel (4)). Included in the general monitoring scheme should be a comparison of measured wind speed with a broad indication of the rotor power at this condition; a wide disparity from known conditions could point to icing taking place (or severe insect build up) requiring remedial action. A time delay would have to be included in the logic to cope with wind perturbations.

Temperature records for gear box oil and hydraulic fluid are amenable to analysis to indicate incipient failure. Other means are necessary to identify particular components (see Chapter 6). Overtemperature relays are included in generators - these could be upgraded to provide diagnostic data.

The logging of run up and run down times often used for aero engine monitoring may not be appropriate for a wind turbine.

4 CONDITION MONITORING IN GEAR BOXES

Wind turbines are quite commonly configured so that rotor shaft power is transmitted to the generator system via a mechanical gearbox and transmission. In most designs, the gearbox is reasonably accessible for inspection and maintenance, but nevertheless is a key element in the sense that unscheduled repairs require the wind turbine to be stopped and taken off line.

It is characteristic of mechanical transmissions that the first indications of an internal fault may be changes in the noise levels emitted by them. This, however, provides no clue as to precisely the nature of the fault which is developing, whether it requires immediate attention or can be left until the next scheduled service, or whether indeed the failure is likely to be catastrophic. Frequently, for example, in gearboxes employing rolling element bearings, increases in noise level may be due to a race which is developing track 'spalling', and would continue to function for some time before requiring replacement. On the other hand a crack developing in a gear tooth would of course require immediate attention. A normal reaction of service engineers when confronted by developing noise in a gearbox is to play safe, and to demand that the gearbox be imemdiately removed and stripped for examination, in some cases leading to the turbine being unnecessarily stopped. In helicopters which frequently employ multistage planetary transmissions, similar approaches were adopted until the recent application of condition monitoring techniques. Adoption of these methods has considerably reduced the incidence of unscheduled main gearbox removals, with corresponding improvements in fleet availability and reductions in strip, repair and qualification testing.

The two major techniques adopted for condition monitoring of helicopter gearboxes are vibration signature monitoring and spectrographic analysis of wear particles present in the lubrication system.

4.1 Vibration Monitoring

The vibrations present in a typical gearbox comprise what at first sight appears a confusing array of frequencies, due, in multimesh arrangements, to the various tooth meshing forcing frequencies and harmonics present. Gear teeth, despite being manufactured to high precision standards, transmit motion with inevitable kinematic inaccuracies, and these give rise to forced vibrations in the shafts, bearing and casing of the gearbox. When these forcing frequencies coincide with the natural modes of the gearbox elements, resonances occur which can produce significant vibration levels and noise at certain running speeds, even in well designed systems.

Vibration monitoring of a gearbox to detect faults is carried out by measuring the vibration signature on a continuous basis and noting the changes in signature which occur over a period of time. Since it is usually impractical to mount transducers directly onto bearing housings, or at other strategic locations in the gearbox, transducers are often mounted directly onto the gearbox casing. To discriminate between elements producing vibration internally, signal processing techniques are employed in conjunction with knowledge of tooth nesting frequencies. These techniques permit changes in power spectrum levels at particular frequencies to be measured and their location within the gearbox to be identified. The application of these techniques in detail ideally exploit previous experiences with similar gearboxes, to identify particular faults from a library data base built up over a period of time.

4.2 Spectrographic Analysis

Spectrographic analysis of particles in the lubrication system, taken from filters or plugs, enables the material of the particles(5) to be identified. In certain circumstances, with a detailed knowledge of the gearbox from which the particles were taken, the material identification provides the clues which may indicate the location within the gearbox where those particles are being shed. In conjunction with the rate and concentration at which debris is occurring in the filter, this technique can sometimes enable the rate at which a component is deteriorating to be followed. With experience, a judgement may be made at which stage repairs should be carried out.

The technique has been particularly successful in detecting incipient rolling element bearing surface fatigue failures in gas turbine engines. Unfortunately, in gearboxes, the large number of ballraces often present makes identification of the location of the failing component within the gearbox sometimes very difficult. Furthermore, the technique has often not been available on site in the past, with resulting delays in receiving information from laboratories processing the debris removed.

However, in certain modern helicopters, the technique is used as an adjunct to vibration monitoring, where the two techniques complement each other. Thus for a wind turbine gearbox, the changes in vibration level confirmed by vibration signature measurement, when combined with the results of a spectrographic analysis, would enable the troublesome component to be identified, it's rate of deterioration monitored and forward planning for removal of the gearbox at a scheduled servicing period to be undertaken with minimum disruption to the operation of the wind turbine.

5   VISUAL INSPECTION

The wind turbine has a sharp division between readily accessible components in the nacelle or tower for example, and those which are in the hostile wind stream.   This colours those methods which may be used and where, and may be size related.

During the design phase, stress investigations will have identified potential high stress regions.   There will be a virtue in ensuring these regions will be readily accessed or instrumented with strain gauges or optical fibres in order to monitor the conditions.

5.1   Nacelle and Tower Housed Components

The main rotating members are housed within these areas (or the tower base in the case of a VAWT) and access must be provided for the appropriate NDT inspection.   There will be sufficient down times occurring during the operations of a turbine because of low winds, these should be used for checking periods rather than disturb operations in a regular manner.   In this way production from the variable energy source can be maximised.   It is anticipated that personnel will be on hand in large installations and multi unit installations who would include inspection as part of the duties.   An accurate recording system will be essential, in view of the slightly irregular service times.

Some inspections may be possible during running, on housing, retention features, etc. Running will highlight loose components and allow cracks to open up and be 'heard' (see Chapter 6).   During braked conditions visual inspection of the shafts, gears, brake discs and couplings, which should be as easy as possible, would include a follow up of faults highlighted during running.   Various crack inspection methods can be employed, dye penetrants viewed directly or with UV light, magnetic particle methods but these latter are likely to be difficult to use in situ (6) because of magnetisation required.   Shafting, coupling outer casings, brake discs, gearbox casings, generator housings would be candidates for these methods, in addition to the mountings of the various units.

Borescopes, devices using illumination and viewing through optical fibres are used extensively in the aero industry and in medicine.   With suitable access port provision, such instruments may be used to inspect gearbox internals, gears, planet carriers and some bearing components.   Rotor and stator sections viewing in the generators and brake disc and pad conditions in enclosed units is possible provided suitable provisions are made.   The difficult hub region on the HAWT which may contain teeter bearings and blade pitch actuating systems may be amenable to inspection using long borescope probe units, overcoming some of the need for external inspections.

MW sized turbines may permit access into the hub region for closer inspections.

There remains the problem of inspection of the invisible, flexible couplings clamped between flanges, bolts and studs, some bearing components.   At the present time these need to be removed for individual inspection and for hole inspection.   The crash of an Alouette helicopter that occurred in June 1983 being used in some North Sea duties, highlights the need for regular inspection of such hidden components and the need to use correct greases in the torque tightening of bolts (7).   The aircraft had not been utilised for a large number of hours so that certain checks were not invoked.   The hidden nature of the failure, however, is pertinant. The event also highlights the need for regular inspection of those items little used which may be components of vital safety systems which may be called on once in a lifetime.

5.2   Rotor

Whether the machine is a HAWT or a VAWT, inspection of the blades is a major problem. These components are in the less hospitable air flow and it will be rare that conditions are warm and calm to provide easy inspection. The blades of a HAWT could be partly visually inspected from windows in the tower and in some cases from service platforms specially provided.   Blades of the straight bladed Musgrove VAWT might be accessed when in a partially furled condition from positions on the cross arm.   Large HAWT and VAWT turbines may permit access into the blades along the main spar, probably the main load carrier. None of these options is straightforward or safe.   A strong attraction of the Vee VAWT is that the blades can be lowered quickly to the ground to give ready access to both sides of the blade and to permit mechanised inspection.   The Darrius VAWT presents special difficulties with much of the blade being away from ready access and parts of the blade being out of sight from the ground.

Possible means of external access to blades of a large turbine could include the following.

*HAWT -'window cleaners cradle' lowered from beneath the nacelle, coupled with a fully variable pitch blade to provide access to both surfaces.

*VAWT (Musgrove) - fully furled blades gives some access to three of the four sides of the blade from the cross arm in addition to the blade actuating mechanism.

*VAWT (Vee) - Release of the tie wire gives ground level access to both sides of the blade.

*VAWT (Darrius) - Access to ends and possibly to parts adjacent to steadying cross members; access to other parts is not easy unless the machine is brought to ground level.   A helicopter survey may be feasible in a farm situation.

*VAWT (Giromill) - access difficult at large scale unless cross arms are substantial.

Binocular access will be viable for general viewing at long range for most types.

Several other methods of external access are conceivable, but the problem of finding personnel to man devices in open conditions at a sea site may be insurmountable.

6 ASSISTED VISUAL INSPECTION

Such methods require surface access, either by direct contact, or by line of sight. These techniques can deal with the undersurface condition of components, cracks within shafts, delamination beneath the gelcoat of a blade for example.

A recent seminar on non destruction testing reported in Flight (8) described some current methods which could be applicable to wind turbines and their components.

The wind turbine shares with large aircraft a large lifting surface and the increasing use of composite materials, which are not so amenable to established means of non destructive testing. Thermography, which is well known for the identification of heat loss sources in buildings has been used in trials for scanning a wing from ground level. Discrimination levels approaching 1°C are sufficient to identify water ingress for example into honeycomb structures as well as identifying internal joins in the structure incurred in the manufacturing process. Coupled with access ports in the tower some of a turbine blade could be inspected. Longer range viewing from ground level may require further investigation, but appears possible. The technique could be used on internal nacelle components to identify gearbox hot spots or leakage in control system, air or fluid, lines. There are likely to be significant electronics in the control systems and thermography will quickly identify overheated components, overheating cables and connections.

Vibration characteristics and their use has been discussed in Chapter 4. However, the operating noise of a machine can reveal useful information, as many will have experienced with a car. With some care it is possible to hear cracks under load and some Canadian work is currently undergoing trials (8).

With many blades now manufactured from composite materials the option of introducing optical fibres into highly stressed areas provide an economical crack detection solution. When illuminated the cracked fibre emits light, even though the crack in the base material has closed up. This method is already being incorporated into highly stressed regions on conventional welded materials.

Ultrasonic testing has been used for some years to identify inclusions in materials and for checking wall thicknesses. Multiple arrays of transducer can provide a 3-D image of damage but this is most effective on simple surfaces. Such methods could expose internal damage within structures exhibiting no external damage. Wind turbines are unlikely to suffer severe impacts on towers or rotating structures, however, assembly in difficult conditions may

give rise to damage which could be inspected on site.

Metallic structures are amenable to use of eddy current techniques and trials have provided useful results. The method appears to require substantial back up equipment at present.

A number of methods based on X-rays for examining clearances within operating test bed aeroengines, for revealing internal cracks and blockages are being used successfully. The support equipment and the need for radiation protection would preclude these from use in the field at present.

Main spars of helicopters have regularly been pressurised, so that when a crack occurs it's presence can be identified by a loss in pressure. This could be used in less accessible areas of rotors and may be particularly useful for the Darrieus VAWT where various sections of the blade could be separately pressurised, perhaps differentially, to provide flow between adjacent areas, isolating the position of the crack.

7 CONDITION MONITORING - THE COST

There must be a right balance between cost and life, cost and safety, these may not always be related. Considerable anguish is experienced by all in the aircraft industry over these balances (9). The presence of a maintenance crew at a multi installation may remove some of the need to automate monitoring on each machine since there are a wide variety of portable units for stress and frequency analysis for example. Single units may have to carry a higher cost of monitoring unless some kind of service agreement is provided.

Turbine size will play a part, the larger being able to absorb the cost of more sophisticated equipment. The margins of safety may also play a part, where an installation adjacent to dwellings may be considered a greater hazard than those in remote sites away from all but specific support staff.

Most wind turbines, except the very small <10kW may have sensors included for other purposes. With few alterations these could be adapted for wider monitoring use. Access for performing the tasks may be more pleasant in larger machines, having better protection within a tubular tower and access therefrom to an enclosed nacelle, smaller machines have towers with outside access and car boot type for access to nacelle components, but with limits in being able to get to specific components.

Additions to blades for later interrogation could include strain gauges and optical fibres which are relatively cheap, one providing stress levels the other giving a ready identifier of cracks in highly loaded areas.

Thus for small machines in particular there may not be large additional expenses in going a long way toward a useful condition monitoring package by using existing equipment. Low cost additions to the blades, which could

Types of Monitoring	LARGE (MW)	SMALL (50-300kW)
Visual	Yes, special equipment	Yes, direct access may be possible
Aided viewing	Yes, special equipment	Yes but, Specialist firm
Vibration	Special equipment	Yes, in emergency shut off systems but could be monitored specially
Performance	Yes, wind speed and power in control system	Yes, wind could be added, power included in many control systems
Temperature	Yes, gearbox and systems	No
Pressure	Yes, hydraulics, air	Maybe, in brake systems
Fluid Contamination	Yes, expert inspection	Unlikely, but could be used

Table 1   Possible types and complications of monitoring

easily be incorporated at manufacture and additions of strain gauges to hub components with appropriate diagnostic connections built in for static checks should hardly affect costs.

An approach is to build with substantial factors of safety, this is likely to be expensive and may give false security since the technology is still young and adequate factoring may be omitted from relevant areas.   It may be impossible to monitor those areas either directly, but a general overview of vibration characteristics for example, may highlight unexpected problem areas.   A utility view (10) requests a simple and robust approach to meet the functional requirements - this along with a strategic monitoring plan would go to achieving the 95% availability required by the utility.

8   DISCUSSION AND CONCLUSIONS

Why not design the WECS fail safe and reduce the need for such extensive inspections?   A concern is the new ground that lies before the wind turbine engineer, and identification of components which need to be designed fail safe.

The recent spate of problems with the established Boeing 747 (11) where fuselage cracks became apparent in older aircraft with many thousands of flights completed, demonstrate the fail safe aspect well, but the perturbation caused 16 aircraft to be withdrawn for inspection by one airline, with 2 cases of cracking identified at overhaul in another airline and 7 further cases being identified by 4 other airlines.   Very significant replacement costs were cited and concern was expressed at what further tear down might reveal. This suggests that fail safe measures themselves ought to be checked regularly and argues for a free access to even 'safe' components.

A facet requiring particular attention is the monitoring of repairs of the faults monitoring has identified.   A recent crash of a Boeing 747 has identified the possibility of deficient repairs to the rear pressure bulkhead as being contributary.   Such in situ repairs should be subject to specific attention.

With the most effective inspection method,

the human element remains a problem, with judgement entering in, even after careful comparative training.   Human intervention alone can cause a decrease in reliability as shown by the wider introduction of magnetic chip detectors into aero engines and thus more breaks into the respective system (12).

At this early stage of current wind exploitation, the knowledge of material response under large number of fatigue cycles is limited. The option to transfer fatigue history from one installation to another may not be justified, or even from one machine to another in an array. An example of low cycle fatigue usage is the Red Arrows squadron (13), where the throttle movements of the leader are compared with that of a team member working hard to maintain position showed the LCF rates different by a factor of thirty eight to one.   This is not wholly removed from a wind farm situation, where a turbine on the windward perimeter will run in less turbulent air than one within the array.   Counts of control surface operation to maintain constant rotational speed say are likely to show large differences.   Monitoring must thus be specific for each machine, although not necessarily equally comprehensive.

There are probably sufficient techniques to move towards long WECS life in a safe fashion, although implementation in the open air is a severe disadvantage compared with hangar conditions for civil aircraft.   The removal times for gearbox and blades proposed in reference (14) should be capable of improvement in the longer run in the light of running at other sites, which will identify specific problem areas to be monitored.   The removal of large components at sea sites using a fleet of barges does depend on a long period of good weather - the prospect at a land site is only slightly better.   Thus detailed monitoring will identify the components really in need of service, rather than the criteria being that of a time interval.

Had the methods of condition monitoring now

available or becoming available been in use at the end of the last war, we might have been celebrating the thirtieth birthday of the Grandpa of large machines sited on Grandpa's Knob, in spite of the wartime welding modifications and sub standard materials. We hope that twenty or thirty years from now, the long life demonstrated by the Gedser machine would be well eclipsed by the twenty year life wind farm.

9 REFERENCES

1 GLASGOW, J.C.CORRIGAN, R.D.MILLER, D.R., The effect of yaw on horizontal axis wind turbine loading and performance, NASA TM-82778, October 1982.

2 PARRY, G., Ice risk to BBC tower, Guardian, 13th February 1986.

3 MORTIMER, A.R., A review of the icing problem for aerogenerators, Wind Engineering, 1980.

4 CHAPPEL, M.S., TEMPLIN, R.J., And the cold winds shall blow.., Wind energy research and development in Canada - Spring 1985, 7th BWEA Wind Energy Conference, Oxford, March 1985.

5 WESTLAND MAGAZINE Article relating to gearbox condition monitoring, 1983.

6 CASTROL BROCHURE Non destructive testing - magnetic particle inspection materials.

7 FLIGHT, Leader and article - The wrong grease led to crash, Flight International, 2nd November 1985.

8 HOPKINS, H., Non-destructive testing, Flight International, 7th December 1985.

9 RAMSDEN, J.M., Affordable safety, Flight International, 25th January 1986.

10 STEVENSON, W.S., A utility's view based on experience, R & D needs in wind energy, Proceedings of BWEA-DEn workshop, September 1985.

11 BBC TV News Report, 12th February 1986.

12 MIDGLEY, R.A., Advanced turboprop and turbofan transmissions, I.Mech.E. Aerospace Industries Division Seminar, 14th November 1985.

13 HURRY, M.F., O'CONNOR, C.M., United Kingdom military engine usage, condition and maintenance systems experience, AIAA/SAE/ ASME 19th Joint Propulsion Conference, June 1983, Seattle Washington.

14 BURTON, A.L., ROBERTS, S.C., The outline design and costing of 100m diameter wind turbines in an offshore area. BWEA7, Ibid.

10 RELATED ARTICLES AND PAPERS

JAMISON, P., HUNTER, C., Analysis of data from Howden 300kW wind turbine and Burgar Hill Orkney. BWEA7, Ibid.

MITCHEL, J.S., Condition monitoring, Mechanical Engineering, December 1985.

STEVENS, K.S., Modal analysis-an old procedure with a new look, Mechanical Engineering, December 1985.

VOICU, G.H., TANTAREANU, C.G., TUDOR, M., Perspectives of wind power usage in Romania, 7th BWEA Wind Energy Conference, Ibid.

WINDWATCH, Comments on composites for wind turbines, Wind Directions Vol.V, No.3, January 1986.

Appendix :

List of delegates to the Conference

Mr. K.A. Abu-Adma	Imperial College,	Dept of Electrical Eng.,	Exhibition Road,	London
Mr. J.C. Addington	Aeolus Ltd.	2 Bladon Close	Oxford	
Mr. J.F. Ainslie	CERL	Applied Physics Branch	Kelvin Avenue,	Leatherhead, Surrey
Mr. L. Andersen	Vestas Energy A/S	Whinneyknowe House	Upper Polmaise	Stirling
Dr. M.B. Anderson	Sir Robert McAlpine & Sons	1 St. Albans Road,	Hemel Hempstead,	Herts.
Mr. Colin Anderson	James Howden & Co. Ltd.	195 Scotland Street,	Glasgow,	
Mr. G.A. Anderson	NOSHEB,	16 Rothesay Terrace,	Edinburgh,	
DR. M.P. Ansell	University of Bath,	School of Materials Science	Claverton Down,	Bath
Mr. S. Arithoppah,	Cranfield Inst. of Tech.	School of Mech Eng.	Cranfield,	Bedford.
Mr. G. Atkinson	Cavendish Laboratory,	Madingley Road,	Cambridge.	
Mr. W. Scott Bannister,	Napier College,	Dept. of Mech & Ind. Eng.	Colinton Road,	Edinburgh
Mr. Peter Barry	B.B.C.	19 Brasley Way,	West Wickham,	Kent.
Mr. Jeremy Hugh Bass	Rutherford Appleton Lab	Chilton,	Didcot,	Oxon.
Mr. E. William Beans	University of Toledo,	Dept. of Mech. Engineering,	Toledo.	Ohio, U.S.A.
Mr. L.A.W. Bedford	ETSU,	Building 156	AERE Harwell	Didcot, Oxon
Mr. S. Bevan	Dept. of Energy,	Thames House South,	Millbank,	London
Mr. J.A.M. Bleijs	Rutherford Appleton Lab.,	Chilton,	Didcot,	Oxon.
Prof. Sir Hermann Bondi	Churchill College	Cambridge		
Mr. Peter Bonfield,	University of Bath,	School of Materials Science,	Claverton Down,	Bath, Avon
Dr. Neil Bose,	University of Glasgow,	Dept. Naval Arch. & Ocean Eng.	Acre Road,	Glasgow.
Dr. Ervin Bossanyi	Rutherford Appleton Lab.	Chilton,	Didcot,	Oxon
Mr. Godfrey Boyle	Open University	Alternative Technology Group	Milton Keynes	
Mr. A. Brown,	James Howden & Co. Ltd.,	Old Govan Road,	Renfrew	
Mr. T.W. Buchanan	Vestas Energy A/S	Whinneyknowe House,	Upper Polmaise	Stirling
Dr. P. Bullen	Kingston Polytechnic	Canbury Park Centre	Canbury Park Road,	Kingston UponThames
Mr. Alan Bullock,	University of Reading,	Whiteknights,	Reading	
Mr. M. Burgess,	I.R.D.,	Fossway,	Newcastle Upon Tyne,	
Dr. A.L. Challis		Classeys,	Low Ham,	Langport, Somerset
Miss Valerie Chapman	Thermax Corp.,	One Mile Street,	Burlington,	Vermont, U.S.A.
Mr. W.V. Chestnutt	Davidson & Co. Ltd.,	Sirocco Eng. Works	Belfast	
Mr. R. Clare	Sir Robert McAlpine & Sons	40 Bernard Street,	London,	
Mr. Lars Clausen	Nordtank A/S	Nyballevej 8,	8444 Balle	Denmark
Dr. B.R. Clayton	University College London	Torrington Place	London.	
Mr. A. Conway	Press	23 The Ridgeway	Kenton,	Harrow, Middx.
Mr. Alun Howard Coonick,	Imperial College Science/Tech	Dept. of Electrical Eng.	Exhibition Road,	London
Mr. D. Corbet,	Gifford Technology Ltd.,	Carlton House,	Ringwood Road,	Woodlands, Southampton
Mr. Robert Coveney	GEC Avionics Ltd.	Elstree Way	Borehamwood	Herts.
Mr. Gunther Cramer,	SMA-Regelsysteme GmbH,	Hannoversche Straße 3,	3501 Niestetal 1/W	Germany
Mr. P. Cray,	Sir Robert McAlpine & Sons Ltd	1 St. Albans Road,	Hemel Hempstead,	Herts.
Mr. K. Diamantaras,	E.E.C. DG 17	Av du Koufec 9	B-1160 Brussels	
Mr. Dobey,	D.T.I.	Thames House South,	Millbank,	London
Prof. G.T.S. Done	The City University,	Dept. of Mech Eng.	Northampton Square,	London.
Mr. Owen Dumpleton	British Shipbuilders	P.O. Box 25,	Pallion,	Sunderland, Tyne & Wear
Prof. Richard Eden,	Cavendish Laboratory,	Cambridge.		
Prof. Sir Sam Edwards	Cavendish Laboratory,	Cambridge University		
Mr. A.J. Eggington	S.E.R.C.	Polaris House,	Swindon.	
Mr. G. Elliot,	National Engineering Lab.,	East Kilbribe,	Glasgow.	
Mr. F.C. Evans	University of St. Andrews	Physics Dept.	North Haugh,	St. Andrews, Fyfe
Mr. M. Falchetta	ENEA-CASACCIA	Dipartimento Fare-Sacco 99	00060 Roma,	Italy
Mr. J.F. Fawkes	Marlec Engineering Co. Ltd.	Unit 5 Pillings Road Ind. Est	Oakham,	Rutland, Leics.
Mr. E.A.N. Feitosa	University of Southampton,	Southampton		
Mr. Paul Feron	Reading University	Dept. of Engineering,	Reading	
Mr. J.E. Foster	Rutherford Appleton Lab.	Chilton,	Didcot,	Oxon
Dr. L. Freris,	Imperial College	Dept. of Elec Eng.,	Prince Consort Road,	London
Mr. E.B. Fuller	A.B. Fuller Ltd.	196 Morland Road,	Croydon	
Dr. N.N. Fulton,	S.R. Drives Ltd.,	Springfield House,	Hyde Terrace,	Leeds
Dr. P. Galbraith,	University of Glasgow,	Glasgow.		

Ms. Jean Galt	James Howden & Co. Ltd.,	Govan Road,	Glasgow.	
Dr. C.R. Gamble,		Abrahams Cottage,	Fordwell, Asthall Leigh,	Nr. Withey, Oxford
Dr. P. Gardner	University of Strathclyde	Energy Studies Unit	Glasgow	
Dr. A. Garrad,	Garrad Hassan & Partners,	10 Northampton Square,	London	
Mr. J. Garside,	Cranfield Institute of Tech.,	Cranfield,	Bedford.	
Mr. T. Gaskell,	Marlec Engineering Co. Ltd.,	Unit 5 Pillings Road Ind Est	Oakham,	Rutland, Leics.
Mr. P. Gipe,	Paul Gipe and Assoc.	P.O. Box 277,	Tehachapi,	California, U.S.A.
Mr. D. Goodfellow	Leicester University	Engineering Dept.,	Leicester University	Leicester
Mr. R.J. Grew	Balfour Beatty Power Const.	P.O. Box 12,	Acornfield Road,	Kirkby, Liverpool
Mr. M.J. Grubb	Cavendish Laboratory,	Energy Research Group,	Madingley Road,	Cambridge.
Mr. R. Haines,	Wind Energy Group,	345 Ruislip Road,	Southall,	Middlesex.
Mr. J.A. Halliday	Rutherford Appleton Lab.	Chilton	Didcot,	Oxon.
Mr. Gordon F. Hancock	Curtis & Green	Arundel Road,	Uxbridge,	Middx.
Mr. M. Hancock,	Gifford Technology Ltd.,	Carlton House,	Ringwood Road,	Woodlands, Southampton
Mr. J.C. Hansen	RISO National Laboratory	P.O. Box 49	DK-4000 Roskilde	Denmark
Dr. R. Harrison	Sunderland Polytechnic	Dept. of Physical Sciences,	Ryhope Road,	Sunderland.
Dr. U. Hassan	Garrad Hassan & Partners,	10 Northampton Square,	London	
Mr. O. Helgasson,	University of Iceland,	Science Institute	Iceland	
Mr. J. Hojstrup	RISO National Laboratory	P.O. Box 49	DK-4000 Roskilde	Denmark
Mr. J. Holt,	C.E.G.B.,	Laud House,	Newgate Street,	London.
Mr. C. Horn,	CAP Industry Ltd.,	Trafalgar House,	Richfield Avenue,	Reading, Berks.
Mr. C.T.J. Hoskin,		Flat B, York Lodge,	The Avenue,	Bushey, Herts.
Mr. F.J. Houghton	Foster Wheeler Power Prod.	R & D Division	Brenda Road, Hartlepool,	Cleveland
Mr. R. Hunter	National Engineering Lab.,	East Kilbride,	Glasgow.	
Mr. B. Hurley,	Dublin Inst. Technology	Maths & Science Dept.,	Bolton Street,	Dublin 1
Dr. David G. Infield,	Rutherford Appleton Lab.	Energy Research Group	Chilton,	Didcot, Oxon
Mr. P.M.W. Ireland,	B.P. Solar Systems Ltd.	Farnbrough Close,	Aylesbury Vale Ind Est.	Stocklake, Aylesbury
Dr. Gary Jenkins	Sunderland Polytechnic,	Dept. of Physical Science	Ryhope Road,	Sunderland
Mr. M.D. Jepson		Ffynnonau,	Penlan,	Brecon, Powys
Philip Jewel		17 Church Road,	Southborne,	Bournemouth
Mr. J. Kawadri	The City University	Dept of Mech Eng.	Northampton Square,	London
Mr. D. Kerr,	Sir Robert McAlpine & Sons Ltd	1 St. Albans Road,	Hemel Hempstead,	Herts.
Mr. J. Kersey	Open University	Walton Hall,	Milton Keynes	
Mr. T.S. Kirby,	National Centre for Alt. Tech.	Machynlleth,	Powys,	Wales
Mr. Paul Kittrick	ASEA Hagglund Ltd.,	Wakefield 41,	Wakefiled,	W. Yorks.
Mr. B.P. Laight		5 Littlemead	Esher,	Surry.
Sir Henry Lawson-Tancred		Aldborough Manor,	Aldborough,	Boroughbridge, N. Yorks
Mr. R. Leicester	James Howden & Co. Ltd.,	Govan Road,	Glasgow.	
Dr. D. Lindley,	Wind Energy Group,	345 Ruislip Road,	Southall,	Middlesex.
Prof. N.H. Lipman	Rutherford Appleton Lab.	Energy Research Group,	Chilton,	Didcot, Oxon
Mr. F. Low,	Royal Aeronautical Society,	4 Hamilton Place,	London	
Mr. Knud Lunde	Inst. of Energy Technology,	Norway.		
Susan Macmillan	Robert Gordon's Inst. Tech.	School of Surveying,	Garthdee Road,	Aberdeen
Mr.G. Manning	S.E.R.C.	Polaris House,	Swindon.	
Mr. C. Maskell	ETSU	Building 156	AERE Harwell,	Didcot, Oxon
Mr. Richard May,	Theta Resource Management	Manor Coach House,	Church Hill,	Aldershot
Dr. R. Mayer,	BP Research,	London		
Dr. I.D. Mays	Sir Robert McAlpine & Sons	1 St. Albans Road,	Hemel Hempstead,	Herts.
Mr. A. Mewburn-Crook	Kingston Polytechnic	Canbury Park Centre	Canbury Park Road,	Kingston Upon Thames
Mr. D.J. Milborrow	CEGB	Laud House,	Newgate Street,	London
Mr. Maurice M. Millar	Aeolus Ltd.	c/o National Wind Turbine Cent	National Eng. Laboratories	East Kilbride
Mr. G.G. Miller	Stork Fans Ltd.	4 Hercies Road,	Hillingdon,	Uxbridge, Middx.
Mr. Andrew Morley,	Vickers Systems Ltd.	P.O. Box 4,	New Lane,	Havant, Hampshire
Mr. C.J.E. Morris	W.S. Atkins & Partners	Woodcote Grove,	Ashley Road,	Epsom, Surrey
Dr. E. Mowforth	University of Surrey,	Guildford,	Surrey.	
Mr. A.T.L. Murray,	N.O.S.H.E.B.	16 Rothesay Terrace,	Edinburgh	
Dr. P.J. Musgrove,	Reading University,	Dept. of Engineering,	Whiteknights,	Reading
Mr. K. McAnulty,	ETSU	Building 156	AERE Harwell	Oxon
Mr. Robert McDonald,		10 Chelwood Gardens,	Richmond,	Surrey,
Mr. A. McDonnell	Davidson & Co.Ltd.,	Sirocco Eng. Works,	Belfast	
Jon G. McGowan	University of Massachusetts	Amherst,	Mass.	U.S.A. 01003
Air Commodore C.T. Nance	c/o Medina Yacht Co.	Mornington House,	Cowes,	Isle of Wight
Mr. L. Naylor,	AERE Harwell,	Didcot,	Oxon,	
Mr. Mark Newham	Information Research	33 Rugby Place,	Brighton	
Mr. J. Newton,	Gifford Technology Ltd.,	Carlton House,	Ringwood Road,	Woodlands, Southampton
Mr. G. Nicholson	Northumbrian Energy Workshop	Tanners Yard	Hexham,	Northumberland
Mr. T. O'Flaherty,	The Agricultural Institute	Kinsealy Research Centre	Malahide Road,	Dublin 17
T.D. Oei	Netherlands Energy Res Unit	P.O. Box 1	1755 ZG Petten,	The Netherlands
Mr. C. Oram,	Cranfield Institute of Tech.,	Cranfield,	Bedford.	
Dr. D.I. Page,	ETSU	Building 156	AERE Harwell,	Didcot, Oxon
Dr. Jean Palutikof,	University of East Anglia,	Climatic Research Unit,	Norwich	
Mr. W. Palz,	E.E.C.,	200 rue de la loi,	1049 Brussels,	Belgium
Mr. W. Parker,	C.E.G.B.,	Burymead House,	Portsmouth Road,	Guildford
Mr. Henry G. Parkinson	Wolfson College,	Oxford		
Prof. B. Maribo Pedersen	Technical University Denmark	Dept of Fluid Mechanics	2800 Lyngby	Denmark
Mr. Penfold	James Howden & Co. Ltd.,	Govan Road,	Glasgow	
Mr. C.M. Peterson,	Windsund Energy Systems Ltd.,	12 Stansby Park	The Reddings,	Cheltenham

312